TP
248.2
.R48
1988

A Revolution in
biotechnology

$44.50   h n

DATE			

A REVOLUTION IN BIOTECHNOLOGY

A Revolution in Biotechnology

Edited by Jean L. Marx

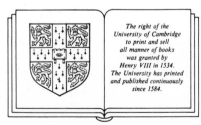

The right of the
University of Cambridge
to print and sell
all manner of books
was granted by
Henry VIII in 1534.
The University has printed
and published continuously
since 1584.

Cambridge University Press
Cambridge
New York   New Rochelle   Melbourne   Sydney

Published on behalf of the ICSU Press by
the Press Syndicate of the University of Cambridge
The Pitt Building, Trumpington Street, Cambridge CB2 1RP
32 East 57th Street, New York, NY 10022, USA
10 Stamford Road, Oakleigh, Melbourne 3166, Australia

First published 1989

Printed in Great Britain by Scotprint Ltd., Musselburgh

British Library cataloguing in publication data

A Revolution in biotechnology.
  1. Biotechnology
  I. Marx, Jean L.
  660″.6

Library of Congress cataloguing in publication data

A Revolution in biotechnology.
  Includes index.
  1. Biotechnology. I. Marx, Jean L. (Jean Landgraf)
TP248.2.R48 1988   660″.6 88—9521

ISBN 0 521 32749 0

MU

Contents

Foreword *by John Kendrew*   vii

Contributors   ix

1   Heredity, genes and DNA
      Jean L. Marx   1

2   Synthesis without cells
      Ethan S. Simon, Alan Akiyama and George Whitesides   15

3   Microorganisms as producers of feedstock chemicals
      Randolph T. Hatch and Ralph Hardy   28

4   Gene cloning opens up a new frontier in health
      L. Patrick Gage   42

5   The microbial production of biochemicals
      Saul L. Neidleman   56

6   Single-cell proteins
      John H. Litchfield   71

7   Bacterial leaching and biomining
      David Woods and Douglas E. Rawlings   82

8   Bacteria and the environment
      Tamar Barkay, Deb Chatterjee, Stephen Cuskey,
      Ronald Walter, Fred Genthren and A.W. Bourquin   94

9   Biological nitrogen fixation
      Andrew W. B. Johnston   103

10   Plant cell and tissue culture
      Edward C. Cocking   119

11   Improving crop plants by the introduction of isolated genes
      Jozef Schell, Bruno Gronenborn and Robert T. Fraley   130

12   Monoclonal antibodies and their applications
      Jean L. Marx   145

13   Site-directed antibodies in biology and medicine
      Thomas M. Shinnick and Richard A. Lerner   159

14   New methods for the diagnosis of genetic diseases
      Yuet Wai Kan   172

15 **The prospect of gene therapy for human hereditary diseases**
    Jean L. Marx   186

16 **Biotechnology, international competition and regulatory strategies**
    Joseph G. Perpich   197

    Glossary   211

    Index   217

Foreword

by John Kendrew
President, International Council of Scientific Unions

This is the first of a series of books on important scientific topics for the general reader that is being sponsored by the International Council of Scientific Unions. ICSU is the world organization representing professional scientists, and an important part of its responsibility is to bring to the notice of the public scientific advances that will significantly affect our lives or that will illuminate our knowledge of the world we live in. One of the ways in which it carries out this task is through the agency of the ICSU Press, which is publishing this book in collaboration with the Cambridge University Press.

Biotechnology is certainly an important scientific topic. In the last few years it has initiated a transformation of many parts of the chemical industry, of agriculture and of medicine – a transformation that has emerged from the laboratory into practical application with quite remarkable speed. The title of the book has been chosen carefully; there has certainly been a revolution in the techniques and practice of biotechnology, but biotechnology is not new.

One of the oldest industries in the world, the brewing of beer, depends on a typical biotechnological process. The breeding of domesticated animals is biotechnology too, if we accept the usual definition of biotechnology as the exploitation – or domestication, depending on the point of view – of other living organisms for the benefit of man. The other organism may be a large mammal, like the cow or pig; a plant, such as wheat or the potato; a microorganism, like the yeast used in brewing beer; or a bacterium, as in many of the processes described in this book.

In general, man is not satisfied with the productivity of other organisms in the wild state. So *breeding* is required to effect a permanent change in the hereditary make-up of the organism – an alteration in the genetic message it passes on to its offspring – to increase the output of the desired product, whether this be alcohol or protein or carbohydrate. Historically, breeding has been the limiting factor in improving biotechnology, because primitive methods, such as some described in the Old Testament, are slow and empirical and proceed by trial and error. Indeed, they have not been very much improved upon in principle even in modern times.

The revolution that gives this book its name originated in the discovery of radically new ways of altering the genetic make-up of microorganisms in a *directed* manner. The promise for the future is that these methods can in practice, as well as in principle, be extended to higher organisms – to plants and animals. And this new power depends on the discoveries of molecular biology: of DNA as the material of heredity; of the genetic code; of the relation between genes and the proteins to which they give rise; of methods of reading the genetic message by sequencing genes; and of the restriction enzymes with which it is possible to cut and splice together sections of DNA in a deliberate fashion. These and the many other elements that make up the subject known as molecular genetics now make it possible to breed microorganisms to order instead of using the hit-or-miss methods of earlier times.

The result has been a vast increase in the potentialities of biotechnology, in effect transforming the whole nature of the subject with many new applications that are becoming important not only in advanced countries, but also in the developing world. Not surprisingly, these new potentialities have brought with them new regulatory and ethical problems.

The new biotechnology is already beginning to affect our lives and in the future its influence will be profound. The purpose of this book is to illuminate the scientific background, to describe what has already been achieved, to discuss the ethical problems, and to suggest what the future is likely to hold. The story it has to tell is as important as it is intellectually exciting.

Contributors

Alan Akiyama
Department of Chemistry,
Harvard University,
12 Oxford Street,
Cambridge, Massachusetts 02138,
USA

Tamar Barkay
US/EPA Environmental Research Laboratory,
Gulf Breeze, Florida 32561,
USA

Al W. Bourquin
US/EPA Environmental Research Laboratory,
Gulf Breeze, Florida 32561,
USA

Deb Chatterjee
US/EPA Environmental Research Laboratory,
Gulf Breeze, Florida 32561,
USA

Edward C. Cocking
Department of Botany,
University of Nottingham,
Nottingham NG7 2RQ,
UK

Stephen Cuskey
US/EPA Environmental Research Laboratory,
Gulf Breeze, Florida 32561,
USA

Robert T. Fraley
Monsanto Company,
700 Chesterfield Village Parkway,
St Louis, Missouri 63198,
USA

L. Patrick Gage
Hoffman-La Roche,
Nutley, New Jersey 07110,
USA

Fred Genthren
US/EPA Environmental Research Laboratory,
Gulf Breeze, Florida 32561,
USA

Bruno Gronenborn
Max-Planck-Institut für Züchtungsforschung,
D-5000 Koln 30,
Federal Republic of Germany

Ralph Hardy
BioTechnica International, Inc.,
85 Bolton Street,
Cambridge, Massachusetts 02140,
USA

Randolph T. Hatch
BioTechnica International, Inc.,
85 Bolton Street,
Cambridge, Massachusetts 02140,
USA

Andrew W.B. Johnston
John Innes Institute,
Colney Lane,
Norwich NR4 7UH,
UK

Richard A. Lerner
Research Institute of Scripps Clinic,
10666 North Torrey Pines Road,
La Jolla, California 92037,
USA

John H. Litchfield
Battelle, Columbus Laboratories,
505 King Street,
Columbus, Ohio 43201,
USA

Jean L. Marx
Research News, *Science*,
1333 H Street, NW,
Washington, DC 20005,
USA

Saul L. Neidleman
Cetus Corporation,
1400 53rd Street,
Emeryville, California 94608,
USA

Joseph G. Perpich
Howard Hughes Medical Institute,
6701 Rockledge Drive,
Bethesda, Maryland 20817,
USA

Douglas E. Rawlings
Microbiology Dept,
Molecular Biology Building,
22 University Avenue,
University of Cape Town,
Rondebosch 7700,
South Africa

Jozef Schell
Max-Planck-Institut für Züchtungsforschung,
D-5000 Koln 30,
Federal Republic of Germany

Thomas M. Shinnick
Hansen's Disease Laboratory,
Centers for Disease Control,
Atlanta, Georgia 30333,
USA

Ethan S. Simon
Department of Chemistry,
Harvard University,
12 Oxford Street,
Cambridge, Massachusetts 02138,
USA

Ronald Walter
US/EPA Environmental Research Laboratory,
Gulf Breeze, Florida 32561,
USA

George Whitesides
Department of Chemistry,
Harvard University,
12 Oxford Street,
Cambridge, Massachusetts 02138,
USA

David Woods
Microbiology Department,
Molecular Biology Building,
22 University Avenue,
University of Cape Town,
Rondebosch 7700,
South Africa

Yuet Wai Kan
Department of Medicine and Howard Hughes
    Medical Institute,
University of California,
San Francisco, California 94143,
USA

1 Heredity, genes and DNA

Jean L. Marx

Biotechnology is the use of living organisms, or of substances obtained from living organisms, to make products of value to man. The origins of biotechnology lie deep in the mists of human history. Our early ancestors originated the science about 10 000 years ago, during the Stone Age when they first began to keep domestic animals and to grow crop plants for food, instead of depending solely on what they could hunt or gather in the wild.

The organisms used today in biotechnology can be as complex as dairy cattle or as simple as the yeasts needed for brewing beer and baking bread. Even single-celled bacteria can be valuable because they provide drugs, including antibiotics such as penicillin, as well as a variety of other complex chemical products that would be difficult to obtain by synthesis in the laboratory.

Farmers and other practitioners of biotechnology have always sought to improve the organisms on which their livelihood depends. Breeding cows for increased milk production is one example. Crop plants can also be bred to achieve such improvements as higher yields, or increased hardiness or disease resistance.

A simple observation – that physical traits are inherited – provides the basis for these efforts. It is easy to see that parents transmit their characteristic features to their progeny. Tracing the lineage of a newborn baby's features is a common family ritual, as in 'She has her father's eyes, but her maternal grandmother's nose'.

Comparable observations of heredity at work can be applied to improve domestic animals and plants. The dairy farmer who wants to improve the milk production of his herd can select and maintain for breeding purposes only those cows that give the most milk. In a similar fashion, the plant breeder also works by crossing plants with the desired characteristics and then selecting for further mating only those progeny that turn out to be superior for the trait or traits in question.

Such plant-breeding efforts have proved highly successful. During the years between 1940 and 1980 the yields per hectare of the major grain crops – wheat, maize and rice – have doubled or even tripled. This 'Green Revolution', as it became known, resulted from the breeding of high-yielding variants of the grains in combination with the application of large quantities of commercial fertilizers, pesticides and herbicides.

Despite these successes, there are limitations to what conventional breeding can achieve. The process can be time-consuming, requiring many years to develop a new, genetically stable plant variant or animal strain. Moreover, not all the crosses that a plant or animal breeder might want to perform are possible. What can be done is restricted by the genetic incompatibility of species.

Strains that are too different genetically either do not produce offspring at all when mated or they produce infertile progeny. Wild rice, for example, may be superior to cultivated strains in such characteristics as tolerance to salty conditions or disease, but those characteristics cannot be transmitted to cultivated rice by sexual crosses.

In addition, current agricultural practices with their dependence on commercial fertilizers and pesticides are expensive and frequently not applicable to the less developed countries where the need for increased food production is the most acute. The practices have also raised concerns about environmental degradation, because phosphate and nitrogen fertilizers can make their way into streams and groundwater, and pesticides and herbicides often have toxic effects on organisms, including humans, beyond their intended insect and weed targets.

The malnutrition and starvation in the world may now be more a matter of problems with distribution than with the actual lack of food. Developed countries, notably including the United States, are often awash with surpluses while people in the less developed nations often fail to produce enough food.

Nevertheless, the world population is still growing. According to figures compiled by the US Census Bureau it was roughly 5 billion in 1986 and is expected to grow to more than 6 billion by the year 2000. Moreover, the growth will be fastest in precisely those regions of the world where food shortages tend to be most severe.

The populations of the developing nations, which currently account for nearly 75 per cent of the world total, are growing nearly three times faster than the populations of the developed countries. These numbers underscore the need for producing new, high-yielding strains of crop plants that can be grown in arid or tropical regions without the large expenditures of fertilizers and pesticides that are now commonly applied in the more developed nations.

The nature of heredity

For the most part, the achievements thus far in plant and animal breeding have not required a detailed understanding of the mechanisms by which heredity operates. Certainly our Stone Age ancestors had no knowledge of the internal elements that direct the growth and development of a plant or animal. In fact, scientists did not begin to understand the biochemical basis of heredity until the 1940s and 1950s. Now we know that the instructions for building a creature, whether this be as small and simple as a bacterium or as large and complex as an elephant, are encoded in the genes. We know what these genes are made of – namely deoxyribonucleic acid (DNA). And we know how they work.

Especially during the past 15 years, scientists have acquired an unprecedented ability to manipulate genes, to isolate and copy them, and even to transfer them into foreign species. These developments, which are the result of the discovery of recombinant DNA technology, have proved invaluable to the basic science of molecular biology. They have allowed researchers to probe the deepest mysteries of gene structure and function.

But more than that, the developments have opened up new approaches for developing economically important strains of plants and animals and thus contributing to efforts to provide more food for a hungry world. For example, scientists have already found that they can produce mice that are twice the normal size by giving them a gene for a hormone that makes the animals grow larger. Similar experiments are under way with farm animals.

The research is also having important medical benefits. Clinicians hope to be able to cure certain hereditary diseases in humans that are caused by defective genes by giving the afflicted individuals new copies of the good genes, although it may be many years before gene replacement therapy becomes a proven clinical option – if it ever does.

Meanwhile, gene transfer on a simpler scale is already providing a new source of hitherto scarce or unavailable products. These include human growth hormone, which is used to treat children who fail to grow properly because their own bodies do not make the hormone, and human insulin, which is needed by diabetes patients who can no longer take bovine insulin because they have become allergic to it. The hormones are made by putting the human genes into bacterial or other cultured cells, which can then serve as living factories for these products. Other human proteins with potential medical value that are being made in this way include the interferons and interleukins, both of which are being tested for cancer therapy.

The contributions of Charles Darwin and Gregor Mendel

Although most of the information that has made these developments possible has been garnered during the past 40 years, the

Fig. 1.1 Charles Darwin, the British naturalist whose extensive observations of the natural world provided the underpinnings for the theory of evolution.

story really began more than 120 years ago with the independent investigations of Charles Darwin (Fig. 1.1) and Gregor Mendel. The contributions of Darwin, whom George Gaylord Simpson credits as being the father of modern biology, received the most immediate recognition, even though that recognition was not always favourable.

Darwin was a naturalist who made extensive and systematic observations of the plant and animal worlds. Working from his own data and from those of others, Darwin came to the conclusion that species are not fixed and unchanging but are capable of evolving over time to produce new species. In addition, Darwin provided a possible explanation for how this evolution could occur.

He had noted that individual members of a given species show a great deal of variation and proposed that some of them would be more fit for the environment in which they found themselves than others. The more fit individuals would thus produce more offspring than those that were less fit. Eventually this process, which Darwin called natural selection, would cause a shift in the characteristics of the population as those traits that favoured survival and reproduction were maintained and propagated, while less favourable traits became less common or died out. In plant or animal breeding something similar happens, although the breeder, rather than nature, provides the selective force by choosing the traits to be maintained. Although scientists are still arguing today about whether evolution occurs by natural selection as Darwin postulated, they do not dispute that species do evolve.

Darwin published his observations and conclusions for the first time in 1859 in the book he entitled *On the Origin of Species by Means of Natural Selection, or the Preservation of Favoured Races in the Struggle for Life*. The book turned out to be a best-seller of its time, although it engendered controversy because of its implication that the human race had evolved from some other form of life.

At the very same time that Darwin was enjoying the acclaim (or notoriety) his work had elicited, the monk Gregor Mendel was labouring obscurely in the gardens of his monastery in Brno, a town in the province of Moravia that then belonged to Austria but is now part of Czechoslovakia. Mendel was laying the foundations of the science of modern genetics, even though it would be another 35 years before the significance of his work was appreciated by the scientific community.

Mendel's research uncovered the basic rules governing heredity. He did this while doing breeding experiments with the garden pea. In one set of experiments he crossed pea plants that normally produced only smooth, round seeds with plants that only produced wrinkled seeds. All of the progeny that resulted from this mating yielded smooth seeds; the wrinkled-seed character had disappeared from the plants of the first generation (Fig. 1.2).

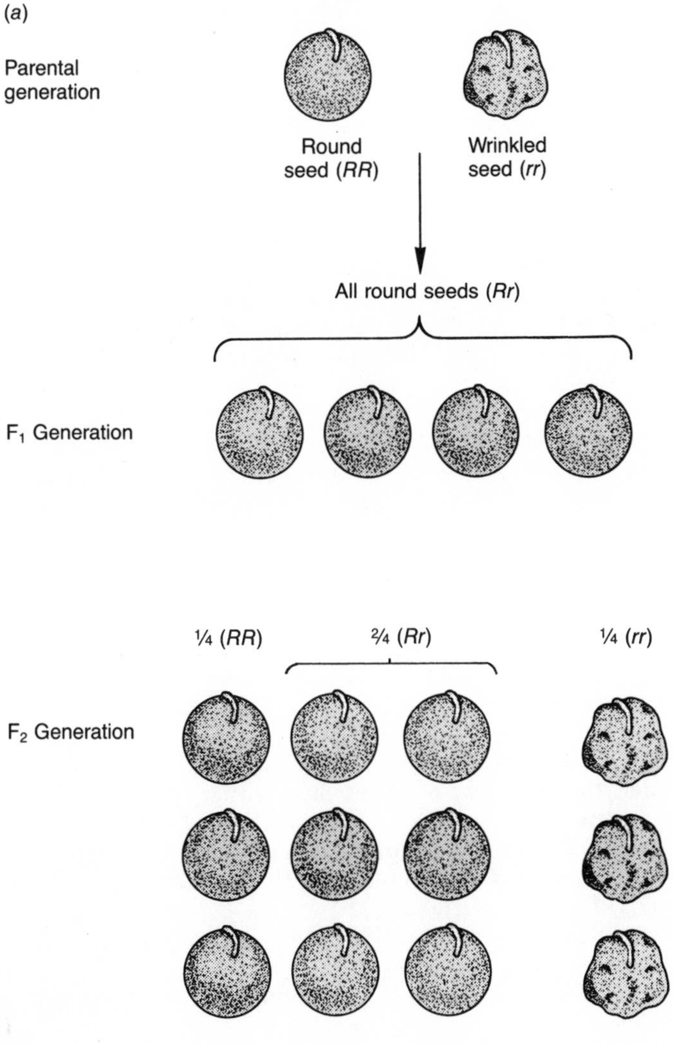

(a)

Parental generation

Round seed (RR)   Wrinkled seed (rr)

All round seeds (Rr)

F₁ Generation

¼ (RR)   ²⁄₄ (Rr)   ¼ (rr)

F₂ Generation

Fig. 1.2   Inheritance in peas. (a) When Mendel mated pea plants that produced only round seeds with plants that produced only wrinkled peas, all the plants in the first (F₁) generation produced round seeds because 'round' is dominant over 'wrinkled'. However, the F₁ pea plants retained the gene for the recessive wrinkled trait. When these plants were mated with one another, one-quarter of the second (F₂) generation plants produced wrinkled seeds.

(b) The behaviour of the genes encoding the dominant round and recessive wrinkled traits. The true-breeding parent plants have either two genes for round (RR) or two genes for wrinkled (rr). When these plants are mated, all the progeny will be Rr and therefore will produce only round seeds. However, when two Rr plants are mated, the genes will combine as indicated to give one RR and one rr plant for every two that are Rr. Three-quarters of these progeny plants will thus produce round seeds, and one-quarter will produce wrinkled seeds.

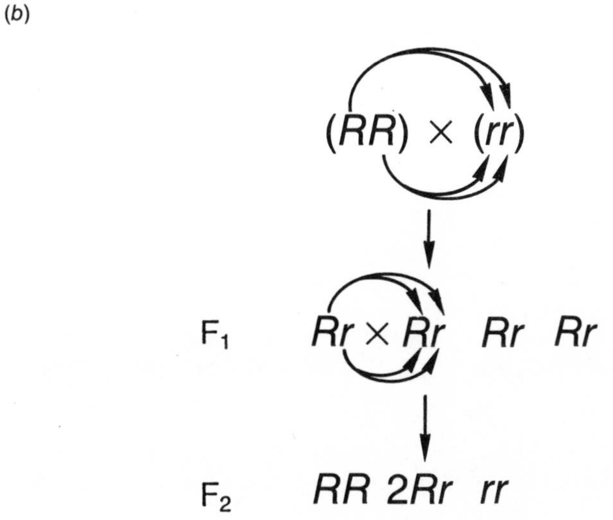

(b)

$$(RR) \times (rr)$$

F₁ $Rr \times Rr$ Rr Rr

F₂   RR 2Rr rr

Mendel then went on to mate the first-generation plants with one another. Three-quarters of the resulting second-generation plants still produced smooth seeds, but the seeds of the remaining quarter were again wrinkled. A trait that could not be seen in the first generation had reappeared in the second. Comparable breeding experiments showed that other characteristics, such as seed colour, behaved the same way,

Moreover, when Mendel followed the inheritance of two different characteristics simultaneously, he found that they could be inherited independently of one another. Mating plants that produced smooth, yellow seeds with those that yielded wrinkled, green seeds produced plants in the first generation that made only smooth, yellow seeds. However, plants of the second generation not only yielded peas with the two initial combinations, but also produced some peas with the novel combinations of wrinkled, yellow and smooth, green.

From these results Mendel deduced that hereditary traits are carried and transmitted to the progeny as discrete units. In essence he originated the concept of the 'gene', although this term was not used until the early 1900s. Mendel's experiments also led him to conclude that each individual carries two units for a given characteristic, but passes just one copy to each progeny. Furthermore, some variants of a particular trait are dominant over others. If they are inherited together the dominant character is expressed, while the other, recessive variant is not seen.

These postulates fully explained the inheritance patterns in the pea plants. For example, the original smooth-seeded individuals had to have two units specifying 'smooth' and the wrinkled-seeded individuals had to have two units for 'wrinkled'. The progeny of the first mating then inherited one of each. The finding that all the peas produced by these plants were smooth meant that this variant was dominant over the wrinkled form. Nevertheless, the unit specifying the wrinkled character was not lost. It was maintained and could again be seen in the progeny of the second mating when two hereditary units for 'wrinkled' were inherited together.

Mendel published his findings in 1865 in the *Journal of the Brno Society of Natural Science*, but they went largely unnoticed. Darwin, in particular, was unaware of them, even though they bore directly on his own research. The hereditary units described by Mendel are the raw material for the variation that is acted on by natural selection. Mendel's research was eventually rediscovered around the turn of the century by Hugo de Vries and Carl Correns who were doing similar breeding experiments. It was about this time that the basic unit of heredity became known as the gene and the science of heredity was given the name 'genetics'.

The identification of the genetic material as DNA

Although Mendel described the essential behaviour of genes, his experiments did not reveal the chemical nature of the hereditary units. That would not happen until the middle of this century. The probable location of the genes became apparent much earlier, however.

In the 1830s Theodor Schwann and Matthias Schleiden had independently recognized that all living creatures are composed of cells. The smallest and most primitive organisms, the bacteria, consist of only one cell. The most complex organisms, such as man, contain billions of cells.

Cells appeared to be simple, fluid-filled sacs when observed with the primitive micriscopes with which they were first discovered. But as microscopes became more powerful, scientists learned that cells have intricate internal structures (Fig. 1.3). The most prominent of these is the nucleus. The cells of all higher organisms, including plants, animals and even some very simple organisms such as yeast and amoeba, have nuclei. The typical plant or animal cell has a diameter of about 20 micrometres. (A micrometre is about 1/25 000 of an inch.) Nuclei are themselves about 5 micrometres in diameter. Organisms composed of nucleated cells are called 'eukaryotes'. The only cells that lack nuclei are those of the bacteria and certain primitive algae. A typical rod-shaped bacterium is about 0.5 micrometres in diameter and a few micrometres long, which makes it somewhat smaller than a nucleus. Organisms made of cells without nuclei are 'prokaryotes'.

Nuclei contain a material that is called chromatin because it becomes intensely coloured when cells are stained with appropriate dyes. The chromatin is organized in thread-like structures named chromosomes, the exact number of which is constant in all the members of a species. The chromosomes come in pairs in every eukaryotic cell except germ cells, which are the eggs and sperm. The cells of human beings have 23 chromosome pairs, or a total of 46 individual chromosomes, while the common laboratory organism, the fruit fly, has but four pairs.

When most cells divide, by a process called mitosis, the chromosomes are first duplicated and then partitioned so that each of the two daughter cells receives a complete set of paired chromosomes, all 46 in the case of the human. However, the cell divisions that produce the germ cells occur in such a way that each egg or sperm receives just one member of every chromosome pair, thereby halving the number of chromosomes. Then, when an egg and sperm fuse to form a new individual, the total chromosome number is restored.

In the early 1880s Wilhelm Roux postulated that this careful preservation of chromosome number might mean that they carried the hereditary material. Once Mendel's work was rediscovered it was easy to see how the behaviour of the chromosomes parallels the behaviour of Mendel's hereditary units. He had proposed that an organism carries two of the units for a given trait, and, correspondingly, ordinary body cells always contain two of each type of chromosome. Moreover, Mendel's theory holds that a parent transmits just one of the paired units for each trait to his or her progeny and germ cells contain one member of each chromosome pair.

Despite the identification of the chromosomes as the probable

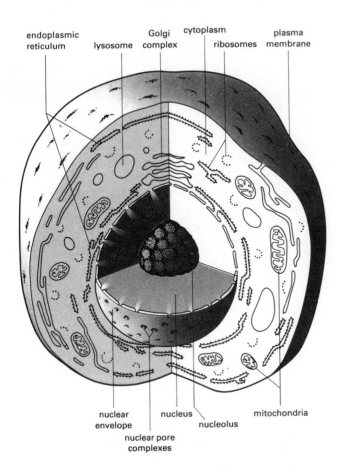

endoplasmic reticulum · lysosome · Golgi complex · cytoplasm · ribosomes · plasma membrane

nuclear envelope · nucleus · nucleolus · mitochondria

nuclear pore complexes

Fig. 1.3 Diagram of a 'typical' animal cell that shows its complex internal structure. The most prominent internal feature is the nucleus, which contains the genetic material. Also seen in the nucleus are one or more structures called nucleoli, where the ribosomal RNA is synthesized and combined with the ribosomal proteins. The nucleus is surrounded by a membranous envelope that has pores to allow the passage of large molecules between the nucleus and cytoplasm.

The ribosomes, which are the sites of protein synthesis, may be either free in the cytoplasm or attached to the membranes of the endoplasmic reticulum. Newly synthesized proteins move in vesicles from the endoplasmic reticulum to the Golgi complex, which appears in electron micrographs to be a stack of flattened membranous discs. Proteins in the Golgi complex undergo a series of maturation reactions, such as the addition of sugars.

Although some energy-producing reactions occur in the cytoplasm, the bulk of the cell's energy is derived from reactions in the mitochondria, which are sometimes called the 'powerhouses of the cell'. The lysosomes contain enzymes for breaking down cellular debris and large molecules, such as proteins, that have been taken into the cell. Surrounding the cell is the plasma membrane, which helps to maintain the normal internal environment by taking in needed nutrients and ions and letting out waste materials. Not shown in the diagram are the various protein filaments and tubules needed for cell movements and for maintaining cell shape.

location of the hereditary material, little or nothing was known about the chemical nature of genes and how the information they carry might be translated into an observable feature, such as pea seed colour.

The discovery of DNA

Chromosomes are composed mainly of proteins and nucleic acids, which were discovered in the 1860s by Friedrich Miescher who isolated the material from cell nuclei. Of the two types of biological substance, proteins seemed to be a much more likely candidate to be the genetic material, an idea that held until nearly 1950 – and even then died fairly hard.

Nucleic acids received very little consideration, primarily because they were thought to be too small and too simple in structure to encode the genetic information. A nucleic acid is built by joining together building blocks called nucleotides. A nucleotide consists of a sugar to which is attached a phosphate group and any of four bases. For the chromosomal nucleic acids the sugar is deoxyribose, hence the name deoxyribonucleic acid (DNA), and the bases are adenine, thymine, cytosine and guanosine (Fig. 1.4).

The size issue eventually resolved itself when molecular biologists realised that the apparent low molecular weights that had been found in early DNA samples had resulted because the material breaks down during isolation. Cellular DNA molecules are in fact very large, having molecular weights in the millions or even billions.

But even as this became clear, another misconception remained to militate against the possibility that genes might be made of DNA. Biochemists generally thought the large nucleic acid molecules were formed simply by repeating over and over again the same sequence of the four nucleotides. A molecule with such a monotonous, invariant structure could hardly carry the information needed to make a mouse or a man.

Proteins, in contrast, have more complicated structures. They are also large molecules, having molecular weights in the tens to hundreds of thousands, and are also constructed by linking together simple building blocks, in this case the amino acids. Proteins contain 20 different amino acids, in contrast to the four nucleotide building blocks of DNA, and the composition of different proteins could be seen to vary.

Moreover, proteins play a very dynamic role in the life of the cell. They are major structural constituents of the various parts of the cellular anatomy. But more than that, they are the enzymes, the biological catalysts that carry out the myriad chemical reactions by which cells obtain energy and grow.

In the mid-1940s researchers found that genes work by directing the synthesis of proteins. Among the main contributors to this development were George Beadle and Edward Tatum, whose research mainly involved simple organisms, such as fruit flies, moulds and bacteria.

Mutations are changes in genes that alter the characteristics of an organism in some detectable way. A mutation might

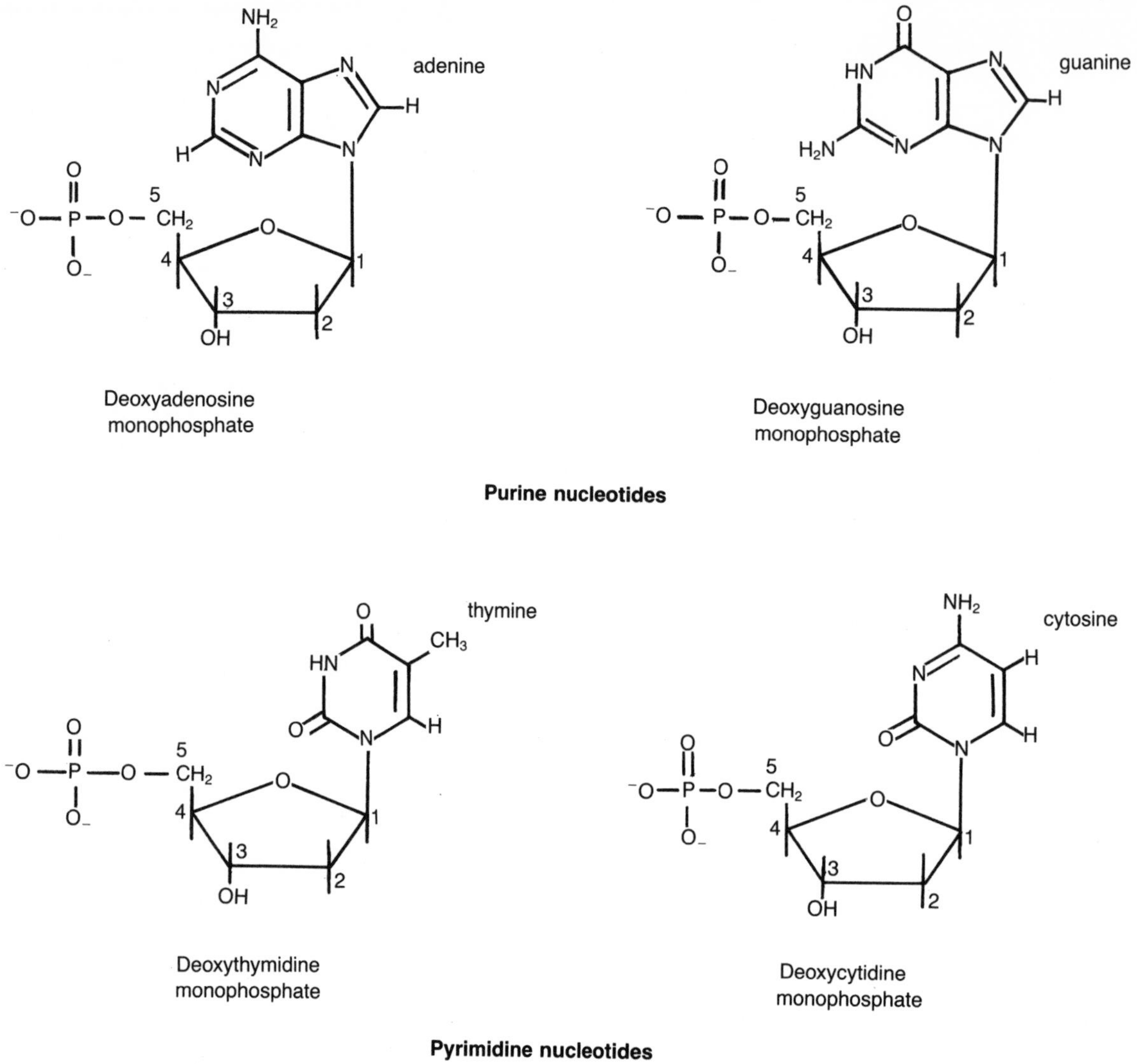

Purine nucleotides

Pyrimidine nucleotides

change the eye colour of a fruit fly, for example. Mutations can also change the nutritional requirements of an organism, perhaps making it dependent on a nutrient not needed by the parent organism. In any event, Beadle and Tatum showed that individual mutations in certain bacteria and fungi are correlated with the loss of particular enzymes by the test organisms.

Another indication that genes direct protein synthesis came from Linus Pauling, who identified the molecular defect that causes sickle cell anaemia, a human hereditary disease in which the red blood cells are very fragile and readily damaged. Red blood cells contain a protein, haemoglobin, that enables them to carry oxygen to the tissues. Pauling and his colleagues showed that the haemoglobin of sickle cell patients is structurally diffe-

Fig. 1.4 The nucleotide building blocks of DNA. Each nucleotide consists of a sugar component, here deoxyribose (shown in ring form with the carbon atoms numbered 1 through 5), that has a phosphate group (a phosphorus to which four oxygens are attached) on carbon 5, and a nitrogen-containing base on carbon 1. The four bases found in DNA include two purines (adenine and guanine) and two pyrimidines (thymine and cytosine). The nucleotide building blocks of ribonucleic acids have similar structures, except that the sugar is ribose, which has a hydroxyl group (–OH) on carbon 2, and thymine is replaced by uracil, another pyrimidine base.

rent from the haemoglobin of normal individuals. In other words, the mutant gene that causes sickle cell anaemia directs the synthesis of an altered protein.

These results gave rise to an interesting philosophical question: If genes direct protein synthesis, can they be proteins themselves? By the mid-1940s, however, research began to point away from proteins as the genetic material – and towards DNA. One of the most convincing pieces of evidence came from Oswald Avery, Colin MacLeod and Maclyn McCarty, who were researchers at the Rockefeller Institute in New York City. (The Institute is now Rockefeller University.)

The groundwork for their research had been laid in the 1920s by Fred Griffith's studies of the bacterium *Streptococcus*

pneumoniae, which causes pneumonia in humans and is quickly lethal in mice. One form of this bacterium, which is highly virulent, produces rough-surfaced colonies when it is grown in culture dishes. A second form, which does not cause pneumonia, produces smooth-surfaced colonies. Griffith had found that mice that were injected with killed, rough-colony bacteria together with living, smooth-colony bacteria got sick, even though the bacteria did not produce illness when injected separately. The conclusion was that some component of the rough-colony bacteria had been transferred to the smooth-col-

Fig. 1.5 The structure of DNA. DNA molecules are built by linking the phosphate on one nucleotide to the oxygen on carbon 3 of the deoxyribose of the next nucleotide in the chain (*a*). This produces a molecule that consists of a backbone of alternating phosphate and deoxyribose groups with the bases protruding. Although the diagram shows a DNA that contains only four nucleotides, the DNA chains of genes can contain hundreds. Molecules of RNA are built in a similar fashion.

In a complete DNA molecule, two chains (one in red and the other in blue) are joined to form a double helix, the common B-form of which is shown in (*b*). The deoxyribose–phosphate backbone of each chain is on the outside and the bases are on the inside of the double helix. The bases of one chain are paired with those of the other according to Chargaff's rules: adenines always pair with thymines and guanines with cytosines. In this view of the DNA, most of the bases are seen edge-on.

(*a*)

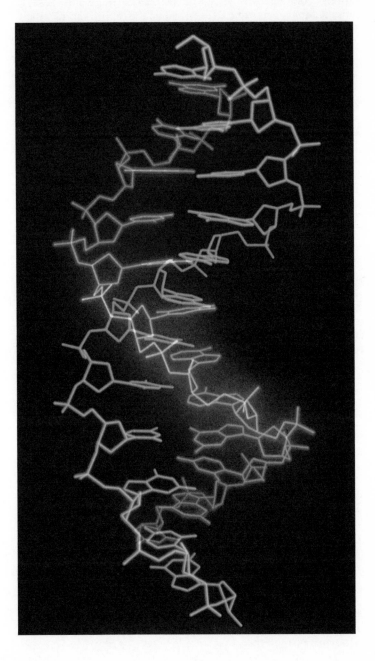

ony bacteria, thus making them virulent. Avery, MacLeod and McCarty demonstrated that the component was DNA. This discovery, that DNA could carry genetic information between bacterial strains, became known as 'Avery's bombshell' because of the explosive impact it had on the then embryonic science of molecular biology.

Then, in the early 1950s, Erwin Chargaff noted some regularities in the base compositions of DNAs from different species that were to provide an essential clue to the structure of the material. In particular, the percentage of adenine always equalled that of thymine and the percentage of guanine always equalled that of cytosine. Nevertheless, the ratio of the adenine–thymine pair to the guanine–cytosine pair varied widely from one species to another.

This latter observation disproved the hypothesis that DNA consists of a monotonous four-nucleotide repeating unit. If that were true, then the four bases adenine, cytosine, guanine and thymine should be present in equal amounts, but this was clearly not what Chargaff had found. The research suggested that DNA's structure might after all have the variety required for it to be the repository of genetic information.

Also in the early 1950s James D. Watson and Francis Crick, who were then working together at the Medical Research Council's Laboratory of Molecular Biology in Cambridge, England, were attempting to solve the three-dimensional structure of the DNA molecule by X-ray crystallographic methods.

In X-ray crystallography a beam of X-rays is shone on a pure crystal of the material being studied. The crystal diffracts or scatters the X-rays in a manner that depends on the three-dimensional structure of the molecule. The pattern of diffracted X-rays can be recorded on a photographic film and from it the structure can be reconstructed by complex mathematical means.

Watson and Crick made a false start or two in their efforts to deduce the DNA structure before they were greatly aided by an X-ray diffraction photograph made by Rosalind Franklin in Maurice Wilkins' laboratory at Cambridge. Using information contained in the diffraction photograph and also Chargaff's rules, Watson and Crick concluded that the DNA molecule consists of two individual chains of nucleotides that are twisted together in a double helix (Fig. 1.5). Each chain consists of a backbone in which the phosphates and deoxyriboses of the nucleotides alternate. The bases protrude from the backbone.

The backbones of the two chains are on the outside of the Watson–Crick double helix with the bases on the inside. The bases of one chain form weak hydrogen bonds with the bases of the other in a very specific fashion. In accordance with Chargaff's rules, adenines always bond to thymines and cytosines to guanines.

The molecular design of the DNA double helix provides a means for maintaining the genetic information and transmitting it to future generations. The molecules are large and can encode a great deal of information in their nucleotide sequences. Moreover, the nucleotide sequence of one chain specifies the sequence of the other.

Before a cell divides, its hereditary material must be duplicated so that each daughter cell inherits an exact and complete copy. To duplicate DNA then, the two strands separate and each one serves as a template or pattern for synthesizing the other (Fig. 1.6). In accordance with the base-pairing rules, adenines match with thymines and guanines with cytosines so that two double-stranded molecules, both of which are replicas of the parent DNA structure, are produced. Each daughter cell inherits one of the two copies.

At first glance there might seem to be a major problem with the idea that DNA is the genetic material. DNA has only four building blocks, but must direct the synthesis of proteins, which are constructed from 20 different amino acids. This is not really a problem, however, because it takes a combination of three nucleotides to specify a given amino acid. Four nuc-

Table 1.1. *The genetic code*

First position	Second position				Third position
	U	C	A	G	
U	Phe	Ser	Tyr	Cys	U
	Phe	Ser	Tyr	Cys	C
	Leu	Ser	Stop	Stop	A
	Leu	Ser	Stop	Trp	G
C	Leu	Pro	His	Arg	U
	Leu	Pro	His	Arg	C
	Leu	Pro	Gln	Arg	A
	Leu	Pro	Gln	Arg	G
A	Ile	Thr	Asn	Ser	U
	Ile	Thr	Asn	Ser	C
	Ile	Thr	Lys	Arg	A
	Met	Thr	Lys	Arg	G
G	Val	Ala	Asp	Gly	U
	Val	Ala	Asp	Gly	C
	Val	Ala	Glu	Gly	A
	Val (Met)	Ala	Glu	Gly	G

A sequence of three nucleotides is required to form the nucleic acid codon for a single amino acid. Four nucleotides can produce 64 different three-nucleotide combinations and all are used in the genetic code. Consequently, all the amino acids except methionine (Met) and tryptophan (Try) have more than one corresponding codon, although GUG, which usually specifies valine (Val) may under some circumstances specify methionine instead. The 'stop' codons UAA, UAG and UGA do not code for amino acids but signal the end of a protein.

The codons are given as they appear in messenger RNA. The four bases in the nucleotides of ribonucleic acids are uracil (U), cytosine (C), adenosine (A) and guanine (G). The amino acids specified by the genetic code are alanine (Ala), arginine (Arg), asparagine (Asn), aspartic acid (Asp), cysteine (Cys), glycine (Gly), glutamine (Gln), glutamic acid (Glu), histidine (His), isoleucine (Ile), leucine (Leu), lysine (Lys), methionine (Met), phenylalanine (Phe), proline (Pro), serine (Ser), threonine (Thr), tryptophan (Trp), tyrosine (Tyr) and valine (Val).

leotides can give 64 combinations when they are grouped in threes, more than enough to specify 20 amino acids.

During the 1960s the genetic code was cracked, largely through the effort of Har Kobind Khorana, Marshall Nirenberg and Severo Ochoa, who found ways of determining which three-nucleotide sequence, or codon as the sequences came to be called, corresponded to which amino acid. Most amino acids proved to be represented by more than one codon, and all but three of the total of 64 codons were assigned to one amino acid or another (Table 1.1). Consequently, if the nucleotide sequence of a gene is known, the amino acid sequence of the corresponding protein can be deduced unambiguously, but the reverse is not true. The amino acid leucine, for example, could be represented in a gene by any of six different codons.

The three codons that do not specify amino acids act as stop signals. They mark the end of the genes, just as a full stop marks the end of a sentence. A gene is thus a linear sequence of three-nucleotide codons, each of which specifies a particular amino acid or, at the gene's end, a stop signal.

Fig. 1.6 Replication of DNA. When DNA replicates, the two strands of the molecule separate and each serves as a template for the synthesis of a new complementary strand, here shown in grey. The obligatory pairing of adenines with thymines and guanines with cytosines ensures that the two daughter molecules will be duplicates of the original.

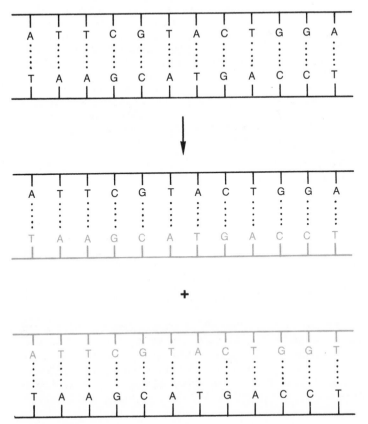

How gene structure is translated into protein structure

Cells have a fairly elaborate machinery for translating the nucleotide sequences of genes into protein structures. The machinery is needed because amino acids do not have a chemical affinity for nucleotides that would enable amino acids to recognise their own codons and line up in the order specified by a gene. This problem is overcome with the help of the ribonucleic acids (RNAs).

RNA structure follows the same overall plan as that of DNA, but there are some significant differences between the two kinds of nucleic acid. RNAs are also large molecules formed by linking together nucleotide building blocks, except that the RNAs contain the sugar ribose and not deoxyribose. In addition, RNAs do not contain thymine, but have the base uracil where DNA would have thymine. Finally, RNA molecules are usually single-stranded, rather than double-stranded, although in some RNAs the single strand folds back on itself, thus producing looped structures.

No less than three different kinds of RNA participate in protein synthesis. When a gene becomes active the first step towards producing the protein it encodes is the copying of the DNA into a molecule of 'messenger RNA' which is so called because it carries the message specifying the protein structure to the part of the cell where the synthesis takes place (Fig. 1.7). In bacteria as well as in nucleated cells, this occurs on the small particles known as ribosomes.

Ribosomes are composed of proteins plus the second type of RNA that participates in protein synthesis (ribosomal RNA). They are located in the cell cytoplasm, where they may either be free or attached to the membranes of the endoplasmic reticulum. This means that in nucleated cells the messenger RNA has to make its way out of the nucleus and into the cytoplasm.

When a messenger RNA is being synthesized the two strands of the double-helical DNA of the gene must separate partially to expose a portion of the nucleotide sequence. The ribose nucleotides that are to be joined to form the messenger RNA line up along the gene sequence according to the familiar rules of base-pairing, except that each adenine in DNA will be paired with a uracil in the RNA being synthesized, rather than with a thymine. As the ribose nucleotides line up in the proper order they are joined by an enzyme called RNA polymerase that moves along the gene until all of it has been copied into a messenger RNA.

The synthesis of messenger RNA is known as transcription. The next step, the actual protein formation, is called translation. At this point, transfer RNA, the third type of RNA that participates in protein synthesis, comes into play. Transfer RNAs serve as adaptors for lining up the amino acids to be joined in the order specified by the gene and its messenger RNA. A transfer RNA recognises the corresponding codon on a messenger RNA by means of a complementary three-base

(a)

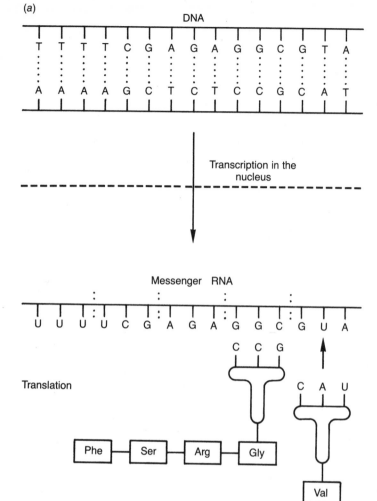

(b)

Fig. 1.7(a) and (b)   Protein synthesis. Protein synthesis requires that the two strands in the DNA molecule separate to allow the copying of one of the strands into messenger RNA. This process, which is known as transcription, takes place in the nucleus where the genes are located. The messenger RNA molecule then moves out of the nucleus into the cytoplasm where protein synthesis occurs.

Meanwhile specific enzymes join each of the amino acids to the appropriate transfer RNA. The messenger RNA and the transfer RNA for the first amino acid of the protein form a complex with the ribosome. Then the ribosome effectively moves along the messenger RNA molecule. As it moves, the 'anticodon' on the transfer RNA for each successive amino acid recognizes the corresponding codon on the messenger RNA and brings the amino acid into position for joining to the growing protein chain. In this way the information encoded in the linear sequence of nucleotides in the DNA of the genes is translated into the linear sequence of amino acids in a protein.

sequence (an 'anticodon') that binds to the codon according to the base-pairing rules.

Every amino acid has at least one transfer RNA and some have more. The amino acids become attached to their transfer RNAs, which then line up on a messenger RNA, with the anticodons recognising and base-pairing with the appropriate codon. Finally, the amino acids are joined enzymatically and synthesis of the protein is completed.

This general picture of gene action was largely complete by the end of the 1960s. Essentially all the work had been done with bacterial genes, however, and towards the end of the 1970s molecular biologists made a surprising discovery about the genes of eukaryotic organisms.

The protein-coding sequences of bacterial genes extend continuously from the beginning of the gene to the end, but the protein-coding segments of almost all the genes of the higher eukaryotes are interrupted by nucleotide sequences that do not code for protein structure. The non-coding regions, which are

called intervening sequences or 'introns' are transcribed into the messenger RNAs for the genes, but are then spliced out before the messenger is translated into protein.

The benefits to the eukaryotic organism of this more complex type of gene organisation are not completely understood. One possibility, which has received some experimental verification, is that genes were assembled during the course of evolution by combining separate DNA sequences that code for protein segments with different functional capabilities. The introns may thus be the relics of the process that brought the different functional segments together.

The early work on gene structure and protein synthesis did little to address the question of how genes are controlled, especially in the cells of higher organisms. Molecular biologists have begun to get to grips with that problem only within the past few years – and this too is part of the revolution that began in the early 1970s with the advent of recombinant DNA technology.

Recombinant DNA

The developments that were to culminate in recombinant DNA technology had their origins early in the 1970s. Recombinant DNA is made by joining, or 'recombining' in molecular biology jargon, DNAs from different sources. The ability to do this grew out of research on the restriction enzymes, which are produced in bacteria as part of the bacterial defences against invading foreign DNAs, such as those of viruses. Restriction enzymes destroy the foreign DNA by cutting it into pieces. Meanwhile, the DNA of the host bacterium is protected by a chemical modification that prevents the enzymes from attacking it.

Restriction enzymes have two properties that make them extremely useful to molecular biologists. The first of these is their specificity. Each enzyme recognizes and cuts only one particular nucleotide sequence in DNA. A few hundred restriction enzymes with different nucleotide specificities have been identified. The enzymes can be used for, among other things, characterising genes.

The structure of two genes can be compared by digesting them with a series of several restriction enzymes. If the genes are identical they will produce exactly the same pattern of fragments with all of the enzymes. If their nucleotide sequences vary, however, they will produce different patterns of restriction fragments. The specificity of restriction enzymes is already being applied as the basis of new techniques for diagnosing hereditary diseases, such as sickle cell anaemia, that are caused by structural changes in genes (see Chapter 14).

The second property that has made restriction enzymes so valuable is the ability of many to produce staggered cuts when they cleave double-stranded DNA. The fragments thus produced are double-stranded but have single-stranded 'sticky ends'. Protruding from the two ends of the fragments are single-stranded tails that have base sequences that can recognize and base-pair with one another (Fig. 1.8). For example, the fragments generated by the enzyme designated EcoRI have the single-stranded sequences 'AATT' on one end and 'TTAA' on the other. (A represents adenine and T represents thymine. When the fragments are brought together under appropriate conditions, their sticky ends will reassociate and then the fragments can be resealed by means of an enzyme (called a ligase) that will join the strands. Even fragments of DNA from two different sources can be joined in this way provided they were both produced originally by the same restriction enzyme.

Paul Berg's group at Stanford University School of Medicine, Palo Alto, California, was the first to join foreign DNAs in a recombinant molecule. They joined genes from the bacterium *Escherichia coli* with DNA from simian virus 40 (SV40). At the time investigators had not yet realised that some restriction enzymes, including the one that they were using, generate sticky ends and the Stanford workers made the ends themselves by adding appropriate bases to the restriction fragments. Letting restriction enzymes do that job is easier.

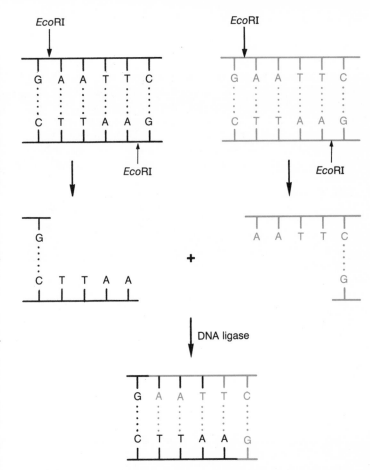

Fig. 1.8   The generation of 'sticky ends'. Restriction enzymes cut at specific nucleotide sequences, often making staggered cuts that produce single-stranded 'sticky ends'. The widely used restriction enzyme *Eco*RI, for example, cuts the sequence shown at the sites marked by the arrows. The single-stranded end of one fragment can therefore recognize and bind to the end of any other fragment produced by the same enzyme, even if the fragments originally came from the DNAs of different species. The two fragments can then be joined by the action of an enzyme called DNA ligase. This procedure forms the basis of recombinant DNA technology.

The ability to make recombinant DNA molecules opened the way to isolating and producing essentially unlimited quantities of a desired gene. Foreign DNAs could be inserted into viral DNAs, as Berg and his colleagues had shown, that could serve as vehicles for introducing the foreign genes into bacteria or other cultured cells. The recombinant DNAs will then reproduce in the cells, thus generating large quantities of the foreign genes.

Plasmids can also be used as vectors for reproduction of foreign genes (Fig. 1.9). Plasmids are circular, double-stranded DNA molecules. They occur naturally in bacteria where they reproduce independently of the bacterial chromosome.

The first plasmid to be developed as a vehicle for introducing foreign DNA into bacteria was pSC101, which was discovered in Stanley Cohen's laboratory at Stanford University School of Medicine. Cohen and Annie Chung, also of Stanford, with Herbert Boyer and Robert Helling of the University of California in San Francisco, inserted a foreign gene into the plasmid and demonstrated that it would still reproduce normally in bacteria.

The reproduction of foreign genes in bacterial or other cells is called gene cloning. A 'clone' refers to a group of entities, whether these be plants, bacterial cells, or as in this case genes, that are all exact copies of one individual.

As a result of the development of recombinant DNA and gene cloning technology, virtually unlimited quantities of specific genes can be produced. The most difficult problem is identifying those cells that make the desired foreign gene, but molecular biologists have devised several strategies for doing this, which are discussed more fully in Chapter 4.

The combination of gene cloning technology with recent advances in methods for analyzing the DNA thus obtained, has made it much easier to determine the nucleotide sequences of genes than the amino acid sequences of proteins. This is especially true for those proteins that are present in cells in low concentrations and would be difficult to separate from contaminating proteins. More often than not, it is the gene cloners who provide the first complete amino acid sequences of important proteins. Once a cloned gene is in hand, the nucleotide sequence, and thus the corresponding amino acid sequence, can be obtained in a matter of weeks.

This relative ease of obtaining gene sequences represents a marked change from the situation in as recent a time as 1970. The sequence of genes, especially those of higher organisms, were then largely thought to be unobtainable. Protein sequences were considered difficult to determine, but possible if given enough time and pure protein sample with which to work. The problem was that relatively large quantities of a protein were needed to obtain a complete amino acid sequence, and many proteins of interest are synthesized in very small quantities. Even today, however, direct protein sequencing still plays an important role in molecular biology. The advent of new auto-mated methods for protein sequence analysis, such as that developed by Michael Hunkapillar and Leroy Hood of the California Institute of Technology in Pasadena, has made it possible to determine at least partial amino acid sequences from very small samples of proteins. Only micrograms or less of material are required. (A microgram is one-millionth of a gram.)

A partial amino acid sequence is often the key to cloning the genes of rare proteins. Probes, which consist of nucleotide sequences corresponding to the amino acid sequence, can be constructed and used to detect the desired cloned gene. In addition, the peptides themselves can be synthesized and used for making antibodies that will react with the intact protein (see Chapter 13). Such antibodies have a variety of applications, including detecting the genetically engineered cells that contain the cloned gene for the protein that the antibody recognizes.

The control of gene expression

Recombinant DNA technology has also greatly facilitated the solution of a major problem concerning protein synthesis that could not be readily addressed at the time the general plan for how genes function was worked out. This problem concerns the question of what causes a gene to be active or inactive in directing the synthesis of the protein it encodes.

The supposition was that at least some of the information needed for regulating gene expression would be encoded in the DNA itself – and that supposition has been largely borne out. Molecular biologists are identifying certain DNA sequences that participate in gene control. The researchers have done this by systematically removing DNA segments from within and around the genes under study and then introducing the altered genes into cells to see how the changes affected the gene activity. The regulatory sequences thus identified include both those that increase gene expression, which include promoters and enhancers, and those that decrease gene activity.

This knowledge is essential for many of the biotechnological applications of recombinant DNA technology. Good expression of a transferred gene in appropriate cells or tissues is a prerequis-ite for many of the applications. This is especially true for the proposed attempts to cure human genetic diseases by gene therapy, but also applies to efforts aimed at improving crop plants by introducing new genes and to the production of hor-mones or other commercially important proteins in foreign cells.

Although some regulatory sequences may act in a broad spectrum of species or cells others may work only in certain cell types. The enhancers of mammalian antibody genes, for example, are active in antibody-producing cells but not in other cell types. Attaching a gene that is to be expressed in a foreign cell to regulatory sequences that work efficiently in that type of cell can then be crucial to the success of a biotechnological application.

Restrictions on recombinant DNA technology

Although the contributions of recombinant DNA technology and gene cloning to basic molecular biology and the developing field of biotechnology are hard to overstate, the research has not been without controversy. The ability to combine DNAs from diverse organisms and to put these recombinant molecules into new species raised concerns that the experiments might inadvertently create virulent new pathogens that might escape

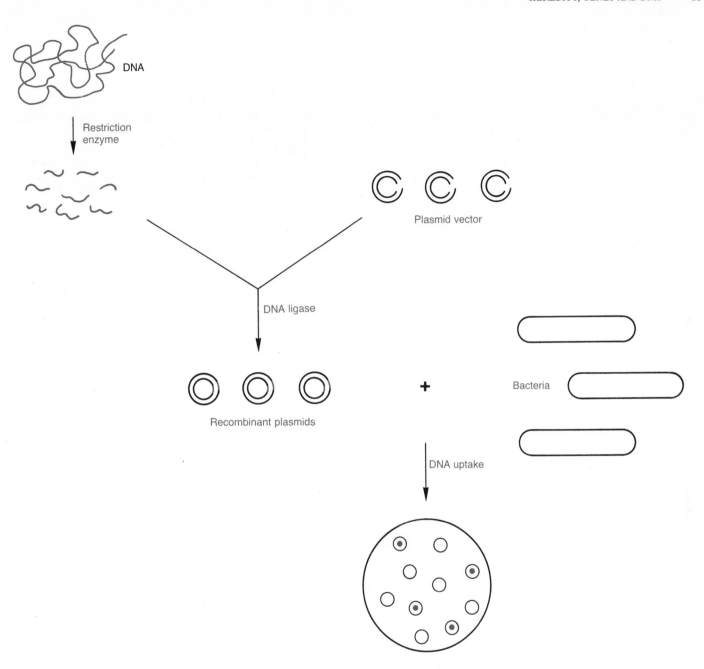

Fig. 1.9 Gene cloning. DNA from the desired source is cut with a restriction enzyme such as *Eco*RI and the fragments are inserted by recombinant DNA methods into a plasmid or other cloning vector that will reproduce in suitable host cells. Bacterial cells have primarily been used in the past, but the cells of higher organisms are coming into greater application. After the recombinant plasmids are introduced into appropriate hosts, the cells are grown in culture so that each colony formed will be a clone consisting of progeny of a single cell.

The cloning vector usually contains a gene coding for some selectable trait such as antibiotic resistance. If this is done, only those cells that acquire the recombinant plasmid will grow in the presence of the antibiotic. The biggest problem with gene cloning is identifying the clones that carry the gene the investigator is seeking to isolate. This can be done in a variety of ways. The protein product may be detected by its activity, if it has a measurable activity, or it may be detected by a specific monoclonal antibody. Once the gene is isolated, essentially unlimited amounts of it, or its product, can be made in the host cells.

from the laboratory. The concern was particularly acute because *E. coli*, the bacterium that was and still is used for many experiments, is a normal inhabitant of the human intestinal tract.

The scientists doing the work were the first to be aware of its potential hazards and they took an unprecedented step. Meeting in 1974 under the aegis of the National Academy of Sciences of the USA they recommended that certain types of potentially hazardous recombinant DNA experiments not be done until an international meeting could be convened to draw up guidelines that would allow the research to proceed safely. The banned experiments included those in which genes for toxins or antibiotic resistance would be put into organisms that did not already have them.

The voluntary moratorium held until February 1975, when the international meeting was held at the Asilomar Conference Center in Pacific Grove, California. The participants recommended that recombinant DNA research be conducted under increasing degrees of physical or biological containment according to the severity of the risks that they might entail. Physical containment means the use of laboratory facilities that will minimize or, at the highest levels prevent, the escape of microorganisms. Biological containment refers to the use of organisms or vectors that have been altered so that they can only survive under specialised laboratory conditions.

Each nation represented at the meeting was to draw up its own guidelines for ranking the hazards of recombinant DNA experiments and defining the conditions under which they could be performed. In the United States this task fell to a committee convened by the National Institutes of Health in Bethesda, Maryland, which at the time supported most of the research in question. The NIH guidelines became available in June 1976. In the intervening years they have been greatly relaxed as increasing experience has failed to demonstrate any hazards of recombinant DNA research in the laboratory.

The NIH guidelines only applied in the first place to research supported by the institutes. The growth of the biotechnology industry has meant that an increasing amount of recombinant DNA work has not been subject to the guidelines, although the companies doing the work have generally followed them anyway. Current concerns centre largely on the safety of deliberately releasing into the environment the genetically altered organisms, including plants and microorganisms, that industry is beginning to produce for agricultural applications. The proposals to treat human genetic diseases by introducing good copies of the affected gene into the patients' cells have also raised concerns. The regulatory and ethical implications of the new biotechnology are considered in Chapters 8 and 16 of this book.

Additional reading

Alberts, B., Bray, D., Lewis, J., Raff, M., Roberts, K. and Watson, J. D. (1983). *Molecular Biology of the Cell*. Garland Publishing, New York and London.

Darnell, J. E. (1985). RNA. *Scientific American*, **253** (October), 68.

Doolittle, R. F. (1985). Proteins. *Scientific American*, **253** (October), 88.

Felsenfeld, G. (1985). DNA. *Scientific American*, **253** (October), 58.

Stent, G. (1971). *Molecular Genetics: An Introductory Narrative*. W. H. Freeman, San Francisco.

Wade, N. (1977). *The Ultimate Experiment: Man-Made Evolution*, Walker and Co., New York.

Weinberg, R. A. (1985). The molecules of life. *Scientific American*, **253** (October), 48.

2 Synthesis without cells

Ethan S. Simon, Alan Akiyama and George Whitesides

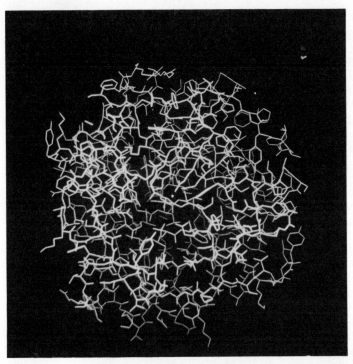

Fig. 2.2

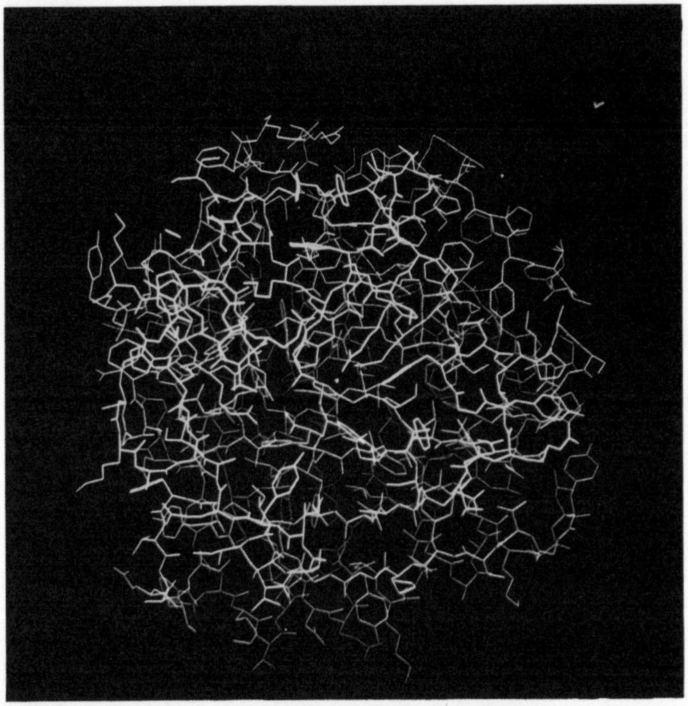

Fig. 2.3

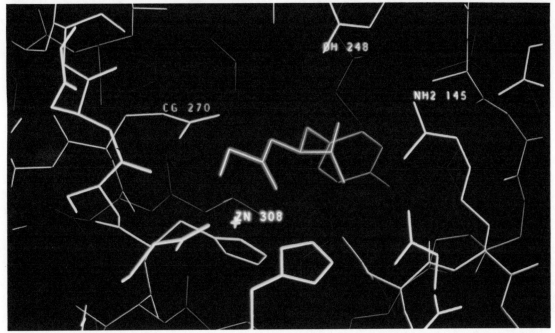

Fig. 2.4

Most of the chemical reactions that occur in living cells would take place too slowly to support life it it were not for the existence of the biological catalysts known as enzymes. These molecules accelerate the rates of the reactions, often by many orders of magnitude, thereby allowing cells to obtain the energy they need, synthesize the molecules of which they are composed, and break down and eliminate their waste products. For example, the first step of glycolysis, the metabolic pathway that converts the sugar glucose into chemical energy, is catalysed by the enzyme hexokinase. Hexokinase accelerates the reaction, the conversion of glucose to glucose-6-phosphate, by a factor of 1 000 000 000.

Enzymes not only make essential contributions to cellular activities, but also find wide application in biotechnology. They are used in the food industry for making cheese, beer, wine and sweeteners, among other things, and in the chemical and pharmaceutical industries for synthesizing amino acids and antibiotics. In addition, they have a small but growing role in medicine. One therapeutic application of increasing importance is the use of the enzyme called tissue plasminogen activator to dissolve blood clots in heart patients.

The biotechnological applications of enzymes depend on their unique properties. Enzymes are proteins that have molecular weights ranging from 10 000 to 500 000. For comparison, water has a molecular weight of 18, and common table salt (sodium chloride) has a molecular weight of 59. The large size of enzymes allows them to adopt well-defined three-dimensional shapes, stabilized by interactions between their constituent amino acids.

These shapes are critical to one of their most important characteristics – the ability to recognize and act specifically on one or a few substances, which are called substrates, even if the substrates are present in a complex mixture. Hexokinase has recognition sites for glucose and adenosine triphosphate (ATP), which bind to the enzyme in close proximity to one another (Fig. 2.1). After binding, a phosphate group is transferred from ATP to the glucose, thereby producing adenosine diphosphate (ADP) and glucose-6-phosphate. The enzyme then releases the two products and is able to catalyse another cycle of reaction between glucose and ATP.

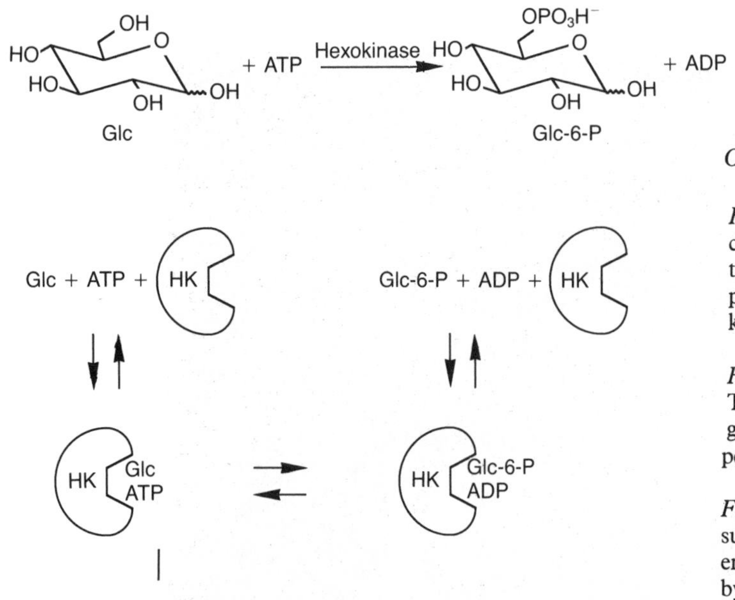

Fig. 2.1 The enzyme hexokinase (HK) accelerates the reaction of glucose (Glc) and adenosine triphosphate (ATP) to form glucose-6-phosphate (Glc-6-P) and adenosine diphosphate (ADP). The hexokinase itself is not consumed in the reaction.

Captions to previous page

Fig. 2.2 Computer-generated model of the active site of the enzyme carboxypeptidase A. The active site, which is the region of the protein to which the substrate binds and where the catalytic activity takes place, forms a small cleft in the surface of the enzyme protein. (By kind permission of David W. Christianson, Harvard University.)

Fig. 2.3 Substrate binding to the active site of carboxypeptidase A. The substrate, here a derivative of benzylpropionic acid (shown in green), fits nearly into the cleft in the enzyme surface. (By kind permission of David W. Christianson, Harvard University.)

Fig. 2.4 Close-up of the active site of carboxypeptidase A. The substrate here is the dipeptide glycyltyrosine. The amino acids of the enzyme that are important for binding are denoted in red. As indicated by the numbers designating their positions in the protein chain, they may be located far apart in the linear sequence of the protein, but are brought close together when the protein folds into its three-dimensional structure. (By kind permission of David W. Christianson, Harvard University.)

A substrate fits into a specific recognition site on an enzyme much as a key fits into a lock (Figs. 2.2 and 2.3), although enzymes are more dynamic than locks in that the shape of the enzyme may change somewhat in response to substrate binding. Only a few of the amino acids of the enzyme are actually in contact with the bound substrate (Fig. 2.4). These amino acids may be widely separated from one another along the linear sequence of the enzyme protein. If the three-dimensional structure that holds them in the correct alignment for substrate binding is disrupted, the enzyme may lose its activity.

The high specificity of enzymes in substrate recognition and catalytic action is the basis for many of their applications. One important type of selectivity that is exhibited by many enzymes is the ability to discriminate between enantiomeric, or mirror-image, molecules (Fig. 2.5). Enantiomers are stereoisomers that have the same chemical composition, but have the atoms arranged differently in space. Enantiomers can be formed by any molecule that has an asymmetric carbon atom; that is, a carbon with four different groups attached to it. Enantiomers are indistinguishable in most of their physical properties and are therefore difficult to separate.

However, enzymes can usually discriminate between them. Exposure of a mixture of the enantiomers *N*-acetyl-L-alanine and *N*-acetyl-D-alanine to the enzyme acylase I results in removal of the acetyl group from the L-enantiomer only. The reaction releases L-alanine, which is the naturally occurring enantiomer of this amino acid. The free amino acid can be easily separated from the unchanged *N*-acetyl-D-alanine. Enantioselective enzymatic hydrolysis thus provides a convenient and widely used technique for separating enantiomers.

The importance of such separation techniques is very great. Enantiomers often behave very differently in biological systems. One may be a useful drug, whereas the other may be highly toxic. The drug thalidomide provides a case in point. The pure R(+) enantiomer is a relatively safe tranquilizer. The serious birth defects that were produced when thalidomide was given to pregnant women resulted from the very small quantities of the S(-) enantiomer that were present as an impurity. Increasing sophistication in drug synthesis and increasing pressure from health regulatory agencies now provide important stimuli for the development of methods for making pharmaceutical agents that are free of contamination by unwanted enantiomers.

Enzyme catalysts have several other important characteristics in addition to their substrate specificity. They are subject to regulation; that is other molecules can modulate their activities – a feature that is especially important for controlling biochemical pathways in the cell. Moreover, enzymes work under mild conditions. Most operate best in water at a neutral pH and at room temperature. For industrial applications, these conditions are attractive with regard to both saving energy and preserving the environment. Accomplishing comparable chemical reactions without enzymes often requires harsh conditions, including high temperatures and very acidic or alkaline pHs. Perhaps most important from an industrial point of view, enzymes can

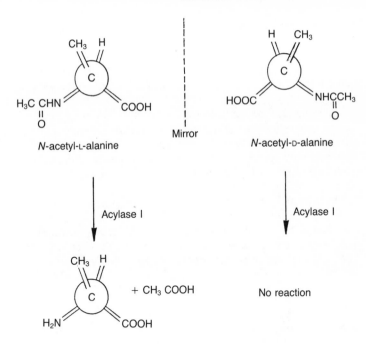

Fig. 2.5 The amino acid alanine has an asymmetric centre. It exists in two forms that are identical except that they are mirror-image isomers of one another. Mirror-image structures such as these are called enantiomers. The L-enantiomer of alanine is the one that occurs in proteins and most other naturally occurring alanine derivatives. D-alanine rarely occurs in nature.

carry out reactions unachievable by any other method. In particular, they may be used for the synthesis of complex biological molecules that cannot be made by other means.

A major disadvantage of enzyme catalysts is that they are relatively fragile in comparison with non-biological catalysts such as platinum or sulphuric acid. Enzymes are rapidly deactivated by temperatures above room temperature, which disrupt their three-dimensional structures. They may also lose activity if exposed to air, to organic solvents, or to acidic or basic conditions. The fragility of enzymes is particularly disadvantageous because they remain expensive relative to most non-enzymatic catalysts, even though methods for producing enzymes are improving.

Nevertheless, this handicap may be overcome by the development of methods for enzyme stabilization. Providing an environment that prolongs the life of an enzyme is a matter of trial and error, but enzyme immobilization is one approach that often works and, at the same time, lowers the cost of using enzymes.

An enzyme can be immobilized by fixing it to, or enclosing it in, a solid support, a procedure which offers a number of practical advantages over using the enzyme in soluble form. Immobilization stabilises enzymes, in part by making them more resistant to shear forces and to attack by proteases, a class of enzymes that destroys other proteins. In addition, an immobilized enzyme can easily be separated from the reaction mixture at the end of the reaction and then re-used. This recovery makes product isolation easier and the reaction less expensive.

Enzymes can be immobilized either as pure entities or as components of whole, but usually dead, cells. Immobilization methods include covalently attaching the enzyme to a solid support, entrapping it in a gel, cross-linking the enzyme molecules to one another, or encapsulating them in small artificial cells (Fig. 2.6).

Once immobilized, an enzyme can be used in any of several types of reactors, each of which has its own advantages and disadvantages. A batch reactor – essentially a tank containing the enzyme and substrate (Fig. 2.7) – is simple to set up, but the mixture requires continuous stirring, which sometimes degrades the enzyme. The most common reactor employed in

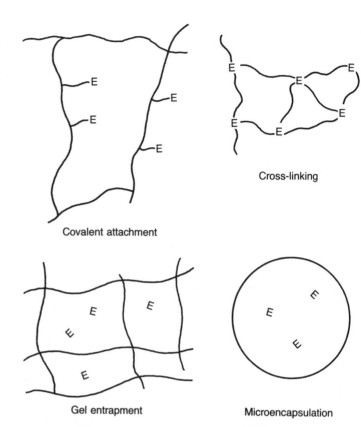

Fig. 2.6 Common methods of immobilizing enzymes (E) include binding them to surfaces (covalent attachment), trapping them in gels, cross-linking the enzyme molecules to one another, and sequestering the enzyme in small artificial cells.

Fig. 2.7 A laboratory-scale reactor that contains an immobilized enzyme.

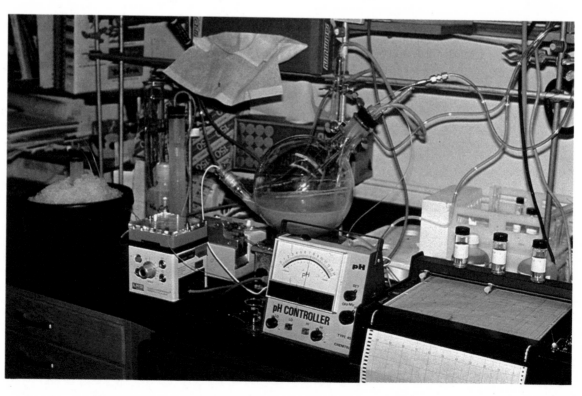

industrial applications, such as the production of high-fructose corn syrup, is the fixed bed reactor. It consists of a column that is packed with an immobilized enzyme through which the substrate flows (Fig. 2.8). The method is efficient and amenable to automation, although plugging of the column may occur with some types of immobilization. The fluidized bed-reactor, in which an upward flow of substrate causes the immobilized enzyme to act like a fluid, solves this problem; however, this procedure is more complicated and costly than other methods.

Fig. 2.8 A laboratory-scale column reactor containing an immobilized enzyme.

Current uses of enzymes

Food and beverages

Enzymes have long been used in the manufacture of cheese, beer and wine, and more recently the commercial application of enzymes for other purposes has grown. In the United States high-fructose corn syrup is the largest volume product made by immobilized enzyme technology. The syrup, which is produced by the enzymatic digestion of cornstarch, is sweeter than the sugar sucrose and is used in the beverage industry to flavour soft drinks and in the commercial baking industry to sweeten biscuits and cakes. A large commercial operation can convert 2 million pounds of cornstarch to high-fructose corn syrup in one day.

The transformation of cornstarch into the syrup requires three enzymes (Fig. 2.9). Treatment with alpha-amylase and glucoamylase first converts the starch to a glucose-containing syrup. Then immobilized glucose isomerase acts on the glucose, producing a mixture of glucose and fructose, which is sweeter, and thus more valuable, than the syrup containing glucose by itself.

Fig. 2.9 High-fructose corn syrup, an important commercial sweetener, is prepared from cornstarch with the aid of the enzymes alpha-amylase, glucoamylase and glucose isomerase. The syrup is an equilibrium mixture of glucose and fructose.

Starch

↓ Alpha-amylase, glucoamylase

Glucose

↕ Glucose isomerase

Fructose

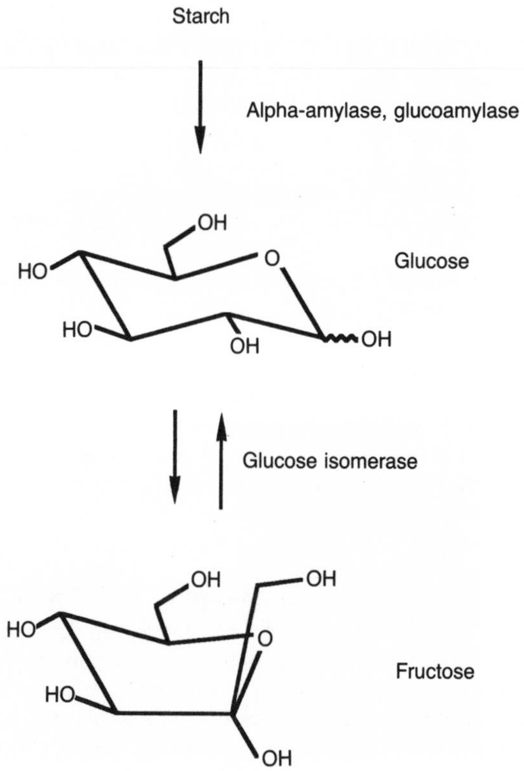

In the dairy industry, enzymes are essential for the production of cheese. Milk contains a group of proteins called caseins. One of the proteins, kappa-casein, prevents milk from coagulating in the presence of calcium ions. The enzyme rennin, which has traditionally been obtained from the lining of calves' stomachs, breaks down casein into the smaller protein called para-casein. Once the kappa-casein is destroyed, coagulation occurs to form a soft solid curd that can be separated from the liquid portion of the milk. The curd is the starting material for the production of cheese.

The fluid remaining after the curd is separated is 'whey'. It contains, among other components, proteins and a non-sweet sugar, lactose. Although these materials are potentially valuable nutrients, no good uses have yet been developed for whey because many individuals cannot digest lactose. If the sugar is present in substantial quantities in food and is not broken down during digestion, the microorganisms in the intestine feed on the lactose, producing gases and causing gastrointestinal distress. Whey has in the past presented a disposal problem.

Current research is exploring methods of using enzymes to convert whey into a material that can be used as an additive to foods. When whey is treated with the enzyme lactase, the lactose is broken down into two sweet, digestible sugars: glucose and galactose. The resulting sweet, protein-rich syrup can be added to certain food products, such as ice-cream.

Because the demand by the cheese industry for rennet, a dry rennin-containing powder, is so large, substitutes for the calf product have been developed. A rennet derived from microorganisms is widely used, but produces cheese of slightly inferior quality because the microbial enzyme is harder to inactivate than the calf product when its action is no longer desired. Scientists at Genentech Inc., South San Francisco, California, are now investigating ways of generating improved microbial rennet. One possibility is to use recombinant DNA techniques to modify the microbial enzyme genetically so that its stability is decreased and it can be deactivated by brief heating at the appropriate time.

In the beverage industry, enzymes are used to chill-proof juices, wines and beer. Juices and wines contain a polysaccharide called pectin that is soluble at room temperature but, when cooled, may form a colloidal suspension that gives the liquid a cloudy appearance. In Europe this cloudiness is considered a sign of high quality in bottled fruit juices, but in the United States it is unacceptable to most consumers. To prevent the haze from forming, enzymes called pectinases are added to the juice or wine. The enzymes degrade the pectin to lower molecular weight, and therefore more soluble, fragments that do not precipitate on cooling.

A similar problem of haze formation occurs during beer manufacture, although in beer the compounds that cause the trouble are proteins and tannins rather than polysaccharides. The proteins and tannins become suspended in the brew and precipitate when the liquid is cooled, thereby making it cloudy. To chill-proof beer, proteases such as pepsin or papain are added to break down the proteins and limit formation of the 'chillhaze' as it is called.

Another application of enzymes is the manufacture of soft-centred sweets. The confections are first made with solid sucrose centres that contain the enzyme invertase. Over a period of three to four weeks, the invertase transforms the sucrose into a liquid mixture of fructose and glucose. In addition to being sweeter than sucrose, fructose has the advantage of retaining more moisture, so that its presence also prevents the sweet from drying out and tasting stale.

Enzymes are also important in packaging foods. For example, the plastic wrapper used for cheese may be coated with glucose and two enzymes, glucose oxidase and peroxidase, to prevent spoilage. The coating slows the development of the rancid flavours that can be produced when oxygen reacts with components of the cheese. Under the influence of the glucose oxidase, the oxygen reacts instead with the glucose in the plastic wrapper to form gluconic acid and hydrogen peroxide, which is in turn converted to water by peroxidase. The products generated in this way have no effect on the taste of the cheese. Similarly, glucose oxidase may be added to mayonnaise to prevent spoilage by oxidation. The amount of enzyme added in these applications is very small.

Large volume chemicals

Although the largest volume transformations carried out by enzymes are in the food industry, as in the production of high-fructose corn syrup, enzymes are also used for synthesizing chemicals. The most important of these chemicals are probably amino acids, which are needed for protein synthesis, as nutritional supplements for hospital patients who are unable to eat normally, as supplements for animal feed, and as chemical intermediates.

Moreover, certain transformations in pharmaceutical synthesis depend on enzymes. For example, the microbially produced antibiotic penicillin G is converted by the enzyme penicillin acylase to 6-aminopenicillinic acid, which is the starting material for production of semi-synthetic penicillins (Fig. 2.10).

Detergents

The best agents available for removing proteinaceous stains such as egg, blood or grass from cloth are the protein-degrading protease enzymes. The first enzyme-containing laundry detergent was introduced in the United States in 1966, but in the early 1970s concerns that the enzymes in detergents might cause adverse effects, such as lung irritation from breathing enzyme dust, on the people who make and use the products caused the removal of the detergents from the market. They remained in use in Europe, however, and proved safe and effective when properly formulated and handled. Enzyme-containing detergents are being reintroduced in the United States where approximately 15 per cent of detergents now have protease additives.

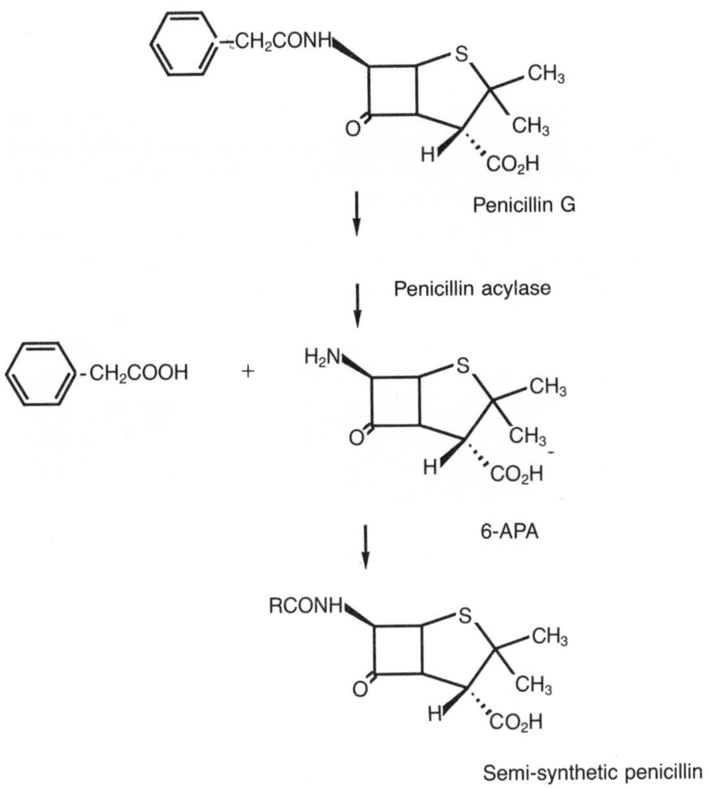

Fig. 2.10 Penicillin acylase selectively hydrolyses the side-chain amide bond in the readily available penicillin G and forms 6-aminopenicillinic acid (6-APA), which is a starting material for synthesizing a variety of semi-synthetic penicillins.

Medicinal applications

Enzymes control and carry out the myriad of chemical reactions occurring in the body. They mediate the digestion of food, build the components of cells, and generate and respond to intracellular messengers such as hormones and the chemical neurotransmitters that carry nerve signals. Enzymes are valuable both for studying these complicated systems and, occasionally, for medical therapies.

In one simple therapeutic application, a crude mixture of pancreatic enzymes called 'pancreatin' is given orally as a digestive aid to people who are deficient in digestive enzymes as a result of genetic disorders, surgical removal of the gall bladder, or advancing age.

A more complex medical application involves the use of the enzyme heparinase for controlling blood clotting. Patients who are undergoing kidney dialysis or certain forms of surgery or who have had heart attacks or strokes are often treated with a polysaccharide called heparin, which acts to decrease the ability of the blood to clot. Although heparin treatment is usually uneventful, it may sometimes be necessary to reverse its action rapidly – if the patient bleeds excessively, for example.

Heparinase selectively cuts heparin apart, thereby destroying its activity. Medical researchers are attempting to use the enzyme, which is immobilised on a filter, to achieve reversal of the anti-clotting effects of heparin. When this becomes necessary, the patient's blood is circulated outside the body through a filter with the immobilized heparinase and then returned to the patient.

A therapeutic enzyme use with possibly enormous potential is treatment with tissue plasminogen activator (TPA) to promote the dissolution of the blood clots that form during heart attacks. If a clot can be dissolved sufficiently early, permanent damage to the heart muscle may be prevented or at least minimized.

In the past, TPA could only be isolated with great difficulty from human uterine tissue and was in too short supply for widespread application. Now, however, it can be produced commercially by recombinant DNA technology and is expected to be the first major new product of the technology to enter the human health care market.

The enzyme has the advantage of being highly selective. It acts only inside the clot and causes little damage elsewhere in the patient. Initial tests of TPA to reopen the heart arteries that have been blocked by clots have proved very promising, provided that the treatment is started soon after the clot forms. Several other enzymes, including urokinase, pro-urokinase and streptokinase, have clot-dissolving activities and may have therapeutic benefits for heart attack victims when administered either alone or in conjunction with TPA.

Enzymes are beginning to be used to modify proteins, an application that will grow in importance as more proteins that have medical or other uses are made available by recombinant DNA technology. Enzymes are the catalysts of choice for this purpose because they provide the selectivity required in this technically demanding area.

The first commercial example of enzyme modification of a protein for human use is the conversion of porcine insulin to human insulin. Diabetes patients who must take insulin to control their disease sometimes become allergic to the bovine or porcine versions of the protein, but may have better tolerance for the human variant, which has been in short supply until recently.

Although porcine and human insulins differ by only one amino acid, chemical methods for converting one to the other are not available. This can be done enzymatically, however. First, an enzyme clips the protein backbone of porcine insulin next to the unwanted amino acid, thus removing it on a short peptide fragment. Then, under different reaction conditions, another enzyme replaces the fragment that has been removed by attaching the corresponding human peptide to the truncated porcine molecule. The human peptide is short enough to be chemically synthesized. Much of the 'human' insulin sold for treating diabetes in Europe is porcine insulin that has been converted to the human amino acid sequence in this way.

A number of efforts have been made to treat cancer with enzymes, but so far there have been no major successes. For example, asparaginase, which converts the amino acid L-asparagine to aspartic acid, has been used to treat leukaemia. Some leukaemic cells seem to have a higher requirement for L-asparagine than normal cells and it was hoped that injection of the enzyme into the blood might reduce the amount of L-asparagine available to the cancer cells and cause them to starve for want of a required nutrient. The treatment proved of little benefit because the tumors often develop resistance to it and the L-asparaginase can have toxic side effects.

Analytical chemistry

Enzymes are extremely valuable analytic tools, for both medical and non-medical applications. Because of their specificity they can be used to assay the amount of a substance, even of another enzyme, in a complex mixture such as blood, urine or other biological fluids. Often a number of enzyme reactions are coupled together in solution so that the sequence of enzyme-catalysed reactions culminates in the conversion of NAD (nicotinamide adenine dinucleotide) to NADH, the reduced form of the molecule. This conversion results in a change in the ability of the sample to absorb light, which can easily be measured by a spectrophotometer. Many of the diagnostic tests carried out by physicians depend on enzymes. Enzymes are routinely used to measure the concentrations of glucose, urea, amino acids, ethanol and lactic acid in biological fluids and to identify proteins and nucleic acids.

Home diagnostic procedures also often depend on enzymes. For example, diabetics must monitor the glucose content of their urine as an indicator of their need for insulin. The glucose analysis is done simply by dipping an analytical test stick in a urine sample. The stick contains the enzymes glucose oxidase and peroxidase as well as a reagent that registers the reaction with glucose by changing colour.

Potential new uses of enzymes

Most of the current applications of enzymes in industry and research involve relatively simple processes, such as the conversion of starch to high-fructose corn syrup. In the future, enzymes may be used as catalysts in more complicated systems. Many biologically important molecules, including polysaccharides, nucleic acids and proteins, are too complex to be easily synthesized by the standard methods of organic chemistry. Because enzymes function well in transformations involving these substances, current research efforts seek to develop enzymatic methods for their synthesis.

Success in these endeavours will represent a large step towards two goals. First, when these complex biological compounds are readily obtainable scientists will be in a better position to understand the molecular basis of life. Second, this new form of biotechnology will provide the tools for making

useful products, not only for biology but also for medicine and agriculture.

A chemist interested in adapting the unique catalytic properties of enzymes views a cell differently from a traditional biologist (Fig. 2.11). The biologist is concerned with the structure and function of the cell and its components and with understanding how the parts work together. The synthetic chemist considers the cell as a bag of enzyme catalysts into which he can place reactants and out of which will come valuable products. Nevertheless, the chemist needs the information garnered by the biologist.

The development of the new enzyme technology requires the design of schemes that rival the intricacy of some of the cells' major metabolic pathways. Enzymes, substrates, plus additional essential molecules called cofactors, will have to be

Fig. 2.11 Chemists and biologists view the cell differently. Chemists concern themselves with molecular details (*a*), whereas biologists study the overall function of the cell (*b*).

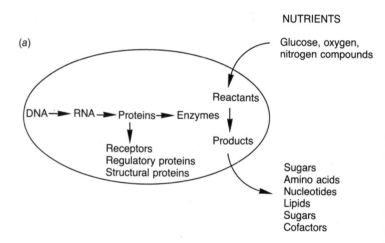

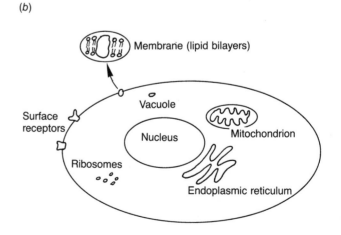

combined in complex cycles. At this stage, no one knows exactly how complicated these systems of artificial metabolism can be while still remaining practical.

Current enzyme research is aimed at developing and improving methods for cofactor regeneration, exploring and extending the range of structures accepted as substrates, synthesizing the larger molecules unattainable by classic synthetic methods, developing systems of artificial metabolism, and genetically engineering new or altered enzymes.

Cofactors and enzyme-catalysed syntheses

Enzymes can be classified in several ways, one of which is by their requirement for cofactors, low molecular weight substances that also participate in the enzyme-catalysed reaction. The cofactor NAD, for example, acts as an oxidizing agent in some enzymatic reactions, whereas its reduced form, NADH, acts as a reducing agent. Another cofactor, ATP, serves as a donor of phosphate groups.

Enzymes that require an added cofactor are in general more difficult to use in large-scale commercial operations than those not requiring added cofactors. Because the cost of cofactors is high – 98% pure NADH costs about $11 000 per pound – their use would be prohibitively expensive in an enzyme-catalysed reaction in which one molecule of cofactor is consumed for every molecule of product formed.

This problem can be solved by coupling a reaction for regenerating the cofactor to the reaction in which it is consumed, a strategy which is similar to that used in nature. Another enzyme and inexpensive starting materials convert the product form of the cofactor back to its original reactant form. A method for the reduction of NAD by formate and the enzyme formate dehydrogenase, which was developed by Zeev Shaked and George Whitesides of Harvard University in Cambridge, Massachusetts, is an example of such a coupled cofactor regeneration system (Fig. 2.12).

To make cofactor regeneration schemes practical, it is necessary to be able to synthesize the ultimate regenerating reagents in an economical way. Practical regeneration schemes are becoming available for the three most important cofactors: ATP, NAD and NADH. Although the schemes can currently provide many kilograms of product, none is yet used on an industrial scale.

Some enzymes have cofactors that are effectively built-in. They are tightly attached to the enzyme protein and regenerate as part of the normal enzyme action. Because no external cofactor regeneration schemes are needed for these enzymes, they are easier to use than those requiring added cofactor. Tyrosinase, which catalyses the formation of the amino acid tyrosine, and tryptophanase, which produces the amino acid tryptophan, are examples of enzymes with built-in cofactors. Different amino acids can be made by varying the substrates fed to these enzymes.

The simplest enzymes to use in organic synthesis do not

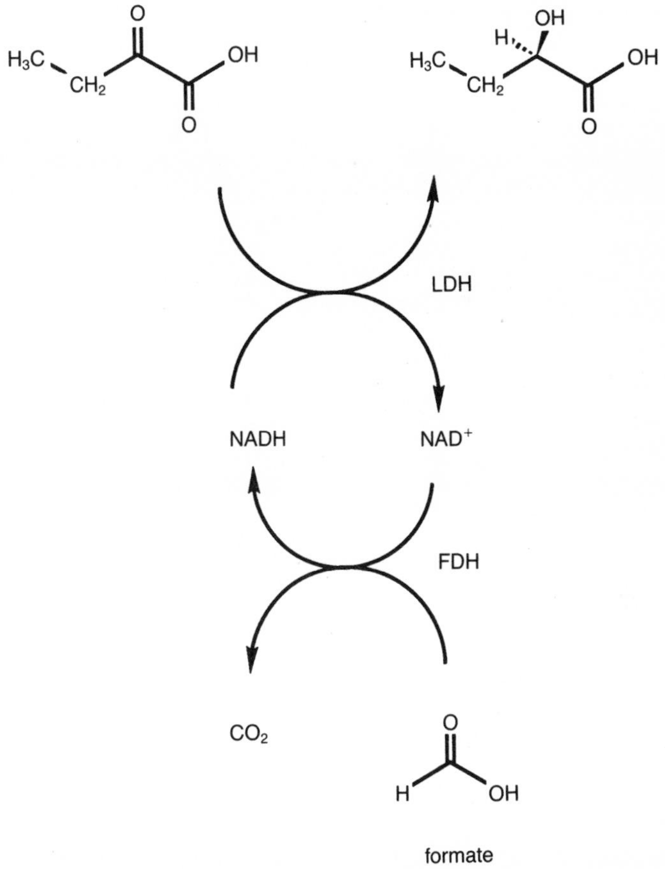

Fig. 2.12 A coupled cofactor regeneration system converts NAD back to its reduced form (NADH). As the enzyme lactate dehydrogenase (LDH) reduces alpha-ketobutyric acid to alpha-hydroxybutyric acid, it uses NADH. The enzyme formate dehydrogenase (FDH) then regenerates the NADH by oxidizing formic acid to carbon dioxide.

require any added cofactors at all. Lipase and aldolase belong to this category. Lipase catalyses the splitting of ester bonds to produce the free acid and alcohol. Its natural substrates are phospholipids, fatty acid esters of glycerol, esters of cholesterol, and related substances. A wide range of esters that do not exist in nature can also serve as substrates for this enzyme.

Lipase is another enzyme that can aid in the isolation of individual enantiomers (Fig. 2.13). If either the acid or alcohol portion of an ester substrate has an asymmetric centre, the enzyme will often split the ester of one enantiomer more rapidly than the ester of the other enantiomer. In this circumstance the lipase-catalysed reaction will yield an alcohol or acid product that is enriched with regard to the more quickly released enantiomer. Such products may be useful in synthetic reactions.

Aldolase catalyses the formation of a carbon – carbon bond

between dihydroxyacetone phosphate and any of a wide variety of aldehydes (Fig. 2.14). The enzyme's specificity towards different aldehydes and the stereospecificity of the reaction are currently under investigation.

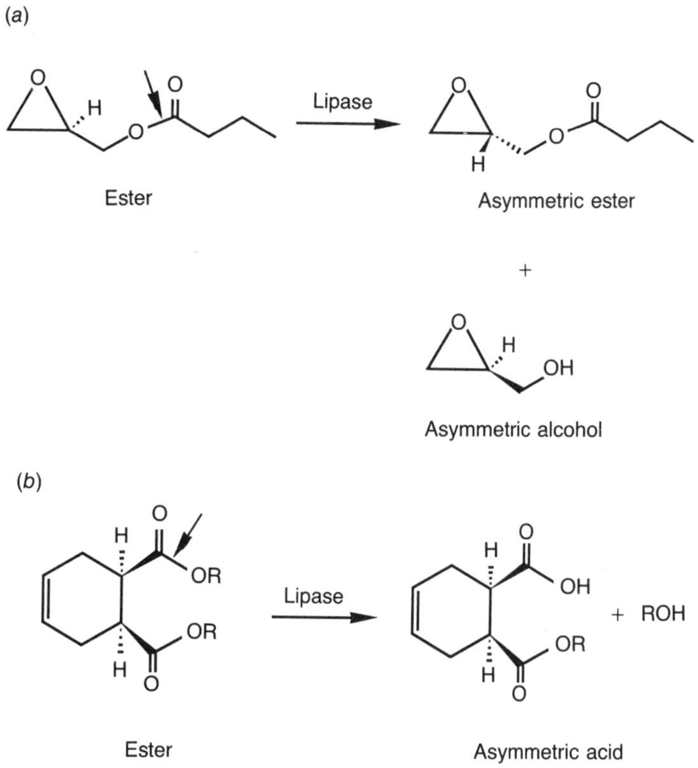

(a)

(b)

Fig. 2.13   The enzyme lipase catalyses the cleavage of esters, releasing the alcohol and acid that were originally joined in the ester molecule. The bonds split by the enzyme are marked with arrows. The alcohol of the ester shown in reaction (a) has an asymmetric carbon and can therefore exist in either of two enantiomeric forms. In such circumstances lipase may split the ester containing one enantiomer either slowly or not at all and can be used to separate the enantiomers. The ester shown in (b) does not contain an asymmetric centre, but lipase creates one in the acid produced by removing one of the two alcohols.

Fig. 2.14   Aldolase catalyses the reaction of an aldehyde and dihydroxyacetone phosphate to yield an aldol product of known stereochemistry.

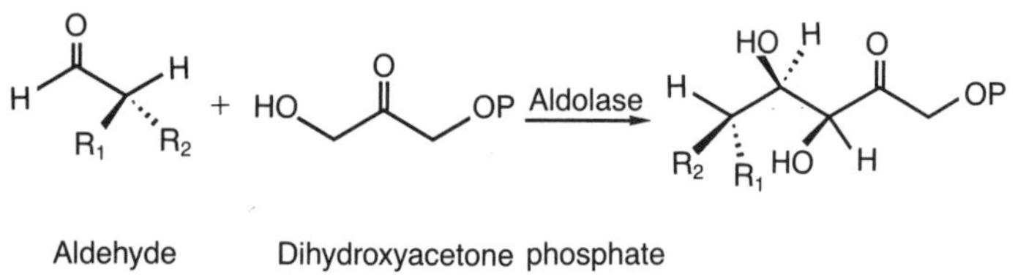

Extending the substrate ranges of enzymes

As already mentioned, the specificity of enzymes for particular substrates is the basis for most of their applications in organic chemistry. The specificities of different enzymes vary, however. Some will accept only one substrate, whereas others accept many. Hexokinase, for example, can act on more than 30 compounds that are structurally related to glucose. Enzymes often accept 'unnatural' substrates that do not occur in nature, in addition to the natural variety.

A typical research programme in enzyme-catalysed synthesis will often be aimed at making a particular molecule, either because it is valuable in itself or because it can serve as the starting point for the synthesis of other compounds. If an enzyme that produces similar molecules is known, the programme can begin by surveying the range of substrates that it recognizes. The survey provides an understanding of how the enzyme works and may allow predictions of other potential substrates. Even if the enzyme does not work as well with the unnatural substrates as with its natural one, the transformation may still be useful – especially if the product cannot be made any other way.

In the next phase, research on a small scale determines the best conditions of temperature, pH, and salt and substrate concentrations for the enzyme-catalysed reaction. If immobilization of the enzyme is desirable and if cofactor regeneration is necessary, appropriate systems are devised. Finally, enough information is collected so that the reaction can be run on the large scale demanded by commercial operations. Such 'scaling-up' can be difficult if problems in plant engineering are not carefully considered.

Modification of enzymes

Known enzyme catalysts do not exist for many of the reactions of interest to synthetic chemists. Sometimes, random screening of previously unexplored enzymes or enzyme sources uncovers the desired catalytic activity. This process is inefficient and very labour-intensive, however. One of the exciting prospects for the future is the modification of existing enzymes to alter or improve their specificity. Even more exciting, is the possibility of designing new enzymes and producing them by recombinant DNA technology, although this goal belongs to the distant future. Accomplishing this objective requires a detailed knowledge of the relations between amino acid sequence, three-dimensional protein structure, and catalytic recognition and action. Virtually none of this information is available now, although researchers are beginning to acquire it.

Nevertheless, it will be some time before chemists can synthesize new enzymes from scratch. Chemical synthesis of an entire enzyme by the stepwise construction of the amino acid chain is technically possible, but not yet practical. One difficulty is in predicting how a particular linear sequence of amino acids will fold. The amino acids of an enzyme's active site, which is the portion of the molecule that contacts the substrates and effects their conversion to products, are frequently distant from one another in the linear protein chain and are only brought into conjunction when the enzyme folds into the correct three-dimensional structure.

Another difficulty concerns determining which amino acids to incorporate in the active site. Computer models can estimate the contribution of particular amino acids to substrate binding and catalysis, but the reliability of these estimates has been checked experimentally in only a few instances and much remains to be learned before the models have firm predictive value.

For the present then, research efforts are aimed at modifying existing enzymes. Once the amino acids in or near the active site have been identified, they can be modified either by direct chemical alteration or by site-specific mutagenesis. Chemical transformations allow one amino acid to be changed into another; or reactive groups may be attached to some amino acids to alter the reactivity of the active site. For example, papain, a proteinase, has been converted by D. Lawrence and E. Kaiser of Rockefeller University into a protein active in oxidation and reduction reactions. They did this by attaching a flavin group to the sulphur of a cysteine residue (Fig. 2.15). Flavins are cofactors for the oxidation-reduction reactions catalysed by a number of enzymes.

Site-specific mutagenesis modifies the enzyme by incorporating a specific mutation in the corresponding gene. The linear

Fig. 2.15 The enzyme papain is chemically modified by attachment of a flavin group that alters the reactivity of the enzyme.

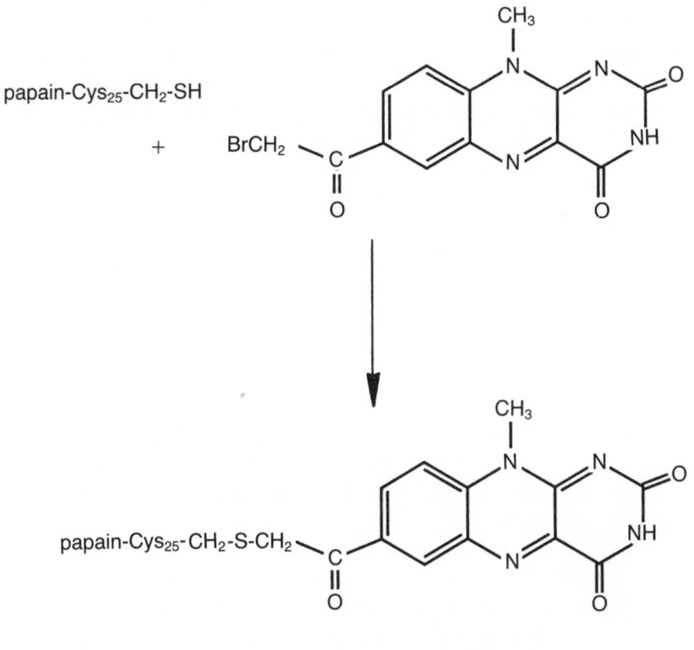

sequence of amino acids in the enzyme protein is determined by the linear sequence of nucleotides in the gene DNA, with three nucleotides constituting a codon for each amino acid (see Chapter 1). If the gene sequence that codes for the amino acids of the enzyme's active site is known, one amino acid can often be replaced by another by changing just one nucleotide in the appropriate codon.

A reliable and predictable method for modifying the gene coding for a protein is 'oligonucleotide-directed mutagenesis'. This method involves synthesizing a short section of DNA (the oligonucleotide) that corresponds to the gene sequence to be mutated except for a change in a single, specific base. When introduced into appropriate cells, this oligonucleotide serves as a primer for DNA replication with the result that the altered sequence becomes incorporated into the gene. Further manipulations permit the generation, identification and isolation of mutant clones that produce the modified enzyme, often with great efficiency.

Alan Fersht of the Imperial College of Science and Technology in London, performed this type of genetic engineering on the gene for tyrosyl-tRNA synthetase, thereby changing a cysteine residue into a serine residue at position 35 of the enzyme. The only difference between the two amino acids is that serine has a hydroxyl ($-OH$) group whereas cysteine has a sulphydryl ($-SH$) group. The mutation resulted in a change in the enzyme's affinity for its substrates because the affected amino acid participates in substrate binding and serine forms stronger hydrogen bonds than cysteine. A number of other amino acids have also been changed in tyrosyl-tRNA synthetase and used to explore the mechanism of action of the enzyme.

Multi-enzyme systems

An important potential use of enzymes is in multi-enzyme systems. Because most enzymes are active under the same conditions – in water at room temperature and a pH of approximately 7 – different enzymes will all exhibit catalytic activity when mixed in the same solution. Moreover, the substrate specificities of the enzymes will allow each to act on its own substrates without interfering with the action of the other enzymes. If the product of each enzymatic reaction is the substrate for another enzyme, the reactions will become coupled into a chain, in much the same way that the metabolic pathways of the cell are composed of sequences of coupled reactions leading from initial reactants to final products.

Laboratory-designed systems of complex sets of coupled enzymatic reactions may be termed 'artificial metabolism'. Such systems are complicated to design and operate, however. Careful control of the enzyme concentrations relative to one another to give optimum concentrations of the intermediates in the sequence and recycling of any necessary enzyme cofactors are required to prevent the build-up of inhibitory intermediates or the depletion of cofactors. Either occurrence could shut the system down.

'Flavo-papain'

At this time, the use of multi-enzyme systems to carry out the total synthesis of complex molecules remains at the stage of laboratory demonstration. One example of such a system, which was developed by Chi-Huey Wong in Whitesides' laboratory at Harvard, uses six cooperating enzymes to convert glucose-6-phosphate and N-acetylglucosamine to lactosamine, a disaccharide occurring as part of the carbohydrate groups that are attached to the surfaces of many proteins.

Multi-enzyme systems for synthesizing complex carbohydrates may eventually be a useful adjunct to recombinant DNA technology. The mammalian proteins made in *Escherichia coli* and other bacteria by recombinant DNA methods do not contain the carbohydrate that they normally would because the bacteria are not capable of adding the material to the proteins. If the carbohydrate turns out to be necessary for the normal function of the proteins, a way to add it will have to be found. Artificial metabolism may help in this regard.

Economic considerations: the future

Economics

Although thousands of enzymes have been identified, a relative few are of commercial importance. Twenty enzymes account for most of the worldwide market. The US Office of Technology Assessment estimated that in 1985 this market produced some 75 metric tons of enzymes, worth about $600 million. In the United States glucose isomerase and the two amylases used for producing high-fructose corn syrup from cornstarch account for about 50% of the market for enzymes. Production of the syrup saved the United States about $1.3 billion in foreign exchange for imported cane and beet sugar in 1980. Other enzymes produced in large volume are rennets for cheese-making, papain for chill-proofing beer, and proteases for detergents.

The short-term potential for growth in the industrial enzyme market is modest. A substantial number of new applications for enzymes are being explored, but the number of processes being actively developed for large-scale use is small. The major opportunities are probably in reactions involving water. The use of a nitrile hydratase to make acrylonitrile and acrylamide for the plastics industry is being actively explored in Japan. Many applications of lipases in the production of fine chemicals, especially for enantiomeric intermediates for the pharmaceutical industry, are very promising. In addition, enzymes are becoming useful for dealing with the important problem of toxic wastes. For example, a commercial enzyme system for destroying cyanide ions in aqueous wastes from chemical plants has recently been developed. Water treatment is an application driven by regulatory and societal pressures.

In general, however, the development of new, enzyme-based technologies for the chemical industry is inhibited by the slow growth and modest profitability of this area. Moreover, the majority of the current products of the chemical industry are water-insoluble and thus not ideal targets for enzymatic transformation.

What then is the future of enzyme-based technologies? In the longer term, it is extremely bright for several reasons.

Recombinant DNA technology provides a new method for lowering production costs for scarce enzymes. The enzymes used until recently have been those that could be prepared inexpensively by classic microbiological methods. Research aimed at developing commercial applications focused on these simply because they were the only ones with which an economical process might be built. The removal of this economic constraint allows current research to explore possible applications, such as chemical synthesis, for enzymes once considered only as biological research tools. Any of these enzymes that are identified as having promise in synthesis will be produced economically by recombinant DNA technology.

Environmental issues – from waste water treatment and industrial waste disposal to the development of chemical manufacturing methods that are perceived by society as safe – are now a major and unavoidable concern of all industries. An increasing awareness of this reality provides a reason for developing enzyme-based technologies for waste treatment and chemical synthesis. Even if these technologies are not quite competitive with traditional technologies, the enzymatic methods may be preferred because they operate in water at low temperatures and often produce few by-products.

The pharmaceutical industry, in particular, requires an entirely new group of technologies to deal with the opportunities afforded by biotechnology. Many of the products emerging from biotechnology are of types with which the pharmaceutical industry has had relatively little experience. Examples are proteins such as the clot-dissolving TPA and the immune-system regulator interleukin-2; polysaccharides such as hyaluronic acid, which is used in eye surgery; and the nucleic acid segments that are used for analysing DNA. All of these classes of compounds are exactly those to which enzymes are most productively applied, and enzymatic synthesis and modification of the substances will clearly be important. The value of the enzymes used for these applications may not be large, but the value added to the product by appropriate enzyme treatment may be very large.

Even for existing classes of pharmaceutical products, enzymes will play an increasingly important role in providing enantiomerically pure intermediates, and in carrying out difficult transformations with high selectivity. Enzymes will always be reserved for 'difficult' problems – those resistant to solution by conventional methods. The sophistication of current activities in drug development is such that difficult problems are commonplace.

The specificity of enzymes will continue to be useful in analytical methods. Continued development and marketing of electrodes that contain immobilized enzymes will make the assay of many aqueous biological substances routine. The specificity of enzymes often allows accurate measurement of one substance in a complex mixture without any sample preparation.

Enzymes will play an increasing role in food processing. The major driving forces in this area are safety and cost reduction. Enzymes provide both in certain areas of application. The food industry is technologically conservative, however, and change will come slowly.

Most of the chemical industry is based on declining petroleum feedstocks. In the long term, and especially in developing nations, there will be economic reasons to use biologically-derived starting materials, including cellulose, starch, lignin and plant proteins, instead of petroleum products. Enzymes will certainly play a role in processing the biological materials.

The pace of the research on the relation between the structure of an enzyme and its catalytic activity is very rapid. Although this problem is a very difficult one, it will eventually – perhaps in 20 to 50 years – be possible to design proteins that will catalyse new types of reactions and produce these unnatural enzymes economically by recombinant DNA technology. The first applications of this type of activity are now appearing from modest programmes that use site-specific mutagenesis to modify the properties of existing enzymes. These programmes will grow more ambitious as time goes on.

Additional reading

Chibata, I. (1978). *Immobilized Enzymes: Research and Development*. Halsted Press, New York.

Commercial Biotechnology: An International Analysis. (1984). Office of Technology Assessment, US Congress, OTA-BA-218, Washington, DC.

Fersht, A. (1985). *Enzyme Structure and Mechanism*, 2nd edn. W. H. Freeman, New York.

Hasselberger, F. X. (1978). *Uses of Enzymes and Immobilized Enzymes*. Nelson-Hall, Chicago.

Kaiser, E. T. and Lawrence, D. S. (1984). Chemical mutation of enzyme active sites. *Science*, **226**, 505.

Kilara, A. and Shahani, K. M. (1979). The use of immobilized enzymes in the food industry: a review. *CRC Critical Reviews in Food Science and Nutrition*, **12**, 161.

Klibanov, A. M. (1983). Immobilized enzymes and cells as practical catalysts. *Science*, **219**, 722.

Mosbach, K. (ed.) (1976). *Methods in Enzymology*, vol. 46, *Immobilized Enzymes*. Academic Press, New York.

Mosher, H. S. and Morrison, J. D. (1983) Current status of asymmetric synthesis. *Science*, **221**, 1013.

Porter, R. and Clark, S. (eds.) (1985) *Enzymes in Organic Synthesis*. CIBA Foundation Symposium 111. Pitman, London.

Ruttenberg, M. A. (1972). Human insulin: facile synthesis by modification of porcine insulin. *Science*, **177**, 623.

Whitesides, G. M. and Wong, C.-H. (1985). Enzymes as catalysts in synthetic organic chemistry. *Angewandte Chemie, International Edition in English*, **24**, 617.

3 Microorganisms as producers of feedstock chemicals

Randolph T. Hatch and Ralph Hardy

Microbial production of industrial chemicals from renewable resources

Feedstock chemicals are the raw materials of the chemical industry. They are used as starting materials for synthesizing a vast array of other chemicals – plastics and rubber, for example; as solvents; and for the manufacture of numerous products, including textiles and paper (Table 3.1).

Industrial chemicals were originally produced by extracting materials such as vegetable oils, starches, cellulose, lignins and waxes from plants. These carbon-containing raw materials are directly renewable by the photosynthetic reduction of carbon dioxide from the atmosphere. They can be chemically converted to other substances, thereby greatly expanding the number of products that can be derived from the plant kingdom.

Many of these conversions are carried out by microorganisms that can use the plant-derived materials as a source of energy for increasing the microbial cell mass, at the same time forming numerous by-products. The role of the microorganisms was not appreciated when they first began to be used for preserving foods and producing alcoholic beverages, however. The Dutch scientist Antony van Leeuwenhoek did not discover bacteria until the seventeenth century, and only later, in the nineteenth century, did the Frenchman, Louis Pasteur, convince the scientific world of the essential role of microorganisms in fermentation. The products of microbial metabolism were then identified at an increasingly rapid rate over the following decades.

At first the products came primarily from the energy-producing metabolic pathways of the microbes. For example, the glycolytic pathway provides ethanol for alcoholic beverages, carbon dioxide for leavening bread, and lactic acid for food preservation (Fig. 3.1). Other organic acids, such as acetic, propionic and citric acids, were identified as products of microbial metabolism by the end of the nineteenth century.

In the early twentieth century several additional organic chemicals were found to be products of microbial energy metabolism. The fermentation method for producing the solvents acetone and butanol was developed during World War I, when acetone was required for the production of the explosive cordite. A similar situation arose during World War II, when supplies of natural rubber from the Far East were disrupted. Researchers then found that 2,3-butanediol could be converted to 1,3-butanediene, which can be used to make synthetic rub-

Table 3.1. *Industrial chemicals produced by fermentation*

Organic chemical	Microbial sources	Selected uses
Formic acid	*Aspergillus*	Textile dyeing, leather treatment, electroplating, rubber manufacture
Ethanol	*Saccharomyces*	Industrial solvent, intermediate for vinegar, esters and ethers, beverages
Acetic acid	*Acetobacter*	Industrial solvent and intermediate for many organic chemicals, food acidulant
Glycolic acid	*Aspergillus*	Textile processing, pH control, adhesives, cleaners
Oxalic acid	*Aspergillus*	Printing and dyeing, bleaching agent, cleaner, reducing agent
Glycerol	*Saccharomyces*	Solvent, plasticizer, sweetener, explosives manufacture, printing, cosmetics, soaps, antifreezes
Propylene glycol	*Bacillus*	Antifreeze, solvent, synthetic resin manufacture, mould inhibitor
Isopropanol	*Clostridium*	Industrial solvent, cosmetic preparations, antifreeze, inks
Acetone	*Clostridium*	Industrial solvent and intermediate for many organic chemicals
Malonic acid	*Penicillium*	Manufacture of barbiturates
Lactic acid	*Lactobacillus, Streptococcus*	Food acidulant, dyeing, intermediate for lactates, leather treatment
Acrylic acid	*Bacillus*	Industrial intermediate for plastics
Butanol	*Clostridium*	Industrial solvent and intermediate for many organic chemicals
Butanoic acid	*Clostridium*	Manufacture of esters
2,3-Butanediol	*Aerobacter, Bacillus*	Intermediate for synthetic rubber manufacture, plastics and antifreeze
Methylethyl ketone	*Chlamydomonas*	Industrial solvent, intermediate for explosives and synthetic resins
Fumaric acid	*Rhizopus*	Intermediate for synthetic resins, dyeing, acidulant, antioxidant
Succinic acid	*Rhizopus*	Manufacture of lacquers, dyes and esters for perfumes
Malic acid	*Aspergillus*	Acidulant
Tartaric acid	*Acetobacter*	Acidulant, tanning, commercial esters for lacquers, printing
Itaconic acid	*Aspergillus*	Textile and paper manufacture, intermediate for plastics

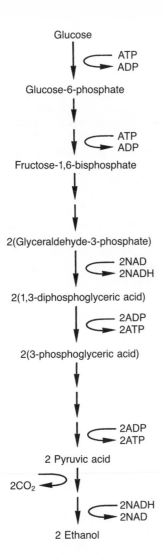

Glucose

→ ATP
→ ADP

Glucose-6-phosphate

→ ATP
→ ADP

Fructose-1,6-bisphosphate

2(Glyceraldehyde-3-phosphate)

→ 2NAD
→ 2NADH

2(1,3-diphosphoglyceric acid)

→ 2ADP
→ 2ATP

2(3-phosphoglyceric acid)

→ 2ADP
→ 2ATP

2 Pyruvic acid

2CO$_2$ ←

→ 2NADH
→ 2NAD

2 Ethanol

Overall balance: Glucose + 2 ADP → 2 Ethanol + 2CO$_2$ + 2ATP

Fig. 3.1 The glycolytic pathway for alcoholic fermentations. In this pathway the sugar glucose is converted in several steps to pyruvic acid, which is then converted to ethanol with a release of carbon dioxide.

ber, and developed a fermentation route for making the 2,3-butanediol.

The modern pharmaceutical industry can trace its origins to 1928 when Alexander Fleming discovered penicillin, which was the first microbial antibiotic to be made commercially. Penicillin is one of the so-called secondary metabolites that are not products of microbial energy metabolism. After World War II the number of products made by microbial fermentations grew rapidly to include an expanding spectrum of antibiotics, amino

acids for animal feed, and enzymes. The wide range of products that can be made by microbial fermentation can be classified as shown in Table 3.2.

Throughout the nineteenth century and the first half of the twentieth, microbial production accounted for the bulk of the world's supply of organic chemicals. Raw materials are usually the key to the economics of commodity products, however, and this is true for organic chemicals as well. In 1920, Standard Oil of New Jersey initiated the production of organic chemicals from petroleum when the company began synthesizing isopropyl alcohol from propylene. Over the following decades, and especially after 1950, the use of petrochemical feedstocks for chemical manufacture grew rapidly because of the ready availability and low cost of petroleum and the rising costs of the plant-derived raw materials.

Despite the existence of numerous fermentation routes to the organic compounds, fermentation as a source of these chemicals declined rapidly although some are still obtained largely by microbial fermentations. These include ethanol, a number of organic acids, and the more complex organic chemicals, such as antibiotics, that require multiple synthetic steps.

However, the oil price shocks of the 1970s not only abruptly changed the relative costs of the raw materials for chemical and fermentation syntheses, but have also exposed the instability of the world petroleum market. As a result agricultural feedstocks once again became more attractive. In the United States, for example, fermentation facilities were rapidly constructed for producing ethanol from glucose that is obtained from cornstarch. Brazil also initiated a major programme for producing ethanol by fermentation methods, in this case using sucrose from molasses.

Table 3.2. *Classification of fermentation products*

	Class	Examples
1.	End-products of energy metabolism	Ethanol, acetic acid, lactic acid butanol, acetone
2.	Energy storage compounds	Glycerol, glycogen, other polysaccharides
3.	Proteins	
	Extracellular enzymes	Amylases, cellulases, proteases, amyloglucosidase, pectinase
	Intracellular enzymes	Glucose isomerase, ligases, glucose oxidase, invertase
	Foreign proteins	Insulin, human growth hormone, bovine growth hormone
4.	Cell structures	Baker's yeast, bioinsecticides, single-cell protein, antigens
5.	Intermediary metabolites	Amino acids, citric acid, vitamins, malic acid
6.	Secondary metabolites	Antibiotics, gibberellins
7.	Chemically modified substances	Steroids, sorbose, glucose

The advent of recombinant DNA technology, which also occurred in the 1970s, has added another new dimension to the manipulation of microbial metabolism. The range of products made by microbial cells can now be expanded; the organisms are being made to function as miniature chemical plants for synthesizing foreign proteins such as human insulin and growth hormone (see Chapter 4). Other genetic modifications may increase the efficiency of current fermentations or give the microbes the ability to grow on previously unusable energy sources.

Industrial solvents

Approximately 80 per cent of the ethanol produced in the world is still obtained from fermentations; the remainder comes largely by synthesis from the petroleum product, ethylene. The alcohol produced in the United States is primarily used in alcoholic beverages (Table 3.3), but this is not always the case elsewhere in the world. Brazil has embarked on a major programme to produce ethanol for fuel and thereby diminish petroleum imports. As of 1984, approximately 7.9 million tons of ethanol were produced by fermentation in Brazil, with sucrose from sugarcane as the carbon source.

The United States is also substantially increasing its fuel alcohol production, originally because of the rapid increase in petroleum costs during the 1970s and the subsequent need for developing alternative energy sources. More recently, as the petrol supply has increased, if only temporarily, the fuel alcohol programme has been sustained by the drastic, federally mandated reduction of the lead content of petrol and the value of ethanol as an octane enhancer.

A number of microorganisms can convert glucose in the absence of oxygen to by-products of energy metabolism, which typically include alcohols such as ethanol, isopropanol and butanol, and short-chained organic acids such as formic, acetic, lactic, propionic and butyric acids. Other products that can be produced are acetoin, 2,3-butanediol and acetone.

Most of the microorganisms that are capable of anaerobic metabolism can produce ethanol, but many are unable to make appreciable quantities because they can not tolerate the alcohol's toxic effects on the cell membrane. Some microorganisms can accumulate high concentrations of ethanol, however. Historically, the most used microbe has been the yeast, *Saccharomyces cerevisiae* which can produce ethanol to give concentrations as high as 18 per cent of the fermentation broth. This yeast can grow both on simple sugars, such as glucose, and on the disaccharide sucrose, which is common table sugar. *Saccharomyces* is also generally recognized as safe as a food additive for human consumption and is therefore ideal for producing alcoholic beverages and for leavening bread.

In the metabolic pathway that produces ethanol, the yeast converts one molecule of glucose to two molecules of the alcohol plus two molecules of carbon dioxide (Fig. 3.1). The microorganism also gains two molecules of the energy intermediate adenosine triphosphate (ATP), which it uses to maintain other cellular activities. If no cell mass were produced, 51 per cent of the glucose would be converted to alcohol. However, the actual yield is closer to 45 per cent because some of the glucose is converted to cell mass and to by-products, such as glycerol.

Although ethanol is primarily produced from glucose and sucrose, a wide range of other sugars can also be used. The sugars can be obtained from a variety of raw materials. Sugarcane and sugar beets are the most common sources of sucrose, while any starch-containing plant material, depending on its cost and starch content, is a potential candidate for conversion to glucose. Fruits are also sources of glucose and fructose, as in the production of fruit wines. In the United States, corn is the preferred source of starch for the production of fuel alcohol, and blends of grains – wheat, corn, rye and barley – are used for making beer and distilled spirits. Molasses is the sugar source for rum production.

Although a great deal of research activity has been directed towards developing cellulose as a raw material for fermentation, it has yet to prove cost-competitive with the sugars and starches. Lactose from whey has also been used as a raw material, again with limited success. Not enough is available for large-scale operations nor are there yeast species that are both capable of using lactose as an energy source and tolerant of high alcohol concentrations. Moreover, whey contains high concentrations of dissolved minerals that further inhibit fermentation.

When corn is used as the starch source for fuel alcohol production, the germ is first separated out so that the valuable corn oil and protein can be recovered. In the corn wet-milling process, for example, the grain is first steeped in water for two days (Fig. 3.2). The softened corn is then broken in a shredding device, the degerminator, and the germ is separated by flotation. The remainder of the corn is wet ground, the fibre and hulls removed by screening, and the starch separated in centrifuges. The stream from the centrifuges, which contains the protein gluten, is dehydrated and the residue dried for sale as an animal feed. The starch is recovered by vacuum filtration and may either be dried before it is sold or further processed into fermentable sugars.

For ethanol production, the starch slurry is liquefied at a

Table 3.3. *US fermentation production of ethanol 1984*

Product	Annual production (million tons, as ethanol)
Beer	2.2
Fuel alcohol	1.6
Distilled spirits	1.1
Wine	0.5
Total	5.4

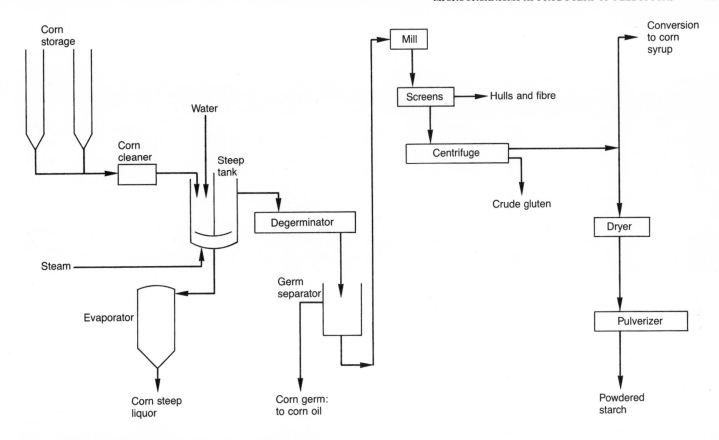

Fig. 3.2 Schematic diagram for the corn wet-milling process. The diagram is explained in the text.

temperature of 90 °C by alpha-amylase enzymes that can withstand this high temperature (Fig. 3.3). The conversion to sugars is completed at lower temperatures using glucoamylase enzymes. Approximately 98 per cent of the starch is converted to fermentable sugars by this treatment. The sugars are then fermented to produce ethanol by industrial strains of *Saccharomyces cerevisiae*.

In a large, efficiently run facility the carbon dioxide produced by the fermentation reaction is recovered for sale as a by-product. After the reaction is complete, the yeast and other solids are separated from the fermentation broth by centrifugation and dried for sale as an animal feed supplement. The clarified broth is sent to a beer still to recover and partially concentrate the ethanol. Further distillation in a rectifying distillation column brings the ethanol concentration to 95 per cent, the maximum that can be achieved by distillation of a water solution. The remaining water is then removed by distillation with another solvent such as benzene, thus producing pure alcohol that can readily be blended into high-octane petrol or used as an industrial solvent.

The acetone–butanol fermentation converts various raw materials into two valuable industrial chemicals with ethanol as a by-product. Hydrogen and carbon dioxide may also be recovered as by-products in a sufficiently large production facility. The bacterial strain most frequently used for acetone and butanol production is the strict anaerobe *Clostridium acetobutylicum*, although *C. butylicum* may be used instead. Consequently, the fermentation can be run only in the complete absence of oxygen.

The acetone–butanol fermentation may start with such raw materials as molasses, starches, cellulose, whey or the sulphite liquor from paper manufacture, as sources of glucose. The bacteria are aided by the availability of complex nitrogen sources that contain proteins and vitamins, in addition to a glucose source. Although corn meal is ideal as a provider of starches, proteins and vitamins, molasses and corn steep are preferred for commercial operations because of their lower prices.

Ethanol production by *Saccharomyces cerevisiae* requires that starches first be hydrolysed enzymatically to glucose. However, *Clostridium* species can do this conversion themselves because during cell growth they produce and secrete amylase enzymes. Once the glucose is released from the raw material by the bacterial enzymes it enters the glycolytic pathway where it produces pyruvate and acetyl CoA, with a release of carbon

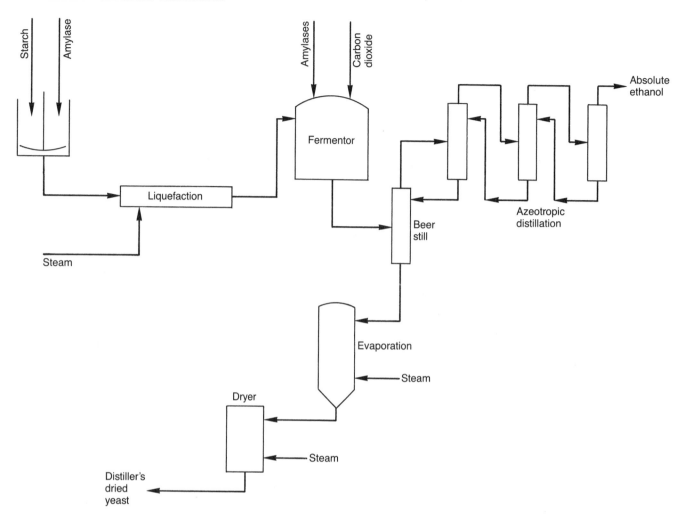

Fig. 3.3   Schematic diagram for the industrial fermentation of sugar to ethanol. The diagram is explained in the text.

dioxide and hydrogen gases (Fig. 3.4). The first phase of the fermentation is complete with the conversion of the pyruvate and acetyl CoA to butyric and acetic acids.

In the second phase, the mixed acids are converted to the mixed solvents butanol, acetone and ethanol, which are present in final ratios of approximately 60 to 30 to 10. The fermentation is over in about 36 hours when the butanol concentration becomes high enough to disrupt the cell membrane and stop the reactions. This occurs at a concentration of approximately 12 grams of butanol per litre of fermentation broth. At the end of the fermentation, conversion of the intermediates to butanol is not complete, but a spectrum of intermediates remains (Table 3.4). Because the residual solids from the fermentation broth are rich in the vitamin riboflavin, it is economic to recover them for sale as a protein-rich supplement for animal feed.

Table 3.4. *End-products of the acetone–butanol fermentation*[a]

Product	C. acetobutylicum	C. butylicum
Butanol	23	24
Acetone	7	—
Ethanol	2	—
Carbon dioxide	54	50
Hydrogen	2	1
Acetic acid	5	6
Butyric acid	2	8
Acetoin	3	—
Isopropanol	—	4

[a]Yields are expressed as grams product formed per 100 grams of glucose consumed.
(Data converted from that of Doelle, H.W. (1975) *Bacterial Metabolism*, 2nd edn. Academic Press, New York.)

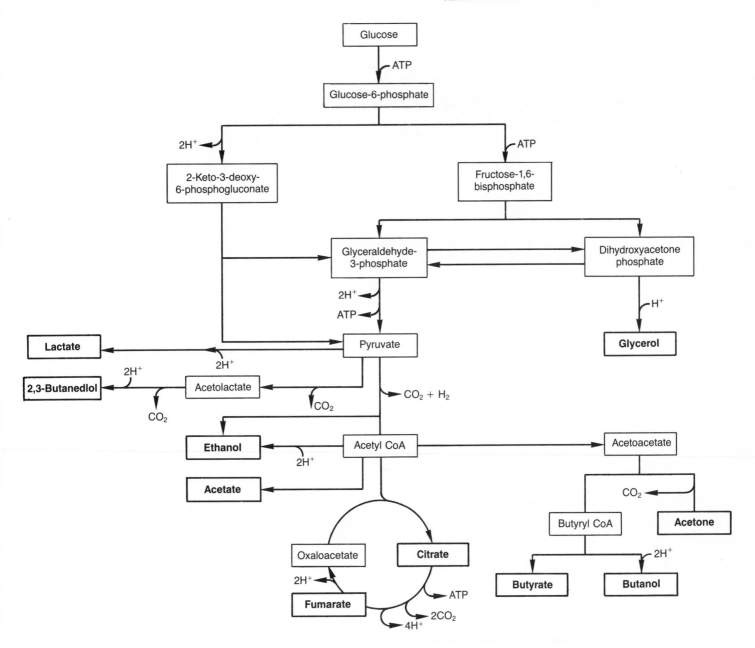

Fig. 3.4 The reactions involved in the fermentation of glucose to butanol and acetone. The glycolytic pathway converts the glucose to pyruvate which is then converted to acetyl CoA. The acetyl CoA in turn is converted to a variety of products, including acetone and butanol, as well as several organic acids.

Organic acids

A wide variety of organic acids can be produced by microbial fermentations. Except for alcohol, these were the earliest fermentation products. Acetic acid in the form of vinegar is the classic example. During the mid-nineteenth century the acid began to be produced industrially for use as a chemical intermediate. This was followed by the development of fermentation processes for making propionic, gluconic, citric and lactic acids, and during the early twentieth century fumaric, malic, itaconic and oxogluconic acids were added to the list of fermentation products. A tartaric acid fermentation process has also been developed but this acid is commercially produced as a by-product of wine fermentations.

Of the other organic acids, the major fermentation product today is citric acid, although significant quantities of lactic,

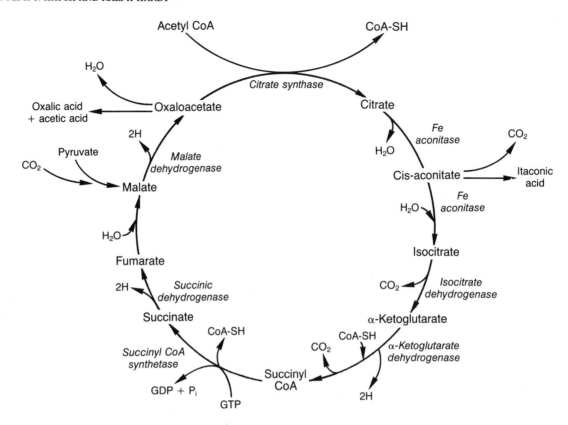

Fig. 3.5 The tricarboxylic acid cycle. This cycle is one of the cell's major pathways for energy production. The first step is the synthesis of citrate from acetyl CoA and oxaloacetate. The next step, the conversion of citrate to *cis*-aconitate, requires the cofactor iron (Fe). In the industrial production of citrate steps are usually taken to remove iron ions from the fermentation mixture to prevent *cis*-aconite formation and thereby improve the yields of citric acid.

gluconic and itaconic acids are also still produced by fermentation. Fumaric and malic acids can be obtained either by fermentation or by organic synthesis, depending on the costs of the feedstocks. Currently, these acids are produced by isomerization and hydration of maleic acid, which is in turn made by the catalytic oxidation of petroleum-based hydrocarbons. Because of regulatory requirements, acetic acid in the form of vinegar is still produced by fermentation, but the industrial grade of the acid is made by catalytic oxidations of low molecular weight petrochemicals or by catalytic carbonylation of methanol.

Citric acid is used widely in foods and beverages as an acidulant and flavour enhancer and industrially as a component that improves detergent action, as a chelator for metals, and as an antioxidant. Until 1930 the acid was produced from citrus fruits, primarily in Italy, although an alternative source, namely fermentation by moulds, had been discovered in the late

nineteenth century. Over the following decades of development of the fermentation process numerous genera of moulds, including *Aspergillus*, *Penicillium*, *Paecilomyces*, *Mucor* and *Ustilina*, were found to produce this acid. More recently, most yeasts and many bacteria were also found to be good producers.

The fermentation production of citric acid was improved by Pfizer, Inc., of Groton, Connecticut, and commercialized in 1923, after which fermentation rapidly became the lowest cost process and dominated the world market. Today, the world demand for citric acid is approximately 250 000 metric tons, approximately one-half of which is still produced by Pfizer. Miles Laboratories, Inc., of Elkhart, Indiana, makes about one-quarter of the total.

Citric acid is synthesized naturally as an intermediate in the tricarboxylic acid cycle, one of the cell's major pathways for energy production (Fig. 3.5). Carbon feeds into the cycle in the form of acetyl CoA from the glycolytic pathway. Consequently, any carbon source that can enter glycolysis can be used to produce citric acid. The most common carbon sources have been low-cost sugars such as those in beet and cane molasses. More recently, hydrocarbons that are oxidized by the *Candida* species of yeast to acetyl CoA have been shown to be economically attractive sources of carbon for the fermentation.

Citric acid was originally produced in tray cultures with the moulds growing on the surface of the fermentation mixture

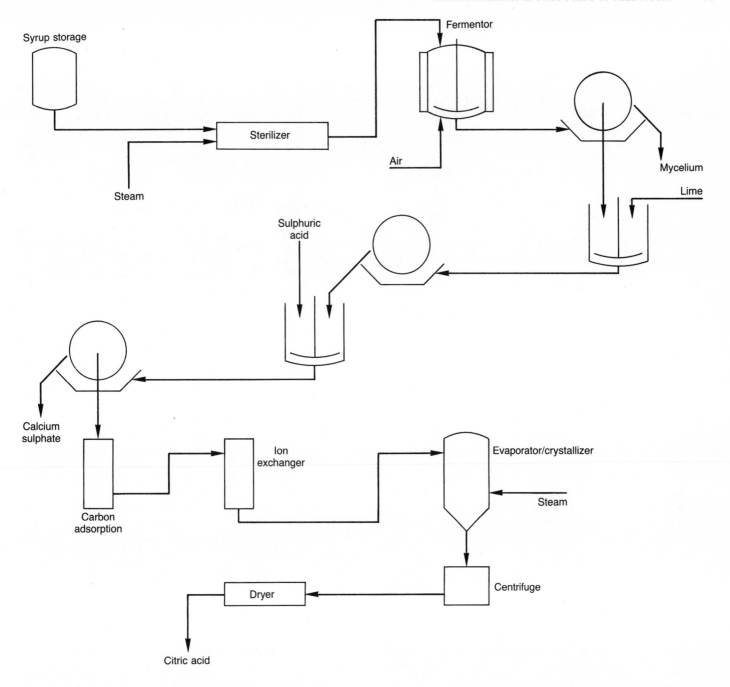

Fig. 3.6 Schematic diagram of citric acid manufacture. The diagram is explained in the text.

and oxygen provided by free diffusion. However, submerging the mould in the broth in an aerated fermentor tank gives higher production rates and is now the preferred method. *Asper-*

gillus niger, which was used from the beginning of citric acid production, is still the mould of choice when the substrate is a sugar.

Early work on the development of the fermentation process showed that pure substrates are required to attain high citric acid production. The metal ions that usually contaminate the substrates have an adverse effect. This is especially true for iron, which stimulates the growth of the microbial cells at the expense of citric acid production. Iron is required for growth, but the optimum concentration for the fermentation is in the range of 0.1 to 10 parts per million, depending on the microbial strain used. Citrate is not normally overproduced and released by the microorganism and the iron deficiency serves to block the tricarboxylic acid cycle at the step in which citrate is converted to *cis*-aconitate. As a result, citrate accumulates. To enhance citric acid yields the iron is removed from the sugars by means of an ion exchange treatment.

Other additives can mitigate the negative effects of iron, however. Potassium ferrocyanide forms precipitates of trace metals and reduces their concentrations in the fermentation medium. The optimum concentration of potassium ferrocyanide for citric acid production is in the range of 0.1 to 1 gram per litre, depending on the initial trace metal content. In addition, alcohols such as methyl, ethyl and isopropyl stimulate yields when used in the concentration range of 1 to 4 per cent. Higher concentrations are sufficiently toxic to the cell growth to reduce the yields. Copper can also counteract the effect of iron, presumably by inhibiting the enzyme that converts citrate to aconitate. A copper concentration of 500 parts per million can counteract up to 10 parts per million of iron.

The sugar-containing syrup to be used in the submerged process of citric acid production is first treated with a cation exchange resin to remove the iron. It is then sterilized and fed to the fermentor, which has already been sterilized (Fig. 3.6). A water solution of nutrients, which has also been sterilized, is introduced into the fermentor to supply the reaction with ammonia, phosphate and other minerals that are missing from the syrup.

The pH in the fermentor is adjusted to the range of 5 to 6 when molasses is the substrate or below 3 if purer forms of sugars are used. After the temperature has been brought up to 30 °C the tank contents are inoculated with *Aspergillus niger*. The fermentation is allowed to proceed for 4 to 5 days, during which time 120 to 150 grams per litre of sugar are converted to 100 to 140 grams per litre of citric acid. To achieve the highest yields sufficient aeration must be maintained to prevent oxygen starvation of the microbe.

At the end of the fermentation the broth is filtered and then treated with lime to raise the pH and precipitate out the calcium salt of citric acid. The precipitate is recovered from the slurry by filtration and redissolved with sulphuric acid, thus producing a calcium sulphate precipitate, which is removed by another filtration step. The acidified citric acid is treated with activated carbon to remove coloured impurities and by ion exchange to remove residual calcium and other cations. The citric acid crystallizes when it is concentrated in an evaporator, after which the crystals are recovered in a centrifugal filter and dried in a rotary kiln drier.

Lactic, itaconic and gluconic acids, the remaining commercial acids that are still made in significant quantities by fermentation, are all produced in a fashion similar to the citric acid fermentation. Species of *Aspergillus* are used except in the case of lactic acid, which is produced instead by species of *Lactobacillus* or *Streptococcus*. The fermentations are run either with molasses as a source of sucrose or starch as a source of glucose. The product concentrations reached in organic acid fermentations generally fall in the range of 5 to 10 per cent by weight. Although lactic acid is produced in the absence of oxygen, the other fermentations all require oxygen for high productivity. Lactic, itaconic and gluconic acids are recovered and purified by methods similar to those used for citric acid. ·

Economic aspects of fermentation

Microorganisms can be considered as microscopic chemical plants that assimilate raw materials and synthesize chemical products while reproducing. The efficiency of the processes within the microbes is often near 100 per cent, with little production of unnecessary by-products.

Over the years the technology that uses these microbial chemical plants has changed a great deal. Through the first half of the twentieth century it was fairly simple. Early fermentations involved the anaerobic production of solvents in large vats. Even today, potable alcohol is frequently produced in 20 000 gallon wood tanks with open tops and no instrumentation to monitor the course of the fermentation reaction. Aerobic fermentations were originally carried out in tray cultures with the microorganisms growing on the surface of a layer of grain or grain products. Acetic and gluconic acids were produced by bringing the fermentation broth into contact with air in beds of wood shavings.

However, the need to produce yeast in large volumes led to the development of large fermentation vessels into which air could be introduced. Once the importance of penicillin was established during World War II, the commercial production of this antibiotic led to the infusion of modern chemical engineering into fermentation technology. Fermentation vessels were then routinely designed for sterilization and control of pH and temperature, although fermentations were still carried out in the 'batch mode' (Fig. 3.7).

In this type of operation the vessel and its nutrient contents are sterilized, and a fresh inoculum of rapidly growing cells is introduced so that the cell volume amounts to 1 to 10 per cent of the total vessel contents. For aerobic fermentations the vessel is agitated and filtered air is introduced. After 2 to 5 days of growth, the fermentation broth is pumped out and filtered to remove the cell mass and solids before the product is recovered and purified. More recently, fermentation operations have been extended to include the continuous type, either with or without recycling of the cells. This permits the processing of more material per vessel with a corresponding reduction in capital costs. The production of high-volume chemicals such as ethanol has been greatly facilitated by the use of the continuous system with cell recycling.

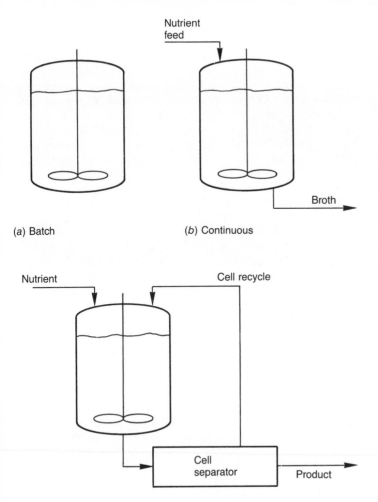

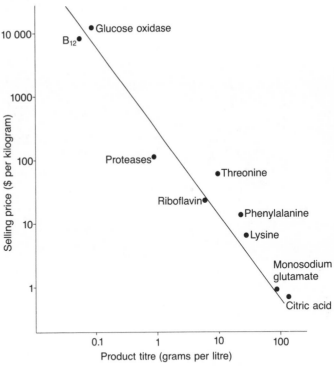

Fig. 3.8 Effect of product titre on the selling price. The higher the concentration of the product in the fermentation broth, the lower the selling price will be.

Fig. 3.7 Three types of fermentor operation. In the batch mode (*a*) the microorganisms and the necessary nutrients are simply mixed in the fermentor until the reaction has reached the desired stage of completion. At that time the contents of the vessel are withdrawn and the product recovered. In the continuous mode, (*b*) a nutrient solution is fed slowly into the fermentor while the broth with the product is slowly withdrawn. Some continuous operations allow for the cells that are removed with the broth to be recycled back into the fermentor (*c*).

The key factors that influence production costs are the fermentation productivity, that is, the product's titre (concentration in the fermentation broth) per cycle time; the yield from the carbon source; and the ease of recovery and purification of the product. Because the cycle times of most fermentations are 2 to 3 days, the productivity is most affected by product titre. Over four log cycles the prices of a wide range of fermentation

products can be directly related to the product concentration in the fermentation broth (Fig. 3.8). The effect is particularly strong for concentrations below 0.1 per cent. Above a 1 per cent concentration the production costs are more dependent on the special requirements of the individual product. An example is the fermentation cost for the production of an amino acid. As the titre increases above 10 grams per litre it has a diminishing effect on the production cost (Fig. 3.9).

A second important cost factor is the yield of product from the carbon source, which typically accounts for 20 to 60 per cent of the total production cost. High yields are therefore imperative to keep those costs down. This is particularly true for the less expensive industrial chemicals. Breakdown of the fermentation costs for a typical organic acid at two different yields illustrates that the cost of the raw material drops significantly as the product yield and titre rise (Table 3.5). The percentage of the total remains almost unchanged, however. The utility costs also drop very significantly because the higher yield permits a large decrease in the fermentor volume.

The recovery and purification costs can be highly variable, depending on the product. Some products, the hydrolytic enzyme amylase, for example, are recovered simply by removing the cell mass from the fermentation broth; further purification

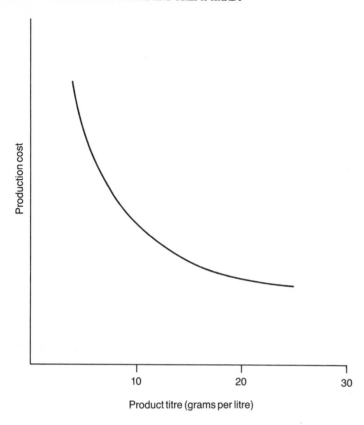

Product cost

Product titre (grams per litre)

10   20   30

Fig. 3.9   Effect of the product titre on the production costs of an amino acid. The price comes down rapidly until the titre reaches 10 grams per litre, after which further increases in the concentration have a diminishing effect on price.

Table 3.5. *Fermentation product cost breakdown*

Annual production: 5 million pounds				
Carbon source: glucose				
Titre:	10 grams/litre		40 grams/litre	
Product yield[a]:	30%		63%	
	$/lb	%	$/lb	%
Raw materials	1.55	45	0.93	43
Utilities	0.61	17	0.20	9
Labour	0.79	23	0.79	36
Capital	0.53	15	0.27	12
Total	3.48		2.19	

[a]Product produced per unit of glucose consumed, as percentage of theoretical maximum.

is not needed. Other products, such as the proteins that are intended for therapeutic use in human patients, require careful purification to remove contaminating agents.

For a fairly simple commodity product, such as an organic acid, the recovery and purification costs amount to approximately one-half of the capital cost investment and about one-half of the operating costs. Other industrial chemicals, including absolute alcohol, which must be distilled, also have approximately one-half of the total cost tied up in the recovery and purification operations. These processes are very sensitive to the size of the facility. Substantial cost savings can therefore be realized as the production capacity increases.

A number of problems have plagued the fermentation industry from the outset. One of the most pervasive and troublesome is microbial contamination. Once a competing microorganism infects the fermentation batch, the product yield drops and the product quality may decline to unacceptable levels. The result may be a complete loss of the contaminated fermentation batch. This could represent up to 1 per cent of the annual production of the product, because there are usually no more than 100 to 150 batches per year. Although many fermentations become contaminated, few result in a total loss.

The bacterial viruses called phages can produce a second type of contamination. Phage infection usually results in a rapid lysis of the entire population of production cells and a complete loss of the fermentation. The bacteria used to produce organic and amino acids are particularly susceptible to this type of contamination and must be protected by operating the fermentation under the strictest of aseptic conditions. Once a contamination occurs, the facility is at increased risk of further problems. The source of the contaminant must be quickly determined and all infected areas sterilized to protect against repeat contaminations.

The mutants that may spontaneously appear in microbial populations constitute a third type of contaminant. The larger the number of generations during the fermentation cycle, the more likely will be the appearance of a less productive mutant subpopulation. This is one of the principal reasons why continuous fermentation processes have not been adopted for many products. The more the producing organism has been modified and selected for high product formation, the greater is the selection pressure for a non-producing subpopulation of the microorganism to take over the batch.

A second production constraint is the need to provide sufficient oxygen transfer rates to aerobic fermentations to maintain growth rates at high cell densities. As the microbial population grows, nutrients may be supplied at high rates simply by using pumps, but the means of oxygen introduction may be less efficient. For rapidly growing bacterial populations in vessels of less than 1000 litre capacity, oxygen availability is the most likely limitation on cell growth. As the vessel capacity is increased beyond 1000 litres, however, heat transfer may become a more significant limiting factor because increased vessel size means that there is less surface area per unit of volume for heat release. In that event, oxygen transfer may be sufficient, but the heat removal insufficient to maintain the correct fermentation temperature.

The viscosity of the fermentation broth constitutes a third possible production constraint. Some of the fungal microorganisms form networks of filaments called mycelia. Although these organisms, if slow-growing, may not be limited by the oxygen supply, the viscosity of the broth may become sufficiently high as the cell population increases for the broth to become unpumpable. As this point the fermentation must be halted, because the broth needs to be transferred by pumps to subsequent processing steps.

A fourth constraint is the need to dispose of the unused components of the fermentation broth. The largest waste component is usually the cell mass itself. Before environmental pollution was identified as a major problem, the cell mass was simply dumped, but once regulatory agencies prohibited this practice, wastes were instead processed through treatment facilities at sites available from the local municipalities. This expense encouraged the development of the solid waste matter as a high-protein additive for animal feed. Today most commercial microorganisms have been approved as feed additives. The exceptions to this include *Escherichia coli*, which is one of the principal bacterial species used for making foreign proteins, and other bacteria that contain toxins. Instead of obtaining a 5 to 10 cent per pound credit for selling the residual cell mass from these organisms, the manufacturer instead has to pay out 10 to 20 cents a pound to dispose of the material in a landfill.

The impact of recombinant DNA technology

The early use of microorganisms for fermentations relied on the selection of specific, naturally occurring strains that produced the desired end-products. These included the alcohol-producing yeasts and the naturally occurring moulds that yielded the organic acids. Although the fundamental metabolic pathways of these organisms are essentially the same, the various species evolved so that they could grow under differing environmental conditions of oxygen concentration, pH and temperature. The organic acids, for example, lower the pH of the surrounding environment to the level preferred by the microorganism, and antibiotics kill foreign bacteria that might be competitors. The production of by-product solvents such as ethanol allows microorganisms to maintain the correct balance between the reduced and oxidized forms of the molecules that participate in energy metabolism. The ability to use different substrates as energy sources also provided competitive advantages for the microbes.

Because the microorganisms evolved to survive and multiply rapidly in specific environments, they became very efficient, producing very little waste. This means, among other things, that the cells maintain very tight control over the numerous

biochemical pathways that they carry out, and that it is therefore necessary to upset these internal controls to increase product yields. Often the control is exerted by blocking expression of the genes that code for the products themselves or for the enzymes needed to make or degrade the products. The cell can also exert a finer level of regulation by blocking the activity of the enzymes after they are made.

One way of disrupting these internal controls is to use various mutating agents, such as chemicals or radiation, that alter the DNA of the cell at random locations. Most frequently, the mutants formed are crippled and of little value. By mutating and screening a large population, however, mutants can be found in which the DNA is disrupted at specific locations that cause the cell to lose its control either over the synthesis of an important enzyme in the pathway that makes a desired product or over the enzyme's activity. As a result, the cell produces increased concentrations of the product and secretes it into the surrounding medium. The process of mutation followed by selection was, and still is, a tedious and imprecise procedure for obtaining a specific deregulated microorganism.

Recombinant DNA technology offers a more direct way of manipulating the cell's metabolism for the production of specific biochemical products. The technology has proved extremely useful for determining the nucleotide sequences of the DNA in and around genes and identifying those segments that are needed for the control of the genes or their products. These segments constitute prime targets that might be specifically mutated or deleted to alter the control of the genes. New genes might also be introduced into bacterial cells to give them novel synthetic capacities.

The first impact of recombinant DNA technology on microbial production of industrial chemicals will most probably be improvement in the yields. Most carbon sources currently consist of starches that must be hydrolysed to simple sugars before they can be used by the microorganism. The hydrolysis does not go to completion but leaves 1 to 2 per cent of the starch as limit dextrins, which are short-chained polysaccharides containing the branch points of the original molecule. The hydrolysed starches also contain unhydrolysed cellulose amounting to as much as 10 per cent of the total carbohydrate.

If new genes that encode enzymes for digesting the limit dextrins and cellulose could be introduced into industrial microorganisms, their carbon source utilization might increase by 2 to 10 per cent. Such a strategy might also obviate the need for the chemical and enzymatic hydrolysis that is now required to convert polysaccharides such as starch into sugars. Introduction of other genes into the microbes might allow carbon sources such as lactose from whey or pentose sugars from the paper and pulp industry to be substituted for hydrolysed starch. By adding new genetic capabilities to commercial microorganisms, additional low-cost raw materials could become available for chemical manufacture.

One of the major goals of fermentation process development is to increase the product yields while reducing the complexity and costs of the raw materials. Using the maximum amount of the raw material for direct production of the end-product with minimal by-product accumulation is most desirable. The primary by-product is the cell mass that accumulates as the microorganisms grow. The best approach to minimizing this accumulation is to uncouple growth from product formation so that the biomass can be used for extended times for the sole purpose of converting the raw materials to the end-product. This has been done for amino acid production by genetically modifying the microorganism to increase the concentrations and activities of the intracellular enzymes that synthesize the product.

For those syntheses that require energy-coupled reactions, a low level of cell growth may be necessary to maintain the oxidation–reduction balance in the cell. The credibility of the strategy of uncoupling growth from product formation is amply demonstrated by such examples as the acetone–butanol fermentation described above, in which the theoretical limit of the product yield from the intermediate organic acid is approached.

Carbon dioxide is a second major by-product of fermentations. The stoichiometry of the reactions within the cell places a limit on the amount of carbon lost to carbon dioxide. This amount can in some cases be reduced by redirecting the carbon flow within the cell. This might be done by adding new genes for alternative pathways that allow different raw materials to be used. This will be particularly effective if the alternative carbon source feeds into the synthetic pathway at a site closer to the end-product than the entry site for the original raw material. The result may be a significant improvement in the yields of traditional industrial chemicals. Modern genetic engineering techniques also permit high-yielding processes for new products to be developed in a small fraction of the time required by the traditional techniques of mutation and selection.

Reducing the length of the fermentation cycle is another approach to improving the biosynthesis of industrial chemicals. Frequently the microorganisms used for fermentations grow slowly and accumulate large quantities of end-products only after a high cell density is reached. The overall process time is typically 2 to 4 days, during which the fermentation is at risk for contamination by phage and foreign or mutant bacteria. Any reduction in cycle time therefore provides two benefits: higher productivity and, as a result, lower capital costs per unit product; and a reduction in the number of fermentation batches lost to contamination.

Growth rates might be increased by switching to a faster-growing microorganism, such as *Escherichia coli*, and engineering into it the capability of producing the desired product. Or, as already mentioned, the growth phase may be uncoupled from the production phase. This should permit the cell to grow unencumbered by the demand to synthesize large quantities of product. Once the proper cell density is reached, the microorganism can be genetically switched from the growth phase to rapid product production by activating a key enzyme within the synthetic pathway. By growing unencumbered, the microor-

ganism should be better able to compete with foreign bacteria and mutants, thereby decreasing the number of batches lost to contamination. When the end-product is toxic to cell growth, this strategy has the additional benefit of allowing the cells to reach a high density before harmful concentrations of product accumulate.

A key constraint in large-scale industrial fermentations is heat generation. Because most commercial fermentations operate near ambient temperatures, heat removal requires the use of refrigerated water to cool the fermentors and large surfaces for heat transfer. The temperature of the operations could be increased, and the problem of heat removal greatly reduced, if strains of producer microorganisms could be developed to withstand high temperatures. This might permit fermentation temperatures to be raised from approximately 35 °C to near 100 °C. Thermophilic microorganisms exist that can grow at such high temperatures.

Protein engineering is another major thrust of research on genetic manipulation. The tools of recombinant DNA techniques can be used to change the nucleotide sequence of a gene and therefore the structure of the corresponding protein. This approach is currently being used to alter the activities of new therapeutic proteins, but it also has great potential for other industrial proteins. Some of the most valuable commodities of the industrial world are fibres and adhesives. Silk, which is all protein, is one of the strongest fibres know to man. It may be possible to use genetic engineering to alter silk's structure to improve its already useful properties. Incorporation of the genes for silk production into alternative organisms may lead to its production at a reduced cost.

The protein-based adhesive made by barnacles is one of the strongest known. Genetic modification of the gene for the adhesive protein could permit the material's application in fields ranging from surgery to manufacturing. Isolation of the gene and its placement in appropriate microorganisms would allow large-scale commercial production of the protein.

In summary, the modern tools of recombinant DNA technology have now opened up an immense potential for manipulating the miniature chemical plants of microorganisms. The first products of the new technology have been therapeutic proteins, such as human insulin and growth hormone, for the health care industry. Improvements in the microbial production of amino acids, industrial organic acids and ethanol will follow, as will further improvements that will allow new raw materials to be used and lower production costs.

Once petroleum prices rise again, the application of recombinant DNA technology to the production of industrial chemicals such as those listed in Table 3.1 will be pursued with even greater vigour. The next area of major impact will be the modification of proteins and polysaccharides to create new products for industrial applications. The large amount of scientific effort now being expended on an international scale virtually guarantees these developments during the remaining years of this century.

Additional reading

Atkinson, B, and Mavituna, F. (1983). *Biochemical Engineering and Biotechnology Handbook*. The Nature Press. New York.

Crueger, W. and Crueger, A. (1982). *Biotechnology: A Textbook of Industrial Microbiology*, ed. T. D. Brock. Science Tech, Inc., Madison, Wisconsin.

Dimmling, W. and Nesemann, G. (1985). Critical assessment of feedstocks for biotechnology. *CRC Critical Reviews in Biotechnology*, 2(3), 233–85.

Eveleigh, D. E. (1981). The microbial production of industrial chemicals. *Scientific American*, 245 (March), 154–78.

Peppler, H. J. and Perlman, D. (1979). *Microbial Technology*, vol. 1, 2nd edn. Academic Press, New York.

Prescott, S. C. and Dunn, C. G. (1959). *Industrial microbiology*, 3rd edn. McGraw-Hill, New York.

Reed, G. (1982). *Prescott & Dunn's Industrial Microbiology*, 4th edn. AVI Publishing Co., Westport, Connecticut.

Rose, A. (1978). *Economic Microbiology*, vol. 2, *Primary Products of Metabolism*. Academic Press, New York.

Schoutens, G. H. and Groot, W. J. (1985). Economic feasibility of the production of iso-propanol–butanol–ethanol fuels from whey permeate. *Process Biochemistry*, 8, 117–21.

Tong, G. E. (1978). Fermentation routes to C_3 and C_4 chemicals. *Chemical Engineering Progress*, 74(4), 70–4.

Tong, G. E. (1976). Industrial chemicals from fermentation, *Enzyme and Microbiological Technology*, 1(3), 173–9.

4 Gene cloning opens up a new frontier in health

L. Patrick Gage

Recent advances in molecular biology have brought novel and powerful methodologies to bear on investigations into the molecular mysteries of life. These tools, which have been provided largely by recombinant DNA and monoclonal antibody technologies, are accelerating the achievement of a biochemical understanding of human physiology. Of particular significance is the impact that these techniques, which have contributed so much to the revolution in biotechnology, are having on the search for the causes of human disease. They are not only promoting the deciphering of the often complex mechanisms that underlie human diseases but, at the same time, are also opening new frontiers in the diagnosis and therapy of those conditions.

Proteins and genes in health research

Human proteins

The deepening understanding of the biochemical basis of life has underlined the prime importance of the role that proteins play in physiology (Fig. 4.1). Proteins, which constitute the most abundant class of molecules in living organisms, make up the principal structural elements of human cells. The structural proteins include the collagen found in skin and connective tissue, the keratin in hair, and the myosin and actin of muscle.

Proteins also play key regulatory roles. The enzymes that catalyse all the biochemical reactions in living cells are proteins. In addition, many of the molecules that transmit regulatory signals from one cell, tissue or organ to another are proteins. Protein hormones such as insulin, glucagon and growth hormone have been known for decades.

More recently discovered proteins that are important in intercellular communications include the numerous cytokines – the interleukins, for example – that cells of the immune system use to communicate with one another, the releasing factors that regulate the release of several pituitary hormones, and a variety of factors that stimulate cell growth. Many additional activities have been ascribed to as yet unisolated and uncharacterized protein factors.

Non-protein organic molecules, including many hormones and chemicals that transmit signals in the nervous system, also

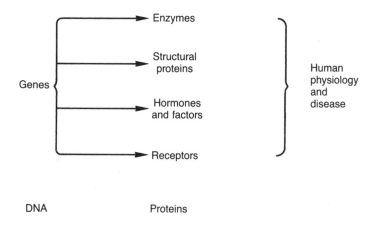

Fig. 4.1   The relation between genes and human physiology. The DNA of the genes encodes all the proteins of the body. If an enzyme, hormone or other protein needed for normal body function is not produced, or is made incorrectly, severe disease can result.

convey intercellular signals. But all bioregulatory molecules, whether protein or non-protein in nature, must interact with specific molecules, called receptors, on the surfaces of the target cells to produce their effects. Because the receptors are themselves proteins, every cell-to-cell regulatory interaction involves specific proteins, either in the role of signal, or receptor, or both.

Biotechnology has provided a set of powerful techniques for the analysis of proteins and for their production and manipulation. In view of the pivotal role that proteins play in all the activities of living cells, the profound effect that biotechnology is having in the production of new therapeutic drugs and diagnostic tests is readily understandable.

Gene cloning and the progress in protein biochemistry

Each and every protein, regardless of whether it is an enzyme, a receptor, a hormone or a structural protein, is a direct representation of a particular gene. The protein-synthesizing machinery that is present in all cells is capable of converting the gene sequence into that of the corresponding protein according to the rules of the genetic code. This code equates a codon consisting of three consecutive nucleotides in the gene DNA to one amino acid in the protein, so that a linear sequence of nucleotide triplets in the DNA becomes expressed as a particular linear sequence of amino acids.

Before the 1970s studies of protein biochemistry were confined to fairly abundant enzymes, structural proteins, and to a few hormones. The techniques of protein isolation and characterization available then were, by today's standards, primitive. As a result, many important bioregulatory proteins, which are generally produced in very small quantities, had either not been discovered or were simply not amenable to study. A full appreciation of the diversity and importance of such rare proteins was not possible.

The discovery of recombinant DNA and the ability to clone genes changed all that. As is perhaps true with the appearance of any breakthrough technology, the availability of recombinant DNA techniques has driven the development of additional methods that are facilitating the achievement of the full potential of the principal technology. The new methods that have sprung up around recombinant DNA technology include techniques for isolating and analysing very small quantities of proteins.

These microscale methods now allow the detection, purification and determination of the partial amino acid sequences of proteins that are present in tissues in minute quantities. Although just 10 to 15 years ago milligrams of a protein were needed for these analyses, they can be done today with 1/10 000 to 1/1 000 000 of those amounts.

Once very small quantities of a protein are purified, biologically characterized, and a partial sequence determined, the focus shifts to gene cloning. The partial amino acid sequence allows the construction of a probe that can be used to identify the gene that encodes the protein in question. When the gene is cloned, its complete sequence can be determined and from that the complete amino acid sequence of the protein can be obtained. The cloned gene can also serve as a probe for identifying any related genes and their protein products.

In addition, the cloned gene can be used to produce large quantities of the corresponding protein, which in nearly all cases would be difficult to obtain by conventional protein purification. The combination of the microscale protein techniques, gene cloning, and protein production from cloned genes has in fact virtually eliminated the large-scale isolation of proteins from natural sources - and laid the foundation for the medical applications of previously rare proteins.

How genes are cloned

Gene cloning is a technology for identifying, isolating and copying a gene for a particular protein with the goal of making the gene available for analysis or of using it to produce the protein. In general, cloning requires three steps: the selection of the protein or protein activity for which the gene is desired; the

identification of a source of genetic material that contains the gene of interest; and the development of an assay for the gene or its product that can used to detect the desired gene clone.

When a protein is present in a cell, tissue or organism it is certain that the gene coding for that protein is present in the cellular DNA – and, for the vast majority of genes, in the DNA of every cell of the organism. To facilitate the identification of particular genes then, investigators frequently use gene 'libraries', which are banks of cloned genes that serve as repositories from which genes can be withdrawn, much as books are withdrawn from traditional libraries.

Gene libraries have been prepared from various microbial, plant and animal sources. For example, Tom Maniatis of Harvard University, Massachusetts, has prepared a widely used library of cloned human genes that represents the total genetic material of a single individual.

To make a complete clone bank of this kind, all the genetic material from a cell or organism is first split into fragments by an appropriate restriction enzyme. The fragments are then inserted individually into self-replicating DNA molecules, which are called vectors and are either plasmids or viral DNAs. The insertion into the vectors involves the formation of recombinant DNA molecules – that is, combinations of DNAs from different sources. Vectors are selected that can replicate indefinitely in a host cell, usually a single-celled organism such as the bacterium *Escherichia coli* (*E. coli*) or yeast, or cultured animal or plant cells.

The host cells, each of which harbours a vector containing a particular foreign DNA fragment, are then grown so that every cell forms a separate colony, that is, a clone. Each colony provides an infinite source of the DNA fragment that was acquired by the colony's parent cell. The clone bank, or library, consists of the pool of all the host cells, each of which harbours a vector containing a particular foreign DNA fragment. In the aggregate, the host cells contain most or all of the genetic information from the original cellular source of the foreign DNA.

Another kind of gene bank contains DNA copies of genes that are actively expressed only in particular cells. Such tissue-specific expression banks, which are made by copying the messenger RNAs in a specific type of cell into DNAs, contain fewer gene clones than banks made from the total DNA of an organism. The copied DNAs are still inserted into vectors that allow replication of the recombinant molecules in suitable host cells, however.

The use of recombinant DNA technology permits the production of large quantities of any gene from any organism in virtually any host cell environment. The challenge is the identification of the desired gene, which is very much like finding a needle in a haystack. The simile is appropriate because the needle (the desired gene) must be present in the haystack (the library of clones) if appropriate cells are selected as the source of the genetic material used to produce the library.

A human clone library that is comprised of gene-sized segments of human chromosomal DNA in *E. coli* cells contains a million or more different clones. Except for the presence of the human genes in the bacteria, the cells are indistinguishable from one another. Unfortunately, the clone library does not come equipped with a Dewey Decimal System for finding the correct gene.

Identification of the desired gene clone

The principal approaches to identifying the gene for a particular protein from a clone bank can be divided into three categories, depending on whether the assay for the gene uses protein sequence information, the biological activity of the gene product, or specific antibodies that can detect the presence of the gene product (Fig. 4.2).

In cases where a partial amino acid sequence is available for the product of the desired gene, radiolabelled DNA probes, containing perhaps 15 to 20 nucleotides, can be constructed so that they have nucleotide sequences that correspond to the amino acid sequence (Fig. 4.3). There will be several possible predicted nucleotide sequences because most amino acids have more than one codon, and consequently several probes will have to be synthesized.

A radiolabelled probe will bind to any gene that has a complementary DNA sequence. The gene thus detected should

Fig. 4.2 Ways of identifying a particular gene in a 'library'. Identifying the desired gene is usually the most difficult part of cloning. A gene library consists of many – perhaps even a million – different clones of cells, each of which contains a DNA fragment from the source material. The desired gene may be present in only one, or at most a few clones and the problem is essentially equivalent to finding a needle in a haystack.

The correct clone may be identified by means of a probe for the gene itself. A probe consists of a short, radioactively labelled segment of DNA that has a sequence corresponding to the known amino acid sequence of a segment of the protein. Alternatively, if the protein product of the gene is made in the cloned cells, it can be detected either by means of a specific antibody or by assaying for the intrinsic biological activity of the protein.

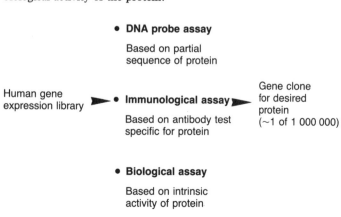

Fig. 4.3 Scheme for cloning a human gene. One method of cloning requires that a small quantity of the protein encoded by the desired gene be isolated and a partial amino acid sequence determined. From this sequence, the nucleotide sequence of the corresponding DNA segment of the gene can be predicted. The DNA segment is then synthesized in radioactive form and used as probe for detecting the bacterial clone in the human gene library that contains the gene itself. Once the correct clone has been identified, the gene can be obtained in sufficient quantity to determine its complete nucleotide sequence, and therefore the complete amino acid sequence of the protein. The gene can also be put in an 'expression vector' that will permit large amounts of the protein to be made in some easily grown cell, such as a bacterium or yeast.

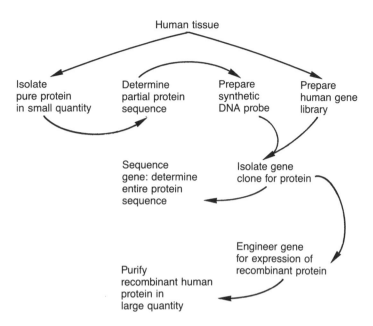

encode the protein sequence on which the probe is based. Under appropriate experimental conditions, probe binding will identify the desired gene from among the millions in the clone bank. This technique has frequently been used to clone genes for proteins that have been isolated only in the small quantities necessary for determining the initial partial amino acid sequence.

When a protein has a biological activity that is easy to detect, the corresponding gene in a clone bank may be identified by screening for expression of the protein. This technique requires that the foreign gene segment produce a functional protein, such as an enzyme or hormone, in the host cells. Any cell that makes the protein must have the desired gene clone. Expression screening has frequently been used for gene cloning because the method has the marked advantage of not requiring that any of the protein be purified first.

The third approach also depends on the ability of the desired gene to produce its protein, but in this method the presence of the protein is detected by a specific antibody, rather than by an assay for a biological activity. The genes coding for proteins for which good antibodies are available are candidates for this method of identification.

Gene characterization and manipulation

Rapid DNA sequencing is another of the valuable techniques that have developed in conjunction with recombinant DNA technology. These methods for the rapid determination of the nucleotide sequences of DNA have been of paramount importance for characterizing cloned genes. They have made the rather tedious process of sequencing of proteins themselves virtually obsolete, because the entire nucleotide sequence of a gene can now be determined in days. The gene sequence can in turn be used to derive the amino acid sequence of the corresponding protein.

Investigators often want to alter the amino acid sequences of proteins, either for the purpose of working out the detailed relation of structure to function or to modify the physiological functions of the proteins in specific ways. Proteins are difficult to alter directly once they have been synthesized, but today genes can be changed in any way desired. The change will be faithfully represented in the protein produced by the gene, and this altered protein can be made indefinitely and in unlimited quantities in cell clones that harbour the altered gene.

The genetic engineering of cells for protein production

A remarkable aspect of the genetic code and protein synthesis is their conservation in virtually all of the diverse life forms on the earth. This commonality in the use of DNA as the substance of genes and of the identical code for the read-out of genes into proteins allows for the use of the cells of any organism to produce proteins from recombinant genes. In fact, it is now a relatively simple exercise in genetic engineering to produce human proteins in bacterial, plant or animal cells by transplanting the appropriate cloned gene into the cells.

For example, the human protein interferon, which is being tested as an anti-cancer and anti-viral agent, was previously obtainable only by extraction from human cells. The small quantities thus produced did not permit the clinical evaluation of interferon in significant numbers of patients. Moreover, humans have several interferon genes and the material isolated from human cells is a complex mixture of structurally different interferons, a situation which was further complicating the efforts to assess its clinical efficacy. The advent of gene cloning has made possible the production in bacterial cells of large quantities of pure interferon from any of the different human genes.

Although the genetic code and the machinery used by cells to make proteins are universal, the actual signals that control the expression of genes differ from one organism to another. Consequently, to obtain high expression of a transferred gene it is nearly always necessary to link the protein-coding sequences of the gene to control sequences from the organism in which the protein is to be produced.

The gene expression and protein synthesis signals are best understood for *E. coli*, a common bacterium that has been intensely studied for many decades. It is not surprising then that *E. coli* is the organism most frequently used for producing recombinant proteins. Foreign genes are usually inserted into a small circular DNA molecule, called a plasmid, that is engineered to contain genetic functions that assure its maintenance in the bacterial cells. The foreign gene is inserted by breaking the plasmid with a restriction enzyme and then using an enzyme known as a ligase to join the ends of the gene to the ends of the plasmid DNA.

Expression is engineered by incorporating appropriate host-cell regulatory signals into the plasmid with the gene (Fig. 4.4). In most instances, extensively studied signals that are derived either from *E. coli* genes or from the genes of viruses that infect the bacterium are used to drive the expression of the foreign genes.

Some recombinant plasmids work so well that when they are introduced into *E. coli*, which already has about 1 000 genes of its own, the foreign protein produced by the plasmid accounts for as much as 60 per cent of the total protein made in the bacterial cells. With the use of this powerful technology, any protein can be obtained in virtually any quantity so long as its gene can be cloned. Human proteins that are manufactured by the fermentation of appropriate recombinant *E. coli* clones have already been approved for clinical use in humans. These recombinant products include the human insulin made by Eli Lilly and Co., in Indianapolis, Indiana; the human growth hormone that is produced by Genentech Inc., of South San Francisco, California, and human interferons that are made by Hoffmann-La Roche, Inc., of Nutley, New Jersey, and Schering-Plough Corp. of Madison, New Jersey.

The yeast *Saccharomyces cerevisiae* is a single-celled fungus that has certain potential advantages over *E. coli* for producing human proteins. The bacterium occurs naturally in the intestines of humans and, under certain circumstances, can cause disease. The yeast is rarely a human pathogen. Moreover, it secretes many proteins into the medium in which it is grown, which means that the proteins can be harvested without having to separate them from yeast cellular proteins. Another advantage of using yeast cells is their ability to add sugar side-chains to the proteins they make. Such side-chains are common features of the proteins of humans and other higher organisms and may be necessary for the normal functioning of the proteins. The side-chains produced in yeast are not quite the same as those made in mammalian cells, however.

Many important human proteins are being produced in yeast, but this organism's advantages over *E. coli* for the production of pharmaceutical proteins have yet to be demonstrated. However, yeast and other fungal cells may be ideal for the large-scale production of bulk proteins, such as industrial enzymes, for which the economics of production are critical.

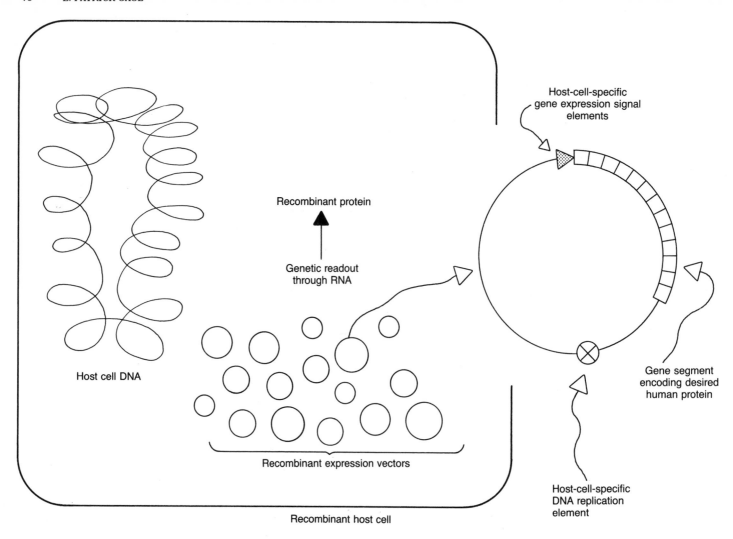

Fig. 4.4   Production of recombinant proteins. The gene encoding the desired protein is inserted into an expression vector (shown enlarged in the right-hand portion of the diagram), which is usually either a plasmid or virus. The expression vector allows the gene product to be made in the desired host cells, whether these be bacterial, yeast, mammalian or plant in origin. The vector must carry the necessary signals for its replication in the host cells, and also the gene control signals needed for transcription of the vector DNA into messenger RNA and the subsequent translation of the messenger RNA into protein.

Biologically important human proteins are often too complex to be synthesized in *E. coli*, which does not add sugar side-chains to proteins. Moreover, several mammalian proteins of therapeutic interest have complex folded structures that are necessary for the biological activity of the proteins and are maintained by cross-bridges of a very specific nature.

These complex proteins include tissue plasminogen activator (TPA), which may be useful for dissolving blood clots, and factor VIII, which is needed for treating haemophilia patients. Proteins such as these are not made in the correct conformation by *E. coli* and are therefore inactive. Mammalian cell systems have been developed that make substantial quantities of active recombinant proteins, which are often secreted into the culture medium.

Advances in understanding disease

The single greatest benefit of biotechnology is the marked acceleration that it is producing in the acquisition of a detailed understanding of human physiology and disease at the molecular level (Table 4.1). In addition to underlining the paramount importance of proteins, the research is providing new information about the genetic aspects of disease, thereby leading to a better understanding of the molecular mechanisms underlying cancer and infectious diseases such as AIDS (acquired immune deficiency syndrome). It is also opening the way to methods for diagnosing, and perhaps treating, genetic diseases.

Table 4.1. *Contributions of biotechnology to basic science*

Advances in biotechnological techniques accelerate the acquisition of fundamental knowledge concerning living systems:

1. By uncovering natural physiological control networks
2. By indicating the molecular causes of disease
3. By revealing key regulatory agents and events
4. By identifying target molecules and molecular interactions for therapeutic intervention

Communication between cells

A very important category of proteins and peptides is those involved in the communication between cells. Intercellular signalling is essential for the appropriate coordination of the functions of the diverse tissues and organs of the body. Often this communication is achieved by regulatory proteins that are secreted by one cell to influence the activity of another. Although hundreds of such regulatory proteins have been identified, researchers are just now beginning to learn about the intricate network of interactions that they control.

This progress is largely the result of gene cloning techniques. For the first time, regulatory proteins such as hormones, growth factors and receptors can be produced in sufficient quantities for analysing their activities and for using them as therapeutic agents, if appropriate.

The interferons are among the earliest of the intercellular signal proteins to be investigated by the techniques of biotechnology. Although progress was being made in purifying two of the main types of interferon, the alpha and beta interferons, before their genes were cloned, the ability to use the cloned genes to make the human proteins in *E. coli* cells has led to an enormous acceleration in interferon investigations.

For example, the cloning showed that there is a family of nearly 20 different genes coding for human alpha interferon. Some of these proteins have now been made and are being biologically and clinically evaluated. The recombinant interferons that were produced by Hoffman-La Roche and Schering were the first recombinant proteins to be approved for human use by the US Food and Drug Administration (FDA) that were not replacing a protein that was already available by purification from natural sources.

Dramatic evidence of the consequences of the availability of the recombinant interferons is the rapid and thorough clinical evaluation of the proteins' potential as anti-viral and anti-cancer agents that has occurred since the cloning. Although interferon's effectiveness as a single agent in cancer therapy has so far been limited to relatively rare malignancies, the evaluation could not have been made without having the recombinant proteins. Now that interferon is available to the clinical oncologist, its potential for use in combination with other therapeutic agents can be investigated.

The routine achievement of the cloning, expression and purification of naturally rare bioregulatory proteins has had a profound impact on the study of cell biology. Previous cell biology studies usually had to be performed with impure mixtures of bioactive proteins. The studies were exceedingly complex and hard to reproduce.

The recent technological advances have resulted in the discovery of many new regulatory proteins and are straightening out some of the complexities that in the past resulted from work with poorly characterized biological molecules. The progress is particularly evident in immunology. Normal functioning of the immune system depends on a complex set of cell-to-cell interactions that are mediated either by secreted proteins such as the lymphokines and cytokines, or by cell surface proteins, such as the T-cell receptor.

Interleukin-2 is one such cytokine. Another is interleukin-1, a potent molecule that is produced by monocytes and other types of immune cell and plays a critical role in the activation of certain immune responses. Researchers had previously suspected that interleukin-1 has many additional activities, not all of which are limited to the immune system. These include fever production and a number of activities that could aid in normal wound healing, but could also contribute to the development of chronic inflammation and conditions such as arthritis.

Because investigators were unable to purify significant quantities of the substances that produce the various effects, they were unable to compare those substances directly. The recent cloning of two distinct, but related, interleukin-1 genes has resolved the issue by permitting the production of the pure recombinant proteins. Studies of these proteins have established that the two interleukins-1 do display a wide range of different biological activities.

Moreover, comparison of recombinant interleukin-1 with cachectin, another monocyte product that has become available in recombinant form, reveals that the activities of these two proteins overlap, but show clear differences. In addition, comparison of cachectin's structure and activity with those of tumour necrosis factor, the gene for which has also been cloned, shows that these two proteins are identical. Because tumour necrosis factor (TNF), a naturally occurring protein, had shown promise as an anti-tumour agent against several types of cancers, clinical trials of the agent in tumour patients had already begun before its identity with cachectin was established.

This finding has considerably broadened the understanding of the molecule's role in human physiology, but may also have diminished its value as an anti-tumour agent. Cachectin, although apparently produced as part of the body's efforts to fight off cancer, causes the severe malnourishment and wasting called cachexia that often afflicts cancer patients. Cachectin-TNF also appears to be the principal mediator of the potentially fatal endotoxic shock that occurs in patients following infection by certain microbial pathogens.

Cancer

Cancer is a complex of diseases that are still incompletely understood despite years of research effort. Cancer cells are normal cells that have been altered in ways that make them proliferate inappropriately. They appear to perform the normal cellular functions of growth and differentiation but do so at the wrong time or place or to an inappropriate degree – they have escaped from control by the body. Past research into the causes of cancer has progressed along several different paths that, until recently, led in different directions and rarely crossed.

One research approach is concerned with chemical carcinogenesis, the study of the carcinogenic agents that induce tumours in animals. This approach has identified classes of chemical carcinogens and suggests that the agents cause cancer by producing gene mutations. Moreover, the studies have indicated that cancer is a multi-stage process that involves independent steps, beginning with the initiating step, which is probably a permanent gene change, and proceeding through a series of promoting steps that convert the altered cell to the fully malignant state.

Careful investigation of the natural course of human cancers and of experimental tumours that have been implanted in animals has similarly shown that cancer development is a multi-stage process. Tumours early on are usually limited to one tissue but in subsequent stages of the disease often spread to additional tissues. Studies of tumour biology have not yet yielded much information about the underlying mechanisms of cancer initiation and spread, however.

Investigations of cancer genetics provide additional evidence for the involvement of genes in cancer development. Several forms of cancer, although mostly rare varieties such as Wilm's tumour and retinoblastoma, occur more frequently in certain families, a circumstance that provides strong evidence for a genetic predisposition to these particular diseases. Moreover there is a strong correlation between the occurrence of these and other cancers and specific chromosomal abnormalities.

The search for cancer viruses has been one of the main themes of cancer research for the past 15 years. This search has turned up some viruses that contribute directly or indirectly to the development of human cancers. These include human T-cell lymphotrophic virus I, which causes leukaemias and lymphomas; certain strains of human papilloma virus, which have been linked to cervical cancer; hepatitis B virus, which is a prominent contributor to the development of liver cancer,

especially in China and other countries of South-east Asia, and the Epstein–Barr virus, which causes a common lymphoma in Africa and has also been linked to nasopharyngeal cancer.

But more than that, the viral work, particularly with a number of viruses that have been found to cause cancers in experimental animals, has contributed a great deal to the understanding of the biochemical changes that transform cells to the malignant state. The studies have implicated particular genes that are carried by the animal tumour viruses as causing the transformation to malignancy. These viral genes, called oncogenes, are closely related to normal cellular genes, called 'proto-oncogenes', that function in growth and differentiation.

Apparently, during the course of infection the viruses picked up these cellular genes, which became inappropriately activated as a result, thereby revealing their oncogenic potential. The work on the cancer viruses culminated with the important observation that the cells of some naturally occurring cancers, including human tumours that are not associated with any known viruses, contain activated oncogenes with transforming ability.

The discovery of oncogenes has helped to reconcile the findings of the disparate approaches to cancer research that were outlined above. The existence of a limited set of genes, which are found in all our chromosomes and are capable of leading to oncogenic transformation when they are inappropriately activated, provides a rational molecular basis for understanding cancer.

One line of investigation that has pointed to the importance of oncogene activation in human cancers involves studies of the chromosomal abnormalities that are present in many cancer cells. For example, Burkitt's lymphoma, an aggressively growing malignancy of the B cells of the immune system, is associated with the specific chromosomal rearrangements in which the *myc* oncogene in the tumour cells is translocated to a new chromosomal location. As a result, the gene apparently escapes from the normal cellular regulation of its expression, thereby causing the cancerous transformation of the cells.

Production of certain animal tumours by chemical carcinogenesis has been shown to correlate well with specific mutations that convert a proto-oncogene to an oncogene. The same mutations have also been found in the cells of some human cancers. In addition, the malignant transformation of some cultured cells appears to involve the activation of two independent oncogenes and therefore serves as a model for the investigation of the multi-stage progression of cancer in humans. The progression of certain human tumours to a highly metastatic and lethal form has also been linked to oncogene activation.

A full understanding of the role of oncogenes in human cancer development has yet to be attained. Other factors, in addition to oncogene activation, may contribute. There are indications, for example, that loss or inactivation of substances that inhibit cell growth may also be involved in cancer aetiology. Nevertheless, the discovery of oncogenes is helping to provide the necessary foundation for the investigation of the fundamental mechanisms of oncogenic transformation. Recombinant

DNA technology and related methodologies have at least vastly accelerated these investigations and perhaps made them possible.

Acquired immune deficiency syndrome

Perhaps the clearest case for the power and value of biotechnology in health is the current battle against acquired immune deficiency syndrome (AIDS). This condition, which was first identified in the United States in 1981, is characterized by the occurrence of unusual opportunistic infections or by a previously rare form of cancer called Kaposi's sarcoma or both. The underlying problem was identified as a severe depression of the immune system that is caused by the nearly complete lack of one class of T lymphocytes: the helper cells that are needed for initiating and maintaining many immune responses.

By 1983 there was sufficient evidence to implicate a newly discovered virus, which infects and kills the helper cells, as the cause of AIDS. The principal routes of transmission of the virus are through sexual contact, blood products and contaminated needles. The virus may also be transmitted from mother to child either in the womb or at birth.

Both the number of AIDS cases and the size of the population infected by the causative virus have been growing geometrically. Except for a modulation in the increase that has resulted from alterations in the life-styles of the individuals at risk, the epidemic continues to grow unabated. By the end of 1986 more than 29 000 cases of AIDS and more than 16 000 deaths had been reported in the United States alone and in excess of one million people were thought to harbour the AIDS virus. How many of those individuals will eventually develop the disease is not known, but estimates now range upwards from 30 per cent. In addition, AIDS is widespread in certain areas of central Africa.

A determined effort to understand AIDS and the virus that causes it is being made. Within one year of the discovery of the virus the complete nucleotide sequence of the viral genome was determined – another accomplishment of recombinant DNA technology – and efforts to characterize the functions of each of its genes were under way. An understanding of how the virus works is critical to combatting its effects in individuals who are already infected and to developing vaccines to prevent further infections.

Because the virus poses a threat to the safety of the blood supply, tests to enable the exclusion of blood donated by individuals who had been exposed to it were rapidly developed. The early tests were crude, but only 3 years after the virus was identified more sensitive and reliable tests are being developed both for detecting the virus and for assessing the degree of progression of AIDS in infected individuals.

Efforts to develop vaccines to protect against infection by the AIDS virus are also under way. Vaccines are normally made from killed or attenuated viruses, but the AIDS virus is so dangerous that immunizing even with the killed whole virus is not being pursued. One approach to vaccines that is being explored for protection against AIDS and other diseases is immunization with proteins from the pathogen in question. The proteins can be made in large quantities in genetically engineered microorganisms into which the appropriate viral genes have been introduced. Both Merck Sharp and Dohme Research Laboratories of Rahway, New Jersey, and Smith Kline and French Laboratories of Philadelphia, Pennsylvania, have used this method to protect against hepatitis B virus infections.

Another approach to making vaccines involves immunizing with vaccinia virus that has been genetically engineered to carry and express antigens from another virus – the AIDS virus for example. Vaccinia virus, which is already widely used to immunize against smallpox, would in this case serve as a vehicle for introducing antigens from another, more dangerous virus that could elicit a protective immune response.

Lastly, recombinant DNA techniques may allow researchers to disable the AIDS virus genetically by removing or altering its genes so that it can infect an individual and generate protective immunity against the virulent form of the virus without causing the disease itself. It is too early to know which, if any, of these approaches to an AIDS vaccine will work – but it is clear that none of them could even be contemplated without the powerful techniques of biotechnology.

Finding effective therapies for AIDS patients and for those who have been infected by the AIDS virus but have not progressed to full-blown AIDS is another important challenge. A drug that would suppress the viral infection could prevent both the progression of the disease in infected individuals and its spread to the uninfected.

In the past there has been little success in finding effective anti-viral therapies. However, the new technologies may provide the means of achieving rapid identification of the viral-encoded activities that are essential for growth of the AIDS virus and are therefore targets for anti-viral therapy. The AIDS virus is rather small and contains only a few genes, which may make it easier to find effective drugs.

The increasing knowledge about the functions of the AIDS virus genes and their proteins holds considerable promise that agents that interfere with them will eventually be identified. Because the proteins' functions are unique to the virus, interfering with their activities should not adversely affect the patient being treated. Success against the AIDS virus, when it comes, will be none too soon.

A new frontier in pharmaceuticals

Biotechnology is not only a tool that has allowed an enormous acceleration in the acquisition of knowledge about human disease, but it is also a vehicle for producing proteins and protein derivatives that play crucial roles in the affected physiological systems. The combination of this flowering of the basic understanding of disease and the availability of potent regulatory proteins has opened an exciting new frontier in pharmaceutical research and human medicine.

Stage I: Natural proteins and peptides as pharmaceuticals

The therapeutic use of recombinant forms of natural human proteins is already beginning to fulfil the promise of biotechnology in health. The first such proteins are recombinant human insulin and human growth hormone, both of which have been approved for human use by the FDA. These two pharmaceuticals are not satisfying some newly discovered need, but are replacing the natural proteins.

Until the genes for the hormones were cloned, the only source of insulin was isolation from cattle and swine pancreases. Human growth hormone could only be obtained from human cadaver pituitaries. Both hormones are used to correct deficiencies: insulin for the type of diabetes in which the pancreas fails to make that hormone, and human growth hormone for the dwarfism that results from a failure of the pituitary gland to release growth hormone.

The development of a recombinant human insulin by Eli Lilly and Company assures an adequate supply of this hormone irrespective of the supply of animal pancreases and will also benefit those diabetes patients who become allergic to the animal insulins. The unlimited supply of recombinant human growth hormone, which is made by Genentech, should allow the optimal treatment of children who suffer from pituitary dwarfism or are short in stature for other reasons. The hormone was previously available only in small quantities and consequently was very expensive. In addition, human growth hormone is being evaluated for use in preventing the loss of protein and wasting that occurs in acute disease and injury states.

Recombinant growth hormone was approved by the FDA at a fortunate time – just when the natural human material had been withdrawn from clinical use in the United States because its administration had been linked to the transmission of Creutzfeldt–Jakob disease, a deadly slow virus disorder that results in the irreversible degeneration of the nervous system. This tragic outcome of the use of a natural product highlights a key safety advantage of recombinant products.

While recombinant growth hormone was being developed, research facilitated by the new technologies identified another regulatory agent from the growth hormone cascade. The existence of a 'releasing factor' that is required for the elaboration of growth hormone from the pituitary had been postulated for some time. With the aid of the advanced methods for purifying, sequencing and synthesizing proteins, the groups of Roger Guillemin and of Wylie Vale at the Salk Institute in La Jolla, California, finally, in 1983, isolated growth hormone releasing factor (GRF), which is produced in miniscule quantities by the hypothalamus of the brain.

The actual biochemical defect underlying most cases of pituitary dwarfism is the absence of GRF. Such afflicted individuals can produce growth hormone normally in the pituitary, but it is not released. Now that the releasing factor has been identified and is easily produced, it may eventually replace growth hormone as the preferred and natural treatment for most growth deficiencies.

Other diseases that result from deficiencies of particular proteins will also benefit from the therapeutic administration of recombinant proteins. An important example is haemophilia, a hereditary defect in blood clotting that is caused by a deficiency of the protein known as factor VIII. To prevent bleeding, haemophiliacs have been treated with natural factor VIII that is prepared from human blood. As a result, many haemophiliacs have become infected with the AIDS virus that was present in contaminated factor VIII preparations. Although recent methods for inactivating the virus have improved the safety of the natural material, the availability of a recombinant factor VIII product will relieve haemophilia patients of the fear of AIDS that so tragically complicates their disease.

The underlying molecular defects in several additional genetic diseases have been identified. Some of these, such as Gaucher's, Tay–Sachs and Fabry's diseases, involve deficiencies of enzymes that result in the abnormal accumulation of materials that would otherwise be broken down by the missing enzymes. Individuals with certain of the diseases may also benefit from injection of recombinant forms of the enzymes. These and other hereditary diseases may also benefit from gene therapy – in which some of the patient's cells are genetically engineered to produce the missing enzyme (see Chapter 15).

The use of recombinant forms of natural bioregulatory proteins to augment the normal physiological responses to diseases is a second category of stage I applications. Recombinant leucocyte interferon was initially approved by the FDA for use against a rare malignancy called hairy cell leukaemia, but the agent is also active against other tumours, including AIDS-related Kaposi's sarcoma, renal cell carcinoma and malignant melanoma. The interferon gives a response rate of about 90 per cent in patients with the leukaemia, whereas previous therapies were largely ineffective.

How interferon causes tumour regression is not known with assurance. The agent inhibits the proliferation of some tumour cell lines in culture and enhances the activity of natural killer cells, which are part of the immune system's defences against tumours. Investigation of the interferons for cancer therapy has largely been an empirical process. Studies in mice or other tumour-bearing animals have not translated well into treatments for human patients.

Leucocyte interferon is not the only recombinant protein being evaluated in cancer therapy. Trials of other agents, either alone or in combination with other forms of therapy, have met with some success and with considerable enthusiasm from clinical oncologists even though the agents' mechanisms of action are incompletely understood.

Throughout the history of pharmaceutical research potent substances have been tried and found of potential value for treating diseases before the relation of their action to the underlying mechanism of the diseases was understood. As has been true for investigations of more traditional pharmaceuticals, clinical cancer research with the recombinant interferons and cytokines should not only identify valuable therapies, but should also help to elucidate the biology and immunology of cancer.

Among the additional agents undergoing investigation is interleukin-2, a protein that is produced by the T cells of the immune system and is an important activator of T-cell responses. Cloning of the human interleukin-2 gene has led to the availability of the recombinant protein for study *in vitro* and in small animals. Moreover, because of its potential as an immune stimulator, it is also being clinically tested for AIDS and cancer therapy. Recombinant interleukin-2 has shown promise for treating certain cancers which have resisted other forms of therapy.

One therapy, which is being developed by Stephen Rosenberg and his colleagues at the National Cancer Institute in Bethesda, Maryland, has shown particular promise in early studies against advanced solid tumours that have proved refractory to other treatments. For this treatment method white blood cells are taken from a patient, activated by exposure to interleukin-2 in culture and then readministered to the patient together with more of the protein.

Tumours have shrunk in nearly 40 per cent of the 55 patients who have received this treatment and five individuals appear to be tumour-free. However, the side effects of the interleukin-2 administration can be severe – a few patients have died as a result – and the procedure is complex and costly. If the early promise of the method is borne out by further investigations, the need for the therapy will drive the development of simpler, less costly procedures with less severe side effects.

Another natural protein that is showing very promising results in clinical trials is tissue plasminogen activator (TPA), a human enzyme that dissolves blood clots by converting plasminogen to plasmin. Plasmin is another enzyme that works by breaking down the protein fibrin, which is a major component of blood clots. The enzymes urokinase and streptokinase are already used clinically for clot dissolution. They also activate plasmin but do so in such a way that both the insoluble fibrin of clots and the soluble fibrinogen of the blood, which produces the fibrin and is required to maintain the integrity of the coagulation system, are destroyed.

TPA activates plasmin predominantly at the fibrin clots and therefore has greater specificity. It has the additional advantage of being amenable to administration by infusion into the veins, whereas urokinase and streptokinase are best administered through a catheter directly to the clot site. If clinical trials provide further evidence of the efficacy of TPA, then its administration may some day revolutionize acute therapy for heart attacks. The enzyme may be the first of a series of enzymes used to augment the body's clot clearing mechanisms.

Stage I research that is directed towards the discovery, biochemical characterization and biological evaluation of natural human proteins for medical purposes is by no means nearing an end. In addition to the few proteins that have been well characterized so far, there still remains a reservoir of poorly understood bioregulatory proteins, critical regulatory enzymes, and receptors that may serve as candidates or targets for therapeutics.

Although recombinant human proteins may have the desired pharmacological effects, they are not necessarily optimally suited for use as pharmaceuticals. In the body the proteins are often made at the site where they are used and they commonly have very short half lives. The need to deliver proteins by injection (they are destroyed in the digestive tract and therefore cannot be given orally) and their short survival times are the two characteristics that makes them less than ideal as pharmaceuticals.

Growth hormone releasing factor provides a case in point. It is synthesized in the hypothalamus and is transported a few centimeters to the pituitary gland where it acts. Although GRF is quickly destroyed in the blood, its concentration in the pituitary is sufficient to regulate growth and the body's nitrogen balance by controlling the release of growth hormone. Also, GRF normally works in concert with negative control factors, such as somatostatin, to maintain growth hormone release at the correct levels. The delicate balance of opposing factors is not achieved when recombinant growth hormone is administered by injection, but is largely achieved with GRF therapy.

Two different approaches are being used to improve therapy with GRF and other recombinant proteins. These are the modification of the agent itself to make it less subject to degradation, and the development of new systems for delivering the agents. The demonstration of the efficacy of protein therapy for more and more diseases will drive the technological innovation that is required to come up with delivery systems that can provide a protein where and when it is needed.

Slow release of material that has been encapsulated in a polymer coating or in the membranous structures known as liposomes is one possibility. Another is the development of computer-controlled instruments for drug delivery. Ultimately the problems of the short half-lives and the need to administer by injection can be overcome.

Stage II: Modified natural proteins and peptides as pharmaceuticals

The traditional approach to drug discovery involves the use of experimental model systems for identifying agents, whether natural or synthetic, that have the desired effects. Most often the molecules identified in this way prove unsuitable for drug use because they are not sufficiently potent or available to living organisms, or because they are toxic or too difficult to synthesize.

Such molecules serve as important leads to the synthetic chemist, however, in that their structures may suggest the specific structural requirements that confer a desired activity on a molecule. The information can be used to construct variants of the original molecule that may, when tested, further define the relation between structure and pharmacological activity. This method has been used many times to progress from an initial compound that has a semblance of a desirable pharmacological activity to an effective and safe drug.

Biotechnology is providing a remarkable parallel to the more traditional approach to drug design. It has produced a new class of leads – the naturally occurring, pharmacologically active proteins. The technologies have also provided a new kind of macromolecular chemistry in which genes can be modified so that they produce altered proteins that may be more suitable for pharmaceutical use than the original molecules (Table 4.2).

Table 4.2. *Stages of biotechnological development*

Stage 1	Making natural proteins, such as: Interferons Interleukins Tissue plasminogen activator Monoclonal antibodies
Stage 2	Making modified natural proteins: Second and third generation tissue plasminogen activator Human–mouse hybrid antibodies
Stage 3	Creating synthetic mimetics of natural proteins

An excellent example of this stage II process of directed protein modification is the effort to develop second-generation TPA molecules. The molecules of strictly regulated enzymes such as TPA are subdivided into functional regions called 'domains' so that intricate regulation can be achieved by means of the binding of other control factors. For example, TPA has an active site domain that catalyses the conversion of plasminogen to plasmin, which then destroys fibrin clots. Another domain of TPA confers on the enzyme its specificity for activation of plasmin at fibrin clots. Still another TPA domain is the binding site for a natural inhibitory substance (or substances).

Because the biological half-life of TPA is very short, substantial doses of the protein are required to achieve effective clot dissolution. Moreover, the complex structure of the molecule requires production in animal cells. In microbial cells, such as those of *E. coli*, the usual choice for production of recombinant proteins, TPA is made correctly with regard to its primary amino acid sequence, but the numerous disulphide cross-bridges are incorrectly formed and the protein does not fold properly.

Because TPA made in animal cells is expensive, it would be desirable to reduce the amounts that have to be given by increasing the stability of the protein and decreasing its sensitivity to the natural inhibitor. A form of TPA with a longer duration of action would also be useful to prevent secondary clot formation, which is a common occurrence after acute therapy with the agent. A great deal of effort is being expended to remodel the natural TPA molecule along these lines to achieve a more suitable pharmaceutical preparation.

Peptides are smaller than proteins and can be made by synthetic chemistry rather than by expression of recombinant genes. Peptides have long been targets for modification and several commercially available pharmaceutical agents are forms of natural peptides that have been altered to achieve an enhanced duration of action.

Peptide systhesis is not limited to the use of natural amino acid as is the production of recombinant proteins by living cells, and many modifications are possible that make peptides stable enough to be given orally. The active sites of some of the larger bioregulatory proteins can even be mimicked by synthetic peptides, a progression that leads to the third stage of the application of biotechnology to pharmacology.

Stage III: Mimetics of natural bioregulatory proteins as pharmaceuticals

Significant efforts are under way to develop the capability to mimic the biological activity of proteins with organic molecules that are free of labile peptide bonds and can be given orally. This work is based on the knowledge that proteins interact with one another and with other molecules through highly specific and space-limited portions of the protein surfaces. The remainder of the protein structure provides the overall scaffolding for the various domains, tying them together into a coherently functioning whole.

The development of suitable protein mimics requires the determination of the detailed three-dimensional structures of bio-active proteins so that their interactions can be understood in atomic detail. In parallel, and contributing to this understanding, are studies of numerous variants of the protein that have altered interactions and biological activities. By combining the results of these investigations a detailed picture of the important structural features of the functional interactions of a protein can be built up.

During the development of recombinant DNA technology there has been remarkable progress in the technologies for determining the three-dimensional structures of large protein molecules. Paramount among these are the advances that have been made in the theory and instrumentation for the determination of protein structure by X-ray crystallography. The use of the new radiation sources, sophisticated detectors that vastly accelerate data collection and powerful computer hardware and software have made it possible to obtain the complete three-dimensional structure of a typical bioregulatory protein in about one year.

The primary limitation of X-ray crystallographic methods remains the need to prepare protein crystals of sufficient quality for analysis, which is still more of an art than a science. Modern technology has provided little help for this problem.

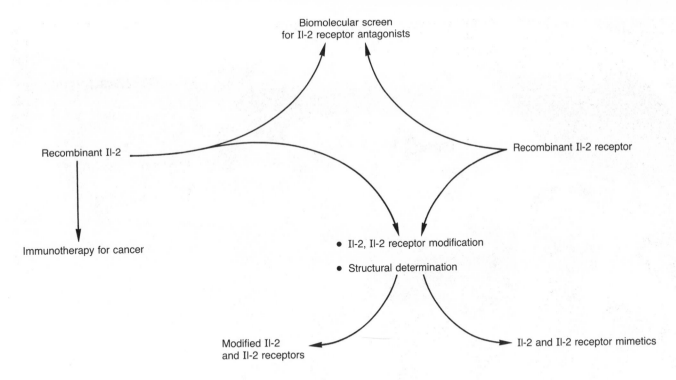

Fig. 4.5 Ways to use recombinant interleukin-2 and the recombinant interleukin-2 receptor. Recombinant interleukin-2 (Il-2) is already being used directly in experimental immunotherapies for cancer. Efforts are also under way to achieve a better understanding of the relation between the structures of the proteins and their activities, with the idea of producing both modified forms of the natural proteins and synthetic compounds that mimic their effects. In this way, it may be possible to tailor new pharmacological agents that are more effective or have a different range of properties from the originals.

The recombinant molecules may also be used indirectly as the basis of an assay for identifying drugs that block the binding of interleukin-2 to its receptor and might therefore be helpful for suppressing immune responses in patients who receive organ transplants or who have autoimmune diseases.

High-field nuclear magnetic resonance spectroscopy, which allows a scientist to determine the structures of molecules in solution, is a second technique that is advancing. This technology is currently limited to molecules that are smaller than most bio-active proteins, but is improving as more powerful instruments are developed.

Lastly, computer methods for predicting the three-dimensional structures of proteins either from their amino acid sequences or from the known structures of related molecules are advancing as more powerful computers become available. This approach, which is at the interface between biotechnology and computer technology, also uses modern molecular graphics techniques that allow scientists to view large molecules in three-

dimensions on a television screen, to modify the structure, and to model the interactions between molecules. The technique will be of enormous value because it is able to display the detailed three-dimensional structures of recombinant proteins.

New tools for pharmaceutical discovery

Although this chapter has concentrated on the direct therapeutic applications of recombinant proteins and their derivatives and mimetics, the indirect use of recombinant forms of natural proteins also offers opportunities for discovering novel pharmacological agents (Fig. 4.5). For example, interleukin-1 and interleukin-2, both of which profoundly stimulate normal immune responses, elicit their effects by interaction with highly specific receptor proteins that are located on the surfaces of the target cells. Agents that block these interactions would suppress immune responses and might be therapeutically valuable for preventing the rejection of organ transplants or treating conditions such as systemic lupus erythematosus and rheumatoid arthritis in which the immune system is overactive.

A new approach to identifying such interleukin antagonists uses recombinant forms of the interleukins and their receptors in highly sensitive and selective assays for detecting agents that block their interactions. Because the assays are computerized and automated, they allow the test of enormous numbers of randomly chosen agents (Fig. 4.6). Synthetic organic compounds or the complex mixtures of materials present in the

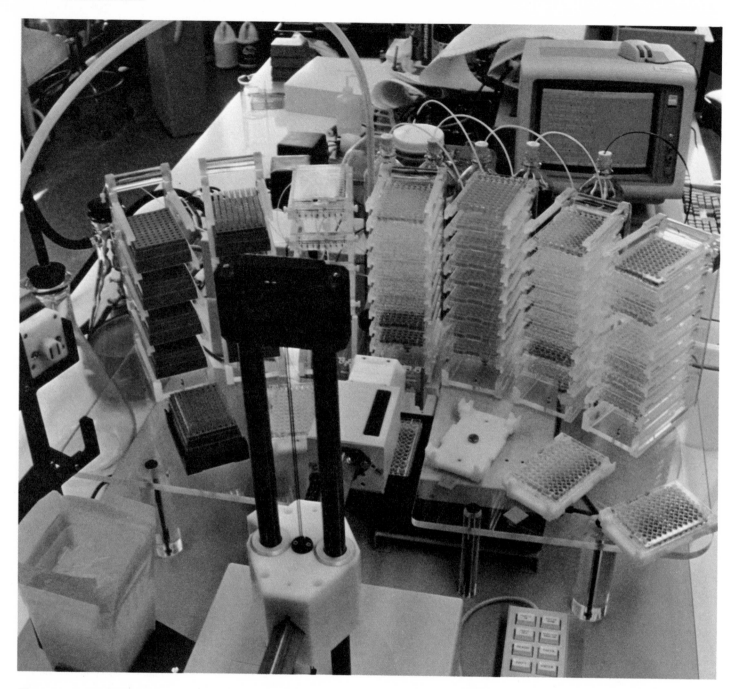

Fig. 4.6 Automated screening for drugs that inhibit the binding of interleukin-1 and interleukin-2 to their receptors. The device is equipped for screening the activity of 400 different samples, in duplicate, in three assays simultaneously. The assays are for interference with the binding of interleukin-1, interleukin-2, and an immunoglobulin to their receptors. The robot that performs the operations is in the foreground with its arm extending backwards to the columns of plates; the computer that controls the procedures is in the background.

The third column from the right contains the plates (amber-coloured) with the samples to be analysed, and the first, second and fourth columns from the right contain plates with the assay materials. Each plate has 96 wells for holding samples or assay materials. The robot hand carries 96 disposable pipette tips, which it picks up from the blue trays on the left and uses to withdraw material from the wells in a sample plate and deposit in plates for each of the three assays.

growth media of microorganisms can be screened for interleukin antagonist activity by the new methods.

Compounds found to have activity could, if not used directly, undergo systematic chemical modification to produce highly specific drugs that are suitable for human use. Because the active molecules are unlikely to be proteins, and therefore to be of manageable structural complexity, this overall approach lends itself to the established procedures of medicinal chemistry.

The elegance of the new screening method resides in the ability to make a rational selection of the components that serve as the basis of the assays employed. Before recombinant DNA technology become available, screens of this kind were not possible either because the underlying mechanisms of diseases were poorly understood or because the specific biological molecules involved in the mechanism could not be obtained in sufficient quantities. Biotechnology, because of its contributions in both regards, is now making it possible to devise more rational procedures for screening for new drugs.

Conclusion

Biotechnology is beginning to have a profound influence on human health. This chapter has focused on a relatively narrow aspect: the impact of biotechnology on the discovery of new therapeutic agents. The accompanying acceleration in the acquisition of knowledge concerning the biology of life was also discussed. The positive impact of this rapidly expanding knowledge based on improved health cannot be overestimated. Although it is possible to predict the kinds of therapeutics that will be derived from the application of new technologies based on today's level of understanding, predicting how the changing knowledge base will alter therapeutics in the long-term future is difficult.

One thing remains certain: biotechnology is a driving change in medicine and health care. This change is occurring at a more rapid pace than in the past. It can therefore be said that biotechnology has opened a new frontier in health.

Additional reading

Croce, C. M. and Klein, G. K. (1985). Chromosome translocations and human cancer. *Scientific American*, **252** (March), 54–60.

Hunter, T. (1984). The proteins of oncogenes. *Scientific American*, **251** (August), 70–9.

Klausner, A. (1986). Researchers probe second-generation TPA. *Bio/Technology*, **4**, 706–11.

Langone, J. (1986). Cancer: cautious optimism. *Discover*, (March), 36–64.

Lawn, R. M. and Vehar, G. A. (1986). The molecular genetics of hemophilia. *Scientific American*, **254** (March), 48–54.

Maranto, G. (1984). Renaissance for vaccines. *Discover* (September), 61–3.

Van Brunt, J. (1986). Neuropeptides: the brain's special messengers. *Bio/Technology*, **4**, 107–12.

Van Brunt, J. (1986). Protein architecture: designing from the ground up. *Bio/Technology*, **4**, 277–83.

5 The microbial production of biochemicals

Saul L. Neidleman

Microorganisms, because of their size, life habits and versatility, have long been mainstays for the production of chemicals. Microbial cells synthesize a tremendous variety of compounds that range from the relatively simple sugars, amino acids and fatty acids, to the more complex antibiotics and pigments, to the most complex proteins and polysaccharides.

The biological activities of the microbial products are equally diverse. They provide several types of drugs, including the antibiotics that have revolutionized the treatment of infectious diseases; agents that inhibit tumour growth and are used in cancer chemotherapy; anti-inflammatory agents for treating arthritis and related diseases; drugs that reduce high blood pressure; and tranquillizers and other drugs that act on the nervous system. Microbes also provide many of the enzymes used in the food, chemical and pharmaceutical industries. Not all of these chemicals are commercially significant, however, although in some cases they could be if the microbial processes for their production were improved.

Table 5.1 illustrates the diversity of microbial products and shows that they can be classified by either their chemical or functional characteristics, although the two categories overlap. For example, many esters have aromas that make them valuable to the food or cosmetic industries and all enzymes are proteins.

Because the production of chemicals by microorganisms exploits the genetic information possessed and expressed by these living cells, the advent of modern genetic engineering has expanded the chances for the commercial development of microbial products. The ability to alter the genetic composition of a microorganism, either by modifying its own genes or by introducing foreign genes, will have an enormous impact on the production of chemicals by microorganisms.

Enzymes: catalysts for biosynthesis and degradation

Enzymes are the catalysts that carry out all the synthetic and degradative reactions of living organisms. Although enzymes are produced by animals and plants as well as by microorganisms, the enzymes from microbial sources are generally the most suitable for commercial applications (Fig. 5.1). One reason for this is that the microbial products can be mass-produced without the limitations that might be imposed by the season of the year or geographic location, as could be the case for a plant-derived enzyme, for example. In addition, microorganisms grow quickly and the production costs are relatively low.

Not surprisingly, in view of the diversity of the syntheses that can be performed by microorganisms, nature has provided them with a vast reservoir of enzymes that carry out a broad range of reactions. Some 25 000 enzymes are estimated to exist in nature. However, most of this reservoir of enzymes remain untapped. Only about 2000 have been isolated and characterized and even fewer have been exploited commercially.

Nevertheless, microorganisms provide enzymes that act on all the major biological molecules (Table 5.2). Many of these enzymes have commercial applications in the food industry (see also Chapter 2). The world market for enzymes was in excess of \$500 million in 1985.

Catalysis by enzymes offers a number of advantages over traditional chemical catalysis. Enzymes have a high catalytic power; they increase the rates of chemical reactions by factors of 10^9 to 10^{12}. Moreover, they work under mild conditions of temperature, pH and pressure, and display high specificity for the substances on which they act. Enzymes frequently distin-

Table 5.1. *Substances synthesized by microorganisms*

Chemical categories	Functional categories
Alkaloids	Antibiotics
Amino acids	Aromas
Carbohydrates	Drugs other than antibiotics
Esters	Enzymes
Lipids	Enzyme inhibitors
Nucleic acids	Flavours
Organic acids and alcohols	Flavour enhancers
Peptides	Hormones
Proteins	Pesticides
	Pigments
	Surfactants
	Vitamins

Table 5.2. *Microbiological production of enzymes by fermentation*

Enzymes	Substrates	Microorganisms
Alpha-amylase	Starch	*Bacillus amyloliquifaciens*
		Bacillus licheniformis
Beta-amylase	Starch	*Bacillus polymyxa*
amyloglucosidase	Dextrins	*Aspergillus niger*
(glucoamylase)		*Rhizopus niveus*
Cellulases	Cellulose	*Phanerochaete chrysosporium*
		Trichoderma reesei
Glucose isomerase	Glucose	*Actinoplanes missouriensis*
		Arthrobacter sp.
		Bacillus coagulans
		Streptomyces olivochromogenes
Glucose-1-oxidase	Glucose	*Aspergillus niger*
Pyranose-2-oxidase	Glucose	*Polyporus obtusus*
Alpha-glucosidase	Maltose	*Aspergillus niger*
		Bacillus amyloliquifaciens
		Saccharomyces cerevisiae
Lipases	Lipids	*Aspergillus niger*
		Candida cylindracae
		Geotrichum candidum
		Rhizopus arrhizus
Pectinesterase	Pectin	*Aspergillus niger*
		Aspergillus oryzae
Proteinase, alkaline serine	Protein	*Aspergillus oryzae*
		Bacillus licheniformis
Proteinase, acid (pepsin-like)	Protein	*Aspergillus oryzae*
		Aspergillus saitoi
Proteinase, acid (rennin-like)	Protein	*Endothia parasitica*
		Mucor miehei
		Mucor pusillus
Proteinase, neutral	Protein	*Bacillus stearothermophilus*
Pullulanase	Amylopectin	*Aerobacter aerogenes*
		Bacillus cereus var. *mycoides*

guish between closely related molecules. Finally, although most enzymes work in water solutions, some can carry out their catalytic activities in organic solvents.

Despite these impressive credentials, enzymes are not without their disadvantages. They may be insufficiently stable for commercial applications and, because they work in water solutions, separating them from the reaction mixture for subsequent re-use is often difficult. Recovering the product is also difficult because both the reaction substrates and products may be present in low concentrations in the water solution. For example, in one reaction for the enzymatic synthesis of halogen-containing alcohols, which are used for making polymers, the product concentrations in the final reaction mixture are between 0.5 and 2.0 per cent.

The reduction or elimination of these handicaps to the commercial exploitation of enzymes is the goal of much current research. Some of the difficulties inherent in the use of soluble enzymes can be overcome by immobilizing the isolated enzymes, thus converting them to insoluble forms. Whole cells, either alive or dead, can also be immobilized and used for carrying our particular reactions. This is especially helpful when the reaction requires the participation of several microbial enzymes. The advantages of immobilizing enzymes or whole cells include increased enzyme stability, reduced enzyme costs, and greater ease of enzyme separation and recovery.

The production of chemicals by microbial fermentation

Fermentation is generally a batch process in which a microorganism that has not been immobilized is grown in a fermentor; this is simply a tank or vat that contains the microbe in a nutrient medium (Fig. 5.2). The growth medium typically contains a carbon source such as glucose, starch hydrolysate or molasses; a protein or nitrogen source such as soybean meal, corn steep liquor or cotton-seed meal; a source of vitamins such as yeast extract; and minerals and miscellaneous nutrients. The fermentor is also equipped with controls for regulating the temperature, pH and oxygenation of the broth in which the cells grow.

As classically defined, fermentations were anaerobic, but today many fermentation reactions require oxygen. The fermentation products may either be contained within the cells or secreted into the medium. The operations required for product recovery and purification will be somewhat simpler for substances that are secreted.

As already mentioned, microbial fermentations provide many of the enzymes that are used commercially. Enzymes, like other proteins, are polymers that are synthesized by linking together amino acid building blocks to form large molecules. Some enzyme molecules are even more complex in structure because they contain added carbohydrate or lipid components.

Microorganisms also provide a number of non-protein polymers (Table 5.3). Almost all of the substances listed are polysaccharides that are built up by joining simple sugars such as glucose or fructose. An exception is poly-beta-hydroxybutyrate, which is synthesized from the simple 4-carbon fatty acid called beta-hydroxybutyrate. Many of the non-protein polymers find application in the food industry as emulsifying agents and as substances that add bulk or texture to food products. They are also used as insoluble matrices for immobilizing enzymes and cells.

The antibiotics are perhaps the most important of the smaller molecules that are produced by microorganisms. Since Alexander Fleming accidentally discovered the antibiotic penicillin in 1928, these natural inhibitors of microbial growth have revolutionized medicine and saved countless lives (Fig. 5.3). Bacterial diseases such as pneumonia, which can now be readily treated with antibiotics, are no longer the scourge that they once were. In addition, some antibiotics – adriamycin, for example – have anti-tumour effects and are being used for cancer chemotherapy.

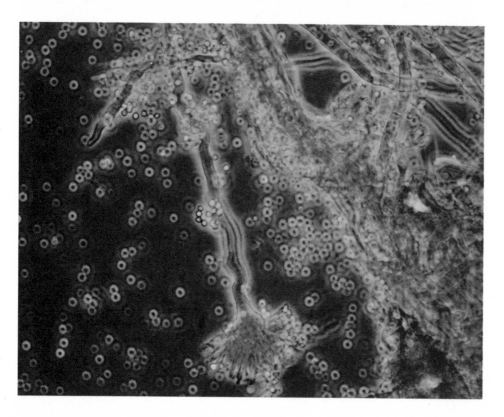

Fig. 5.1   Micrographs of *Aspergillus*. Moulds of the genus *Aspergillus* provide a variety of products, including several enzymes that are used by the food industry and chemicals such as citric and gluconic acids. The moulds are also used for the production of fermented oriental foodstuffs, including sake, shoyu and miso.

Aspergillus sends up stalks topped by the round spore cases called conidiophores (*a*). Micrograph (*b*) shows a higher-magnification view of a conidiophore surrounded by the small spherical spores (the conidia) of *Aspergillus*. Staining with a dye gives the red colour. (Micrographs (*a*) and (*b*) are by kind permission of Robert Toso and Angela Belt, respectively, both at Cetus Corporation in Emeryville, California.)

Fig. 5.2 A laboratory-scale fermentor. The fermentation tank, which has a capacity of 14 litres, is being used for the production of 2-ketogluconic acid, a food-grade organic acid that is produced by the fungus *Acetobacter*. The fermentor has controls and monitors for regulating temperature, pH, oxygenation and nutrient intake.

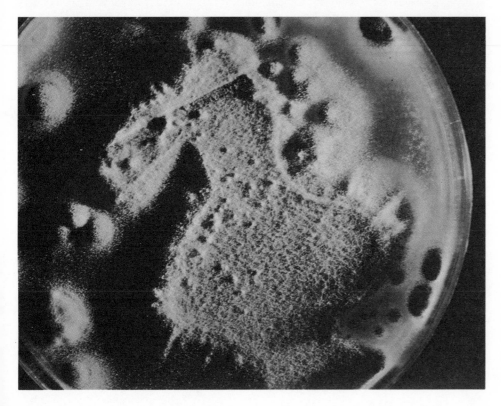

Fig. 5.3 Photograph of a *Penicillium* mould growing in a laboratory dish. Moulds of this genus are the source of the antibiotics of the penicillin family. (By kind permission of Angela Belt, Cetus Corporation.)

60   SAUL L. NEIDLEMAN

Table 5.3. *Microbiological production of polymers by fermentation*

Polymers	Microorganisms
Alginate	*Azotobacter vinelandii*
	Pseudomonas aeruginosa
Cellulose	*Acetobacter* sp.
Curdlan	*Agrobacterium* sp.
Dextran	*Acetobacter* sp.
D-Fructose homopolymer	*Zymomonas mobilis*
Levan	*Bacillus* sp.
	Leuconostoc mesenteroides
	Pseudomonas sp.
	Serratia marscens
Phosphomannan	*Hansenula capsulata*
	Hansenula holstii
	Physarum polycephalum
	Rhizobium meliloti
Poly-beta-hydroxybutyrate	*Alcaligenes eutrophus*
	Methylobacterium organophilum
Scleroglucan	*Sclerotium glucanicum*
Xanthan	*Xanthomonas campestris*

Table 5.4. *Microbiological production of antibiotics by fermentation*

Antibiotics	Chemical classes	Microorganisms
Amphotericin B	Tetraene	*Streptomyces nodusus*
Cephalosporin C	Beta-lactam	*Cephalosporium acremonium*
Chloramphenicol	Nitroaromatic	*Streptomyces venezuelae*
Chlortetracycline	Tetracene	*Streptomyces aureofaciens*
Clavulanic acid	Beta-lactam	*Streptomyces clavuligerus*
Demeclocycline	Tetracene	*Streptomyces aureofaciens*
Erythromycin	Macrolide	*Streptomyces erythreus*
Gentamicin C	Aminoglucoside	*Micromonospora purpurea*
Gramicidins	Polypeptide	*Bacillus brevis*
Griseofulvin	Coumarone + hydroaromatic ring	*Penicillium griseofulvum*
Hydroxytetracycline	Tetracene	*Streptomyces rimosus*
Kanamycin A	Aminoglycoside	*Streptomyces kanamyceticus*
Lincomycin	Amino acid + carbohydrate	*Streptomyces lincolnensis*
Neomycins	Aminoglycoside	*Streptomyces fradiae*
Nystatin	Heptaene	*Streptomyces noursei*
Oleandomycin	Macrolide	*Streptomyces antibioticus*
Paromomycin	Aminoglycoside	*Streptomyces rimosus* forma *paromomycinus*
Penicillins	Beta-Lactam	*Penicillium chrysogenum*
Rifamycins	Aliphatic + aromatic	*Nocardia mediterranei*
Streptomycin	Aminoglycoside	*Streptomyces griseus*
Sulfazecin	Beta-Lactam (monobactam)	*Pseudomonas acidophilus*
Tetracycline	Tetracene	*Streptomyces aureofaciens*
Tyrocidines	Polypeptide	*Bacillus brevis*

Even within the general antibiotic grouping there is a great diversity of chemical structures. More than 2000 antibiotic substances, a few of which are listed in Table 5.4, have been identified. These molecules include peptides, carbohydrates, hydrocarbons, benzene derivatives and a host of other chemical species. The function of these compounds within the microorganisms that produce them is still shrouded in mystery. They probably help the producer defend its territory by killing off other, potentially competing microbes, but this seems unlikely to be their only function.

Having such a broad range of antibiotics available is advantageous from a medical point of view. For one, pathogenic microorganisms vary in their susceptibilities to the various antibiotics. Having a wide selection of the compounds to choose from improves the chances of finding one that is effective against a particular microbe. For another, bacteria can become resistant to an antibiotic if they are exposed to the agent over a period of time. Using a variety of antibiotics can help prevent the emergence of resistant strains.

Over the years antibiotics have been a major source of profit for pharmaceutical companies, thereby providing a compelling demonstration that microorganisms can be a source of valuable commercial products. In the United States alone, antibiotic sales have been variously estimated to have amounted to $1.5 to $3.0 billion in 1986. Finally, antibiotics have also contributed to the fermentation industry because their manufacture has served as a proving ground for the development of new fermentation equipment and also of improved isolation and purification techniques.

The other small molecules that are produced by microbial fermentations include amino acids and organic acids (Fig. 5.4). Most of the 20 or so amino acids that occur in proteins can be produced in this way (Table 5.5). The amino acids are used as

Table 5.5. *Production of amino acids by microbial fermentation*

Amino acids	Microorganisms
DL-Alanine	*Brevibacterium flavum*
L-Arginine	*Brevibacterium flavum*
L-Citrulline	*Bacillus subtilis*
L-Glutamic acid	*Brevibacterium flavum*
L-Histidine	*Corynebacterium glutamicum*
L-Isoleucine	*Brevibacterium flavum*
L-Leucine	*Brevibacterium lactofermentum*
L-Lysine	*Corynebacterium glutamicum*
L-Methionine	*Brevibacterium flavum*
L-Ornithine	*Microbacterium ammoniaphilum*
L-Phenylalanine	*Brevibacterium lactoifermentum*
L-Proline	*Corynebacterium glutamicum*
L-Serine	*Corynebacterium hydrocarboclastus*
L-Threonine	*Corynebacterium glutamicum*
L-Tryptophan	*Brevibacterium flavum*
L-Tyrosine	*Corynebacterium glutamicum*
L-Valine	*Brevibacterium lactofermentum*

nutritional supplements for man and animals. In addition, they serve as precursors for the synthesis of commercial products such as the sweetener aspartame, which is widely used in sugar-free soft-drinks and other low-calorie products. Aspartame is a dipeptide containing phenylalanine and aspartic acid.

Arachidonic and eicosapentaenoic acids are long-chained fatty acids that have important nutritional roles. Arachidonic acid, which contains 20 carbon atoms and 4 carbon-to-carbon double bonds, is one of the essential fatty acids that cannot be synthesized by higher animals and must therefore be acquired in the diet. It is the starting point for the synthesis of a number of extremely potent, physiologically active agents, including the prostaglandins and leukotrienes. These compounds influence a wide range of physiological activities. Some make blood platelets less sticky and may therefore help to prevent abnormal clot formation. Others influence the contractility of smooth muscles, including those in the blood vessels, or regulate kidney function. The effects on blood vessel contractility and kidney function are important components of blood pressure regulation. Still other of the prostaglandins and leukotrienes participate in inflammatory and allergic reactions. Almost all have potential medical applications either as drugs or as targets of drugs.

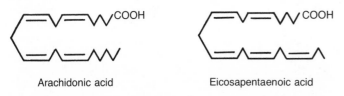

Arachidonic acid   Eicosapentaenoic acid

Eicosapentaenoic acid is an 'omega-3' fatty acid, which means that it has a double bond between carbons 3 and 4. It has recently attracted a great deal of attention as a possible preventer of heart attacks because of its positive effects on blood cholesterol concentrations. Arachidonic and eicosapentaenoic acids occur naturally in the oils of certain fish, including salmon and tuna. The alga *Porphyridium centrum* also produces the fatty acids and could become a commercially important source.

Some of the most fascinating compounds that are produced by microorganisms have pharmacological properties that would appear to be of no significance to the producing organism itself Table 5.6). These include compounds that influence immune responses (immunomodulators), for example, and agents that relieve depression (anti-depressants), reduce the blood pressure (hypotensives), or have anti-inflammatory effects – all activities that would not seem to be of any benefit to the microbes themselves.

The existence of these agents may be nothing more than a pure – albeit fortunate – coincidence. Their pharmacological activities in higher animals may be unrelated to their true biological functions in the producing organisms. Nevertheless, the coincidence may reflect the universality of nature. The molecular configurations at the sites of the agents' actions in mammals and microbes may be similar even though the biological consequences of the actions are presumably divergent in the two groups of organisms.

Also unexpected to some degree is the production by microorganisms of a large number of chemicals that have pleasant aromas or flavours (Table 5.7). Some of these are under investigation as replacements for more expensive flavouring or aromatic agents in foods and perfumes.

The wide range of miscellaneous compounds produced by microbial fermentations includes pigments, enzyme inhibitors, surfactants, plant hormones, herbicides and insecticides (Table 5.8). The normal biological functions of some of these agents is clear. For example, the surfactants act to emulsify water-insoluble foods that would otherwise be inaccessible to the microbes that feed on them. Surfactants also help the producing organisms to adsorb to and release from the surfaces on which they grow. The normal role of others of the miscellaneous microbial products, such as the plant hormones and insecticides, is just as unclear as those of the pharmacological agents mentioned previously.

Not all the materials made by microorganisms are beneficial to man. One example of a harmful microbial product is aflatoxin, a potent carcinogen produced by a fungus that grows on peanuts and grains. In addition, the fungus *Claviceps purpura*, which grows on cereal grains and some grasses, produces a group of compounds known as the ergot alkaloids. These

Table 5.6. *Microbiological production of pharmacologically active compounds*

Compounds	Pharmacological activities	Microorganisms
Amastatin	Immunomodulator	*Streptomyces* sp.
Ascofuranone	Antilipidaemic	*Ascophyta viciae*
Bestatin	Immunomodulator	*Streptomyces olivoreticuli*
1,3-Diphenethylurea	Anti-depressant	*Streptomyces* sp.
Dopastin	Hypotensive	*Pseudomonas* sp.
Esterastin	Immunomodulator	*Streptomyces lavendulae*
Forphenicine	Immunomodulator	*Streptomyces fulvoviridis*
Fusaric acid	Hypotensive	*Fusarium* sp.
Griseofulvin	Anti-inflammatory	*Streptomyces griseofulvum*
N-acetylmuramyl tripeptide	Immunomodulator	*Bacillus cereus*
Naematolin	Coronary vasodilator	*Naematoloma fasciculare*
Oosponal	Hypotensive	*Gloeophyllum striatum*
Oudenone	Hypotensive	*Oudemansiella radicata*
Phialocin	Anticoagulant	*Phialocephala repens*
Slaframine	Salivation inducer	*Rhizoctonia leguminicola*
Zearalenone	Oestrogenic	*Gibberella zeae*

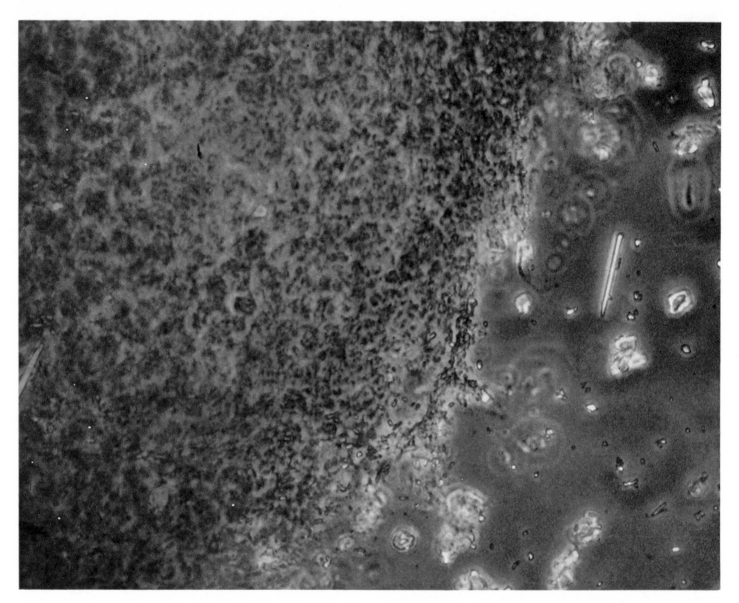

alkaloids have complex structures that include a lysergic acid component:

Fig. 5.4 Micrograph of an *Acetobacter* mould. This mould is used for making a variety of products including 2-ketogluconic acid. (By kind permission of Richard Toso, Cetus Corporation.)

When eaten by man or animals they act on the nervous system and cause convulsions (St Vitus' dance), gangrene and death. The lysergic acid derivative lysergic acid diethylamide is the well-known hallucinogen LSD. Despite the untoward effects of the ergot alkaloids, they are nonetheless valuable, when properly used, for controlling uterine contractions during childbirth and for controlling haemorrhage.

Whatever the explanation for the existence of the various microbial chemicals, fermentations are proving to be a major source of unexpected compounds that may be of value in human medicine. Many of the pharmacologically active substances are undergoing scientific investigation in Japan as potential

Table 5.7. *Microbiological production of aroma and flavour chemicals*

Compounds	Aroma/flavour notes	Microorganisms
Anisaldehyde	Anise-like	*Trametes sauvolens*
Benzaldehyde	Almond-like	*Trametes sauvolens*
Benzyl alcohol	Fruity	*Phellinus igniarius* *Phellinus laevigatus* *Phellinus tremulus*
Cinnamic acid methyl ester	Fruity, jasmine	*Inocybe corydalina*
Citronellol	Rose-like	*Ceratocystis variospora* *Trametes odorata*
Citronellyl acetate	Fruity, rose-like	*Ceratocystis variospora*
Gamma-decalactone	Peach	*Ceratocystis moniliformis* *Sporobolomyces odorus*
Diacetyl	Buttery	*Streptococcus diacetylactis*
3,6-Dihydro-3-methylacetophenone	Fruity	*Mycoacia uda*
p-Alpha-dimethyl-phenyl alcohol	Grassy	*Mycoacia uda*
Ethyl benzoate	Fruity	*Phellinus igniarius*
Ethyl butyrate	Fruity	*Lactobacillus casei* *Pseudomonas fragi* *Streptococcus diacetylactis*
Geranial	Rose-like	*Ceratocystis variospora*
Geraniol	Rose-like	*Ceratocystis variospora*
Geranyl acetate	Rose, lavender	*Ceratocystis variospora*
Linalool	Floral	*Ceratocystis variospora* *Phellinus igniarius* *Phellinus tremulus*
Methyl anisate	Anise-like	*Trametes sauvolens*
Methyl benzoate	Fruity	*Phellinus igniarius* *Phellinus laevigatus*
Methyl benzoate	Fruity	*Phellinus tremulus*
p-Methylacetophenone	Fruity, floral	*Mycoacia uda*
p-Methylbenzyl alcohol	Hyacinth, gardenia	*Mycoacia uda*
Alpha,4-methyl-cyclohex-3-ene ethyl alcohol	Fruity	*Mycoacia uda*
Methyl-p-methoxy-phenylacetate	Anise-like	*Trametes odorata*
Methylphenylacetate	Honey-like	*Trametes odorata*
Methyl salicylate	Wintergreen	*Phellinus igniarius* *Phellinus laevigatus* *Phellinus tremulus*
Neral	Rose-like	*Ceratocystis variospora*
Nerol	Rose-like	*Trametes odorata*
Phenylethyl alcohol	Rose-like	*Phellinus igniarius* *Phellinus tremulus*
6-Pentyl-alpha-pyrone	Coconut-like	*Trichoderma viride*
Tetramethylpyrazines	Nutty	*Corynebacterium glutamicum*
p-Tolualdehyde	Almond-like	*Mycoacia uda*

Table 5.8. *Production of miscellaneous compounds by microbiological fermentation*

Compounds	Properties	Microorganisms
Antipain	Protease inhibitor	Various *Streptomyces*
Astaxanthin	Pigment	*Phaffia rhodozyma*
Avermectin	Anthelminthic	*Streptomyces avermitilis*
Carotenoids	Pigments	*Dunaliella bardarwil* (alga) *Trentepohlia* (alga)
Chymostatin	Protease inhibitor	Various *Streptomyces*
Dopastin	Dopamine-beta-hydroxylase inhibitor	*Pseudomonas* sp.
Elastatinal	Elastase inhibitor	Various *Streptomyces*
Emulsan	Surfactant	*Acinetobacter calcoaceticus*
Fusaric acid	Dopamine-beta-hydroxylase inhibitor	*Fusarium* sp.
Gibberellins	Plant hormones	*Gibberella fujikuroi*
Guanosine	Flavour enhancer precursor	*Bacillus subtilis*
Herbicidin	Herbicide	*Streptomyces saganonensis*
Inosine	Flavour enhancer precursor	*Bacillus subtilis* *Brevibacterium ammoniagenes*
Leupeptin	Protease inhibitor	Various *Streptomyces*
Oudenone	Tyrosine hydroxylase inhibitor	*Oudemansiella radicata*
Pepstatin	Protease inhibitor	Various *Streptomyces*
Phosphatidyl ethanolamine	Surfactant	*Rhodococcus* sp.
Piericidin	Insecticide	*Streptomyces mobaraensis*
Polyketides	Pigments	*Monascus purpureus*
Polymyxin B	Surfactant, antibiotic	*Bacillus polymyxa*
Rhamnolipids	Surfactant	*Pseudomonas aeruginos*
Sophorolipids	Surfactant	*Sorulopsis* sp.
Surfactin	Surfactant	*Bacillus subtilis*
Tetranactin	Miticide	*Streptomyces aureus*
Trehalose lipids	Surfactants	*Rhodococcus erythropolis*
Vitamin B_{12}	Vitamin	*Bacillus* sp. *Propionobacterium shermanii* *Pseudomonas* sp.

therapeutic drugs and as models for the elucidation of pharmacological phenomena. The application of microbial products in a diverse range of additional uses can also be expected to increase.

Products of chemicals by immobilized microorganisms

The use of immobilized cells for biochemical synthesis may have several advantages over the use of fermentation procedures. Higher cell densities may be attainable with immobilized cells than in traditional fermentations and the immobilized cells may give greater yields of product for a given reaction volume. This increased productivity means that there will be less pollution produced per unit of product. Moreover, continuous operation of the process will be simplified.

Immobilized cells may also have advantages over purified and immobilized enzymes. Using immobilized cells can reduce enzyme costs. It eliminates the need for enzyme extraction and purification, and the enzymes in the cells may be more stable during the operations required for the product synthesis than are purified enzymes that have been immobilized.

In addition, more complicated synthesis may be possible with immobilized cells, especially when living cells are used. In most of the early work with immobilized cells only a single enzyme was required to convert the substrates to products. Dead cells suffice for this purpose, as long as they contain the active enzyme. More recently, however, the chemical conversions have involved several enzymatic steps, which are better carried out by living cells. Regeneration of non-enzymatic cofactors, which may be necessary for continued enzyme function in some reaction sequences, can also be performed by living cells.

Cells can be immobilized for synthetic purposes by any of six different methods. The most common is to entrap the cells in a polymer matrix. The polymer may be synthetic, such as polyacrylamide, polyvinylchloride or polyurethane, which are perhaps better known for their uses in plastics and insulation materials. Alternatively, the polymer may be of natural origin. Agar, calcium alginate and K-carrageenan, all of which are produced by algae, are among the natural materials that have been used.

A second technique for immobilizing cells involves cross-linking them with chemical reagents. In a third procedure they can be chemically bonded to some of the same natural or synthetic polymers that are used for entrapping the cells or to inorganic materials such as glass or stainless steel beads, sand, or metal oxides. A fourth immobilization method is absorption of the cells to ion exchange matrices such as diethylaminoethyl cellulose, Sephadex or carboxymethyl cellulose. A fifth process involves encapsulating the cells in microcapsules, in the artificial membranes known as liposomes, or in hollow fibres. The last procedure is cell flocculation through physical, rather than chemical, cross-linking of the microorganisms.

In selecting a matrix a number of factors must be considered. The immobilizing material should be stable and compatible with the microbe used and with the substrates and products of the reaction. The enzyme that catalyses the reaction must not leach from the support nor should the support leach into the product.

Immobilized microorganisms produce many of the same types of compounds as are synthesized in fermentation processes. However, in addition to the advantages already mentioned for the immobilized organisms, immobilization techniques frequently allow the exploitation of microbes that are not ordinarily employed in fermentation reactions. This is true for the synthesis of amino acids and organic acids, for example.

Moreover, immobilized microorganisms can carry out reactions and make products that are not attainable by the more traditional fermentations. Whereas antibiotic production by fermentations usually involves the total synthesis of the compounds, three different types of processes are carried out by immobilized cells. They, too, can perform total antibiotic synthesis as in the production of the widely used compounds bacitracin, penicillin G and streptomycin. But immobilized cells also make precursors for use in the synthesis of a variety of antibiotics and convert the precursors to the final products. An example is the conversion of penicillin G, which is obtained by a fermentation reaction, first to the precursor 6-aminopenicillanic acid and then, in the presence of phenylglycine methyl ester, to the product ampicillin (Fig. 5.5).

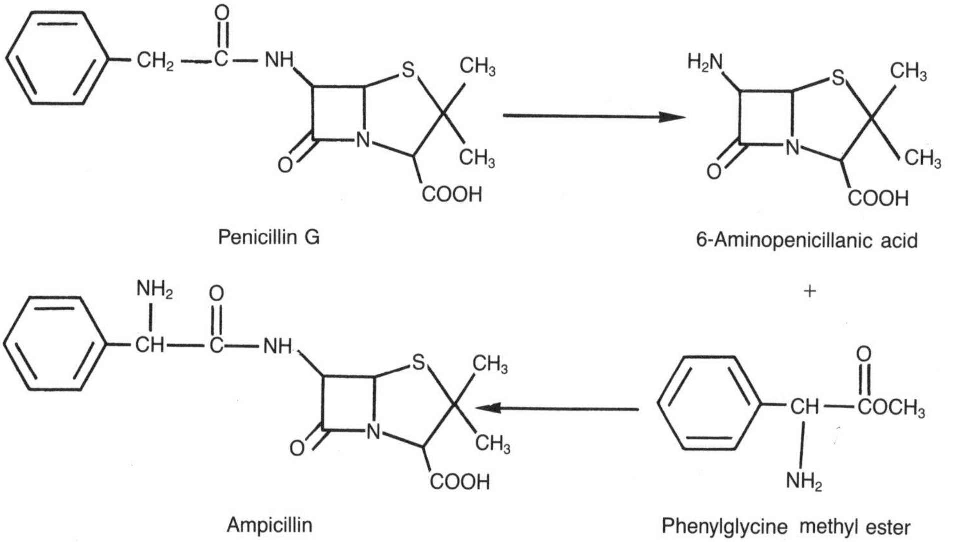

Fig. 5.5 Synthesis of ampicillin. Penicillin G, which is produced by a *Penicillium* mould, is converted to 6-aminopenicillanic acid by any of a number of microorganisms. Then, in a second step, the 6-aminopenicillanic acid and phenylglycine methyl ester are converted microbially to the widely used antibiotic ampicillin.

Immobilized cells also catalyse some commercially significant carbohydrate modifications, most of which would not be feasible in fermentation methods because the desired products would undergo further reactions. Among these carbohydrate modifications is the production of high-fructose syrup, which is frequently used as a sweetener for soft drinks and other products, from either the simple sugar glucose or the polysaccharide inulin.

Finally, the miscellaneous products of immobilized cells include a diverse group of substances such as cofactors for enzyme reactions, precursors for making plastics and other polymers, and the plant hormone gibberellic acid. Although some of these materials, including gibberellic acid, can be prepared by fermentation methods, the preferred methods for making them use immobilized cell technology.

The use of immobilized cells does have potential disadvantages, however. The multiplicity of enzymes in the cells may result in undesirable side reactions, bacterial contamination can occur, and the catalytic activity of the cellular enzymes may be lower than that of immobilized, purified enzymes. In addition, the insoluble matrix may impede the transport of materials into or out of the cells. This can hinder removal of high molecular weight products and the entry of oxygen needed by the cells. It can also have a detrimental effect on the pH of the cells if acidic or basic reaction products cannot diffuse out.

A number of factors determine whether it is better to use immobilized cells or a traditional fermentation method for making a particular product. These factors incude the costs of growing and preparing the immobilized cells; the enzyme activity and stability of the cells; the ability to re-use the cells; the overall productivity of the system; the requirements for capital investment; and the costs of clean-up and pollution control. Each case must be analysed individually because there is no universal solution for the problems encountered in the microbial production of chemicals.

Synthesis of foreign proteins by microorganisms

Numerous foreign proteins can now be synthesized in microorganisms, including bacteria and yeast (Fig. 5.6), as a result of recombinant DNA technology (Table 5.9 and Chapter 4). This has a number of advantages. Previously rare proteins, such as human growth hormone and the interferons, are now available in much greater quantities for study and for possible therapeutic use. Nevertheless, there are a number of problems that still require solution.

Many of the foreign proteins that are produced in microorganisms are fragile molecules. In addition, recombinant DNA molecules may be genetically unstable and often require genetic manipulations to correct the alterations they undergo. Consequently, the integrity and purity of the microbially produced proteins must be constantly tested and assured, especially if

Table 5.9. *Production of foreign proteins by microorganisms*

Proteins	Microorganisms
Aspergillus glucoamylase	*Saccharomyces cerevisiae*
Bacillus subtilis endo-beta-1,3, -1,4-glucanase	*Escherichia coli* *Saccharomyces cerevisiae*
Bacillus subtilis penicillinase	*Escherichia coli*
Bovine growth hormone	*Escherichia coli*
Calf prochymosin	*Escherichia coli* *Saccharomyces cerevisiae*
Cellulomonas fimi cellulase	*Escherichia coli*
Beta-endorphin	*Saccharomyces cerevisiae*
Erwinia chysanthemi pectinases	*Escherichia coli*
Escherichia coli beta-galactosidase	*Saccharomyces cerevisiae*
Growth hormone releasing factor	*Saccharomyces cerevisiae*
Hepatitis B surface antigen	*Saccharomyces cerevisiae*
Human epidermal growth factor	*Saccharomyces cerevisiae*
Human growth hormone	*Escherichia coli*
Human immunoglobulin E chain	*Escherichia coli*
Human alpha-interferon	*Bacillus subtilis* *Escherichia coli*
Human beta-interferon	*Escherichia coli*
Human gamma-interferon	*Escherichia coli*
Human interleukin-2	*Escherichia coli* *Saccharomyces cerevisae*
Human proinsulin	*Bacillus subtilis*
Human tumour necrosis factor	*Escherichia coli*
Insulin-like growth factor I and II	*Saccharomyces cerevisiae*
Invertase	*Saccharomyces cerevisiae*
Mouse immunoglobulin E	*Saccharomyces cerevisiae*
Mouse interleukin-2	*Escherichia coli* *Saccharomyces cerevisiae*
Mouse interleukin-3	*Saccharomyces cerevisiae*
Rat proinsulin	*Escherichia coli*
Semliki Forest virus protein E1	*Bacillus subtilis*
Serratia marcescens nuclease	*Escherichia coli*
Staphylococcus aureus protein A	*Bacillus subtilis*
Staphylococcus aureus nuclease	*Bacillus subtilis*
Streptococcus equisimilis streptokinase	*Escherichia coli*
Thermomonospora cellulase	*Escherichia coli*
Wheat alpha-amylase	*Saccharomyces cerevisiae*

the proteins are intended for therapeutic use in human patients. Government regulations, many of which are aimed at recombinant DNA products, constitute an additional burden on the companies that make the products.

The ultimate marvels of biosynthesis by genetically altered microorganisms are in the future. The new technologies are not limited to producing the proteins that nature has designed. Chemical and genetic methods are available for constructing novel variants of proteins in the hope of improving their properties. This has already been done for the enzyme subtilisin, for example. Protein engineering studies at Genex Corporation in Rockville, Maryland, have improved the heat stability of the

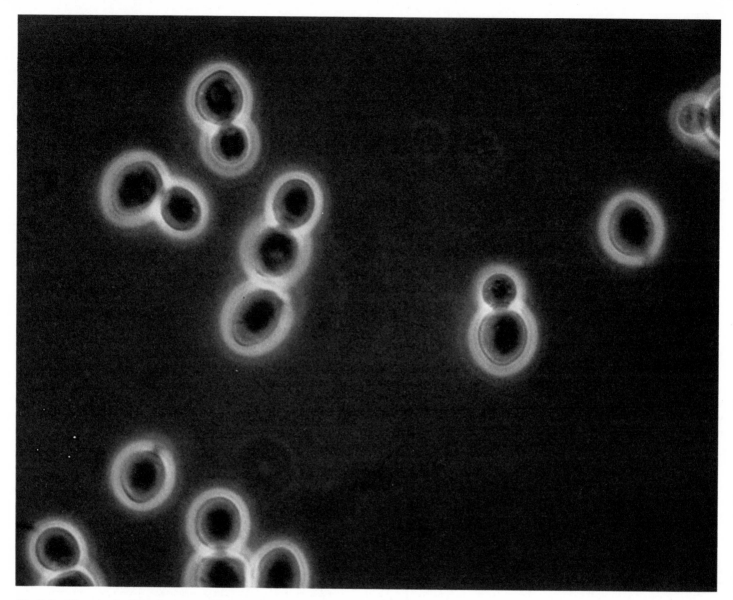

Fig. 5.6   Micrograph of *Saccharomyces*. The yeast *Saccharomyces* has long been applied in the production of bread and beer. In a more modern application the organism serves as a genetically engineered host for synthesizing foreign proteins, including human interleukin-2. This is being tested clinically as a treatment for cancer and acquired immune deficiency syndrome (AIDS). (By kind of permission of Angela Belt, Cetus Corporation.)

enzyme, while researchers at Genencor in South San Francisco have synthesized mutant subtilisins with increased bleach resistance and catalytic efficiency.

Meanwhile, the production of human proteins in microorganisms raises an intriguing question regarding the effects that substances such as human growth hormone, interferon or interleukin might have on the physiology of the producing organism. A study of such effects might yield some basic insights into the universality of the biochemistry of living organisms, on the one hand, and the strategies of evolution, on the other.

Steroid transformations by microorganisms

Microorganisms are used to modify the structures of a diverse group of substances, including antibiotics, alkaloids, cyclic and linear hydrocarbons, pesticides, terpenes and steroids. Many of these reactions would be difficult, if not impossible, to achieve by standard chemical means.

One of the landmark uses of the specific catalytic capacities of immobilized cells and enzymes is in steroid transformations – an application that differs conceptually from most of those described thus far. The previous examples largely dealt with the synthesis of chemicals that are normally made by microorganisms, with the notable exception of the foreign proteins that are produced by genetically engineered bacteria and other microbes. The steroid transformations are also foreign to microorganisms, which nonetheless have enzymes that can use steroids as substrates.

All steroids have the same core structure consisting of one 5-carbon and three 6-carbon rings. Steroid compounds are widely distributed in nature, occurring in both plants and animals.

The steroid core

One of the most important steroids in animals is cholesterol. Although this compound generally has a bad reputation, primarily because the consumption of excess amounts in the diet has been linked to an increased risk of heart attacks, it is nevertheless an essential component of cell membranes and is also the starting point for the synthesis of the body's steroid hormones. These include the male sex hormone testosterone;

the female sex hormones oestrogen and progesterone; the hormones of the adrenal cortex, which are needed for regulating salt and water balance and glucose metabolism; and the major bile salt, glycolate, which is necessary for absorption of dietary fats from the small intestine.

Many steroids have proved valuable as drugs. They are used to treat individuals who have hormonal deficiencies. In addition, patients who have autoimmune diseases or who have had organ transplants may be treated with steroids that suppress immune responses. Steroid drugs may also find application in cancer chemotherapy. But the most widely used steroids are the synthetic oestrogens and progesterones that go into birth control pills. The steroid drugs would not be as widely available as they are today without the chemical and microbial methods that have been developed for their synthesis.

These syntheses often start with plant steroids, which are relatively cheap and easy to obtain. By the early 1950s chemical methods had been developed for converting plant steroids to progesterone, an intermediate both in the body's conversion of cholesterol to the various steroid hormones and in the chemical methods for producing steroid drugs.

Steroid synthesis often requires the specific addition of a hydroxyl group to one of the 17 carbons in the steroid ring of the compound undergoing the reaction. For example, the conversion of progesterone to the glucocorticoids, which include the adrenal cortical hormones and their derivatives, requires the hydroxylation of carbon-11 of progesterone.

Such specific hydroxylations are very difficult to achieve by strictly chemical means, but in 1950 H. C. Murray and D. H. Peterson of the Upjohn Company in Kalamazoo, Michigan, showed that the fungus *Rhizopus stolonifer* (Fig. 5.7) could

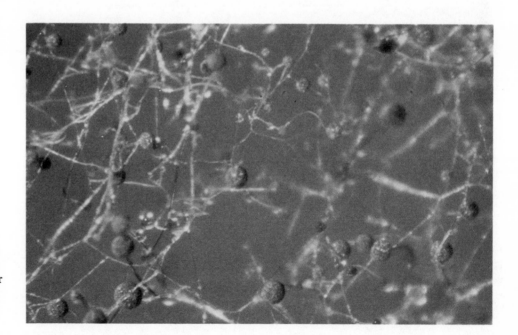

Fig. 5.7 Micrograph of a fungus of the *Rhizopus* genus. *R. stolonifer* performs a key reaction in the synthesis of steroids of the glucocorticoid family, which may be used for suppressing the immune systems of patients who have received organ transplants and for treating cancer.

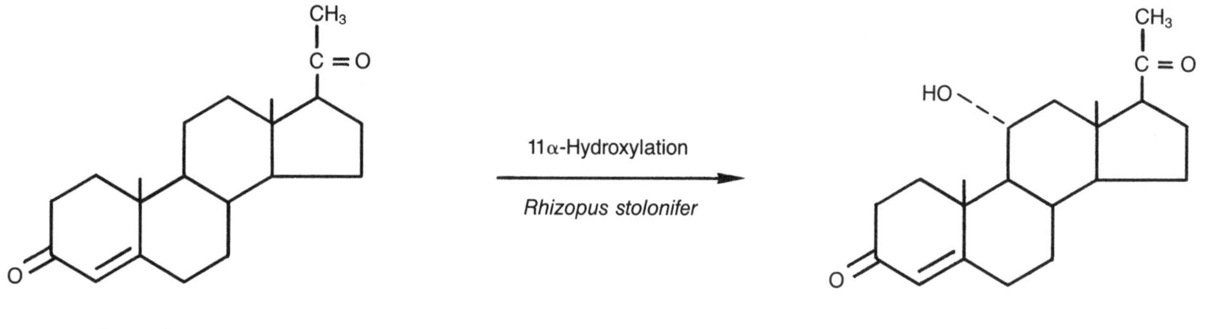

11α-Hydroxylation

Rhizopus stolonifer

Progesterone

11α-OH-Progesterone

perform this hydroxylation with a commercially useful efficiency.

Since then microorganisms have been identified that can specifically add hydroxyl groups to almost all of the carbons of the steroid ring and some of these conversions have achieved industrial utility. The microbiological hydroxylation of steroids remains one of the best examples of the superiority of biological catalysis over chemical catalysis. In addition to hydroxylating the steroid ring microbes can effect other transformations of steroids, including the removal of hydrogens from specific carbons. For example, the commercial conversion of cortexolone to prednisolone, which is used as an anti-inflammatory drug, involves first a hydroxylation reaction and then a dehydrogenation, both of which are carried out by microorganisms.

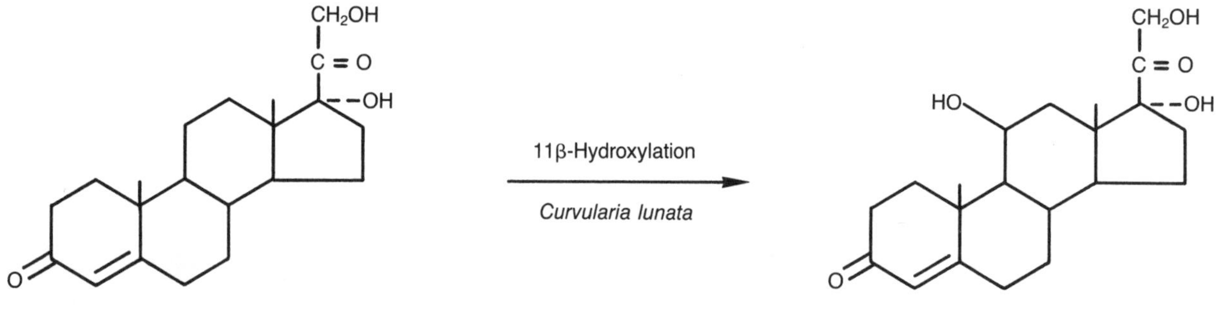

11β-Hydroxylation

Curvularia lunata

Cortexolone

Hydrocortisone

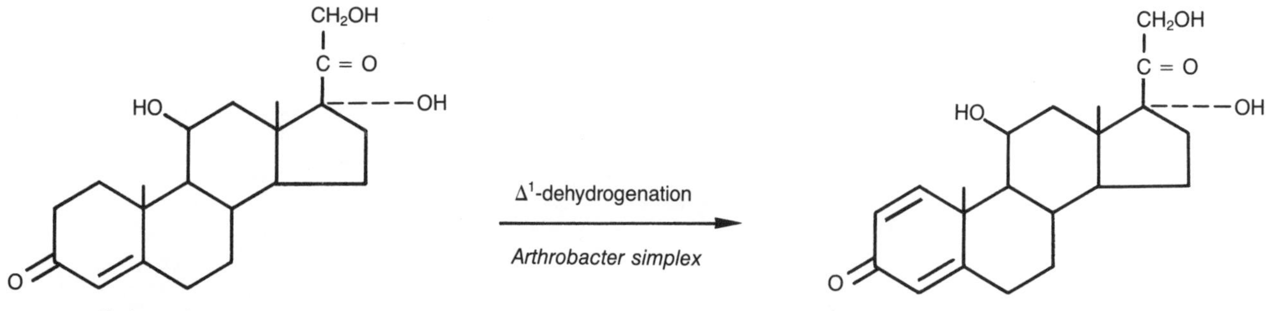

Δ^1-dehydrogenation

Arthrobacter simplex

Hydrocortisone

Prednisolone

Effects of substrate composition and environment on the structure of chemicals produced by microorganisms

The natural synthetic abilities of microorganisms, which are already extensive, can be expanded by applying conditions that deviate from those usually confronting the microbe. Sometimes this is done by varying the composition of the substrate on which the microorganism is grown. A prime example is the production of wax esters by the bacterium *Acinetobacter* sp. H01-N.

Wax esters consist of long-chain fatty acids and alcohols that are joined by an ester linkage. Studies by J. Geiger of Cetus Corporation in Emeryville, California, have shown that when *Acinetobacter* sp. H01-N is grown on the simple straight-chain hydrocarbon hexadecane, which has 16 carbon atoms, the main wax esters each contain 32 carbons. This is because the alcohols and acids produced by the bacterium from hexadecane all contain 16 carbons and the esters formed by the joining 16-carbon alcohols and acids will necessarily contain 32 carbons.

$$CH_3(CH_2)_{14}CH_3 \longrightarrow CH_3(CH_2)_{14}CO_2H + HO(CH_2)_{15}CH_3$$

Hexadecane   Hexadecanoic acid   Hexadecanol

$$\underset{\text{O}}{\overset{\text{O}}{CH_3(CH_2)_{14}CO(CH_2)_{15}CH_3}}$$

C-32 wax ester

In a similar fashion, growing the bacterium on octadecane, which has 18 carbons, will lead to the production of wax esters with 36 carbons. If eicosane, a straight-chain hydrocarbon with 20 carbons, is the substrate, the result will be a mixture of esters, primarily those containing 40, 38 and 36 carbons. This occurs because the 20-carbon hydrocarbon is converted to ester precursors with 20 and 18 carbons. Finally, if the 22-carbon hydrocarbon called docosane is used as the substrate for *Acinetobacter* sp. H01-N, the main wax esters will contain 44, 42 and 40 carbons.

Researchers are interested in altering the chain length of the wax esters produced by *Acinetobacter* because the organism may provide economical replacements for the wax esters found in certain natural oils, including those of the partially protected sperm whale, the orange roughy fish and the jojoba plant. These oils, each of which has a different composition (Table 5.10), are used as lubricants and, in the case of jojoba oil, in cosmetics. The *Acinetobacter* work has shown that appropriate substrate selection can enable the bacterium to produce oils with compositions mimicking those of the more complex organisms.

Table 5.10. *Distribution of major wax esters from various biological sources*

Source	No. of carbon atoms in wax esters										
	28	30	32	33	34	35	36	38	40	42	44
Orange roughy					+		+	M^a	+	+	
Sperm whale	+	+	M		+		+				
Jojoba bean								+	+	M	+
Acinetobacter H01-N											
N-Hexadecane			M								
N-Octadecane					+		M				
N-Eicosane							+	M	+		
N-Docosane									+	M	+
Acetic acid			+		M		+				
Ethanol			+		M		+				
Propionic acid			+	+	M	+	+				
Propanol			+	+	M	+	+				

^{a}M, main ester.

A second, and more subtle method of altering the carbon-chain length of wax esters is by growing an organism on propionic acid, which has three carbons. The principal starting material for the synthesis of long-chain fatty acids and alcohols is the 2-carbon acid, acetic acid. Because the fatty acids and alcohols are ordinarily built up in 2-carbon units, they and their esters have even numbers of carbon atoms. But with propionic acid as the substrate for *Acinetobacter* sp. H01-N the fatty acids and alcohols are built up in 3-carbon as well as 2-carbon units, thereby leading to the production of wax esters with both odd and even numbers of carbons.

Varying the temperature can also alter the chemical structure of the *Acinetobacter* wax esters. In nature there is a common and inverse relation between temperature and the number of double bonds in lipids such as the fatty acids. When an organism such as *Acinetobacter* is maintained at high temperatures the lipids it produces are largely 'saturated' which means that they do not contain double bonds. As temperatures are lowered, however, the lipids become increasingly 'unsaturated' that is, the number of double bonds increases.

The long-chain fatty acids are prominent components of the lipids of cell membranes. Unsaturated lipids melt at lower temperatures than saturated ones. Consequently, the alteration in lipid saturation that occurs as the temperature changes helps to preserve lipid fluidity and membrane function. The alterations in lipid composition can be substantial. Lowering the temperature of *Acinetobacter* from 30 °C to 17 °C results in an increase in the percentage of wax esters with two double bonds that it produces from 9 to 72 per cent, while the percentage of saturated wax esters decreases from 66 to 10 per cent.

The future

The realization of the full potential of the microbial production of biochemicals remains for the future. Many of the microbial reactions described in this chapter will require improvement. They have shown their commercial promise, but can still be better.

The improvements needed extend through every phase of the microbial processes. Better conditions for growing the organisms will have to be defined, and methods of immobilizing either the microbes or the enzymes they produce will have to be made more effective. More economical methods of producing the chemical feedstocks for the reactions are also desirable.

Bioreactor design is another target for improvement. This could allow better contact of an immobilized biocatalyst with the feedstock chemicals. A search for more favourable conditions in the bioreactor could help increase product yields as could better methods of separating the product from by-products and waste materials. In particular, the methods for isolating products from dilute solutions need to be made more efficient. Finally, the recycling of unused reaction materials and the disposal of wastes will have to be integrated into the overall process.

The microorganisms themselves will be the targets of much research. This will include efforts to improve those already in use, possibly through genetic engineering by recombinant DNA techniques, and the identification of new microbes for commercial development. Microbial strains that produce their products efficiently and can also resist such environmental stresses as high temperatures and salt concentrations, variations in pH and exposure to solvents are highly desirable. Once such strains of microorganisms are identified the problems often encountered in scaling-up from laboratory to commercial production will have to be solved. These problems vary from process to process and often require a great deal of ingenuity to overcome.

Ultimately the research may lead to the replacement of chemical reactions by bioreactions. The Japanese have been especially active in this regard. The scientific literature of Japan is filled with reports indicating a very active effort to use microbial methods to make a wide variety of compounds, including alcohols, fatty acids and sugars, that are now made by chemical methods.

Microorganisms have already shown their promise in this regard and their potential in the future is enormous. It remains for the biotechnologists to convert this dream to reality.

Additional reading

Chibata, I. and Tosa, T. (1983). Immobilized cells: historical background. *Applied Biochemistry and Bioengineering*, 4, 1–9.

Chibata, I., Tosa, T. and Sato, T. (1985). Immobilized biocatalysts to produce amino acids and other organic compounds. In *Enzymes and Immobilized Cells in Biotechnology*, ed. A. I. Laskin, pp. 37–70. Benjamin/Cummings Publishing Co., Menlo Park, California.

Hasegawa, M. (1985). Biotechnology and chemical industry. *Chemical Economy and Engineering Review*, 17, 26–35.

Layton, D. (1985). Isolation and recovery of recombinant DNA products: an overview. *The World Biotech Report (Asia)*, 3, 265–70.

Linko, P. (1985). Fuels and industrial chemicals through biotechnology. *Biotechnological Advances*, 3, 39–63.

Linko, P. and Linko, Y. Y. (1983). Applications of immobilized microbial cells. *Applied Biochemistry and Bioengineering*, 4, 53–151.

Neidleman, S. (1984). Applications of biocatalysis to biotechnology. *Biotechnology and Genetic Engineering Review*, 1, 1–38.

Neidleman, S. (1986). Enzymology and food processing. In *Biotechnology in Food Processing*, ed. S. K. Harlander and T. P. Labuza, pp. 37–56. Noyes Publications, Park Ridge, New Jersey.

Neidleman, S. L. and Geigert, J. (1984). Biotechnology and oleochemicals: changing patterns. *Journal of the American Oil Chemists' Society*, 61, 290–7.

Nicaud, J.-M., Mackman, N. and Holland, I. B. (1986). Current status of secretion of foreign proteins by microorganisms. *Journal of Biotechnology*, 3, 255–70.

Pace, G. W. (1985). Scale-up of fermentation processes. *The World Biotech Report (Asia)*, 3, 249–63.

Vezina, C., Rakkit, S. and Medarvar, G. (1984). Microbial transformation of steroids. In *CRC Handbook of Microbiology*, 2nd edn, ed. A. I. Laskin, vol. 7, pp. 65–466. CRC Press, Boca Raton, Florida.

Woodruff, H. B. (1980). Natural products from microorganisms. *Science*, 208, 1225–9.

6 Single-cell proteins

John H. Litchfield

Single-cell proteins are the dried cells of microorganisms such as algae, certain bacteria, yeasts, moulds and higher fungi that are grown in large-scale culture systems for use as protein for human or animal consumption. The products also contain other nutrients, including carbohydrates, fats, vitamins and minerals.

The consumption of microorganisms as part of the human diet is not a new event (Tables 6.1–6.3); people have eaten them in one form or another since ancient times. For example, yeast cells are a component of leavened bread; lactic acid bacteria are present in cheeses, fermented milks such as yoghurts, and fermented sausage; and moulds are the agents used in preparing oriental fermented soybean and fish products. The ancient Aztecs of Mexico harvested algae of the genus *Spirulina* from alkaline ponds and consumed them in their diet. People of the Lake Chad region in Africa eat dried *Spirulina* even at the present time.

Modern technology for single-cell protein production originated in Great Britain in 1879 with the introduction of aerated vats for producing baker's yeast (*Saccharomyces cerevisiae*). The centrifuge was first used about 1900 in the United States for separating baker's yeast cells from the medium in which they grow.

Table 6.1. *Developments in single-cell protein production: ancient times to 1900*

Period	Organism	Technical development
2500 BC	*Saccharomyces cerevisiae*	Top fermenting yeast recovered for baking
1781–2	*S. cerevisiae*	Compressed yeast prepared from brewer's yeast (UK, Netherlands, Germany)
1860	*S. cerevisiae*	Vienna process aerated, malted grain mash substrate (Austria)
1868	*S. cerevisiae*	Introduction of compressed yeast manufacturing into US (Fleischmann)
1879	*S. cerevisiae*	Continuous aeration (UK)
1900	*S. cerevisiae*	Centrifuge used for yeast separation (US)

Table 6.2. *Developments in single-cell protein production: 1900–1945*

Period	Organism	Technical development
1914–18	*Saccharomyces cerevisiae*	Incremental feeding, molasses, ammonium salts (Germany)
1918–19	*Endomyces vernalis*	Fat production from sulphite liquor (Germany)
1920	*Aspergillus fumigatus*	Growth on straw and inorganic N for animal feed (Germany)
1936	*S. cerevisiae*	Heiskenskjold process using sulphite liquor (Finland, Germany)
1936	*S. cerevisiae*	Scholler–Tornesch process for fodder yeast from wood sugar (Germany)
1941–5	*Candida utilis*	Production of food yeast from sulphite liquor and wood sugar (Germany)
	Geotrichum candidum (*Oidium lactis*)	Production of fat (Germany)

In Germany during World War I, baker's yeast was produced for consumption as a protein supplement. Molasses served as the carbon and energy source for growing the yeast and ammonium salts were used as the nitrogen source. Then, during World War II, the Germans cultured torula yeast (*Candida utilis*) as a protein source for humans and animals. The raw materials in this case were the sulphite waste liquor from pulp and paper manufacture and wood sugar that was obtained by the acid hydrolysis of wood.

In more recent years, advances in scientific knowledge regarding the physiology, nutrition and genetics of microorganisms have led to significant improvements in single-cell protein production from a wide range of microorganisms and raw materials. For example, bacteria with high protein contents – 72 per cent or greater – can be produced continuously with methanol as a raw material, and food-grade yeasts can be grown in high cell concentrations to minimize the energy costs of drying.

Table 6.3. *Developments in single-cell protein production: 1945 to present*

Period	Organism	Technical development
1946–54	*Candida utilis*	Continuous process yeast from sulphite liquor – Waldhof fermentors (US)
1948–53	*Chlorella* sp.	Production of algae in open circulation systems (Japan)
1959	*Saccharomyces cerevisiae*	Continuous production of baker's yeast on commercial scale (UK)
1954–63	*Morchella* sp.	Submerged culture of mushroom mycelium (US)
1958–64	*Kluyveromyces fragilis*	Fragilis food yeast from cheese whey (US)
1959–72	*C. lipolytica*, *C. tropicalis*	Feed yeast from hydrocarbons, *n*-paraffins, gas oil, air lift fermentors (UK, France, Japan, USSR)
1963–74	Fungi	Fungi from spent sulphite liquor, Pekilo process (Finland)
1970–74	*C. utilis*	Food yeast from ethanol (US)
1971–5	*K. fragilis*	Continuous production of fragilis yeast and/or ethanol from cheese whey
1979–80	*Methylophilus methylotrophus*	Continuous production of bacterial single-cell protein from methanol on a commercial scale (UK)
1983–5	*C. utilis, K. fragilis, S. cerevisiae*	High cell density, direct dry process for single-cell protein production from ethanol and carbohydrates (US)

Both photosynthetic and non-photosynthetic microorganisms have been used for single-cell protein production. At a minimum, these organisms require a carbon and energy source, a nitrogen source, and supplies of other nutrient elements, including phosphorus, sulphur, iron, calcium, magnesium, manganese, sodium, potassium and trace elements, for growth in a water environment. Some organisms cannot synthesize amino acids, vitamins and other cellular constituents from simple carbon and nitrogen sources, and in that event these substances must also be supplied for the organism to grow.

Photosynthetic organisms in single-cell protein production

Both algae and bacteria are among the photosynthetic microorganisms that are used for single-cell protein production. The photosynthetic growth of the algae of interest, which include *Chlorella*, *Scenedesmus* and *Spirulina* (Table 6.4), can be represented as follows:

Carbon dioxide + water + ammonia or nitrate + minerals
(carbon source)   (nitrogen source)

$$\xrightarrow[\text{energy}]{\text{light}} \text{algal cells} + \text{oxygen}$$
(single cell protein)

The carbon dioxide concentration in the air is about 0.03 per cent, which is inadequate to support algal growth at the high rates needed for single-cell protein production. Additional amounts of carbon dioxide can be provided by the carbonates or bicarbonates that are present in alkaline ponds, by the gases produced during combustion, or by the decomposition of organic matter in sewage and industrial wastes. The concentration of carbon dioxide in combustion gases ranges from 0.5 to 5 per cent, for example.

The nitrogen sources for algal production include ammonium salts, nitrates, or the organic nitrogen that is present in sewage oxidation ponds. Phosphorous and other mineral nutrients are usually present in natural waters and waste waters in concentrations sufficient for algal growth. The algal 'bloom' problem that occurs in many reservoirs in midsummer is evidence for the adequacy of the concentrations of these nutrients.

Light intensity and temperature are important factors for algal growth. Artificial lighting is too costly to use for producing algal single-cell proteins as an animal feedstuff. For large-scale cultivation of the microorganisms to be economically feasible then, the skies at the culture sites should be relatively clear and there should be minimal variations in light intensity during the year. In addition, temperatures should be above 20 °C for most of the year. Consequently, artificial outdoor ponds in semitropical, tropical or arid regions are the most suitable systems for algal cultivation. Materials for constructing the ponds include concrete, plastics or glass fibre laminates.

Large pond areas are needed because the algal growth occurs

Table 6.4. *Selected processes for making algal single cell protein*

Organism	Raw material	Production	Producer or developer
Chlorella sp.	CO_2 (photosynthetic); cane syrup, molasses (non-photosynthetic)	2 metric tons/day	Taiwan Chlorella Manufacture Co. Ltd, Taipei
Scenedesmus acutus	CO_2, urea (photosynthetic)	20 grams/square metre/day	Central Food Technological Research Institute, Mysore, India
Spirulina maxima	CO_2, or $NaHCO_3$–Na_2CO_3 (photosynthetic)	320 metric tons/year	Sosa Texcoco, SA Mexico City

mainly in the top 20 or 30 centimetres where the light intensity is greatest. Usually the pond waters are agitated either continuously or intermittently by pumps, paddle wheels or windmills to prevent the algae from settling. This ensures that the cells are uniformly exposed to light and nutrients.

Algae are generally cultivated under non-sterile, mixed culture conditions in locations where the environmental conditions favour the dominance of the desired algal species over competing species. In Mexico the blue-green alga *Spirulina maxima* has been produced on a demonstration basis in the naturally alkaline waters of Lake Texcoco, which have a pH in the range of 9 to 10. The Spirulina cells float to the surface in clumps and are skimmed off. A pilot production facility at this site has produced 1 ton per day of dry algal single-cell protein for sale as a health food.

Spirulina is also grown on a small commercial scale in Hawaii, Thailand, Israel and Taiwan. In Hawaii, Cyanotech Corporation uses two ponds to produce about 625 kilograms of dry product per month for sale in health food stores at about $18 per kilogram. At this price the product is definitely for affluent customers, not for feeding the world's hungry millions.

In Japan and Taiwan, *Chlorella* species are grown either photosynthetically in ponds or in indoor tanks with sugar syrups or molasses as carbon sources. Dried algae and algae tablets are sold in these countries as health foods.

The Indian and West German governments, cooperating in the Indo-German Algal Project, have instituted a cooperative programme at the Central Food Technological Institute in Mysore, India, for culturing *Scenedesmus* species in artificial ponds. This programme has resulted in projects in Egypt, India, Peru and Thailand. In addition, studies in Israel and Argentina have shown that salt-tolerant algae of the genus *Dunaliella* can be grown in saline waters to yield single-cell protein, glycerol and beta-carotene as co-products.

Growing the algae on sewage serves the dual purpose of cleaning up potential environmental pollution while at the same time producing valuable protein. Researchers at the Israel Institute of Technology in Haifa, for example, have conducted extensive studies in that city on treating sewage in algal ponds while simultaneously producing single-cell protein for animal feed. Haifa appears to be highly suitable for algal cultivation because it has a high light intensity throughout the year.

The ponds there consist of shallow, meandering canals that are provided with equipment for mixing and aeration. A mixed culture of algae and bacteria grow in the ponds, with algae of the *Chlorella*, *Euglena*, *Micractinium* and *Scenedesmus* genera predominating. The system is a symbiotic one in which the algae produce oxygen that is used by the bacteria.

The outdoor cultivation of algae for single-cell protein production involves a coupling of agricultural technology and industrial microbiology. These systems depend upon, and are limited by, climate and the availability of water, light, carbon dioxide and nutrients. They can be operated in an optimum manner by controlling the carbon dioxide supply, the fluid flow and the agitation of the system. Practical productivities are in the range of 35 metric tons of dry algal protein per hectare of pond surface per year.

Because the algal cell concentrations in the culture fluids amount to only about 1 to 2 grams of dry material per litre, large quantities of water must be handled in the harvesting operations. This contributes significantly to the production costs of algal single-cell proteins. The cells must first be concentrated by any of a variety of methods, including flotation, filtration, centrifugation, and flocculation followed by sedimentation, and are then dried in drum dryers. Thermal drying in drum dryers is costly, but such heat treatment is essential to destroy potentially pathogenic microorganisms that may be contaminants of algae grown in sewage treatment ponds.

The photosynthetic bacteria that are used for single-cell protein production include bacteria of the genus *Rhodopseudomonas*, which have been grown on sewage or industrial waste in Japan for use as an animal feed. These bacteria grow in mixed cultures with nitrogen-fixing and other aerobic bacteria. They must be supplied with organic substances as carbon and energy sources, and will not grow using only carbon dioxide and light as the algae do. The culture densities of the bacteria are in the range of 1 to 2 grams of dry material per litre, and the problems of separation and concentration that occur with the algae are also present in this system.

The non-photosynthetic production of single-cell protein

The non-photosynthetic microorganisms that are grown to produce single-cell proteins include bacteria and moulds, yeasts and other fungi. These organisms are aerobic and must have oxygen to grow. They also require an organic carbon and energy source, together with sources of nitrogen, phosphorus, sulphur, and the mineral elements that were previously mentioned as necessary for algal growth.

The conversion of organic compounds to single-cell protein by non-photosynthetic microorganisms can be represented by the following equation:

$$\text{Organic carbon} + \text{nitrogen} + \text{mineral nutrients} + \text{oxygen} \rightarrow$$

$$\text{Single-cell protein} + \text{carbon dioxide} + \text{water} + \text{heat}$$

The bacteria

Many species of bacteria have been investigated for use in single-cell protein production. Among the characteristics that make bacteria suitable for this application is their rapid growth. Their generation times are short; most can double their cell mass in 20 minutes to 2 hours. The comparable times are 2 to

3 hours for yeasts and 4 to 16 hours for moulds and higher fungi.

Bacteria are also capable of growing on a variety of raw materials, ranging from carbohydrates such as starch and sugars, to gaseous and liquid hydrocarbons such as methane and petroleum fractions, to petrochemicals such as methanol and ethanol. Suitable nitrogen sources for bacterial growth include ammonia, ammonium salts, urea, nitrates, and the organic nitrogen in wastes. A mineral nutrient supplement must be added to the bacterial culture medium to furnish nutrients that may not be present in natural waters in concentrations sufficient to support growth.

The bacterial species most likely to be used for single-cell protein production grow best in slightly acid to neutral pHs in the range 5 to 7. The bacteria should also be able to tolerate temperatures in the 35 to 45 °C range, because heat is released during the bacterial growth. The use of temperature-tolerant strains will minimize the need for refrigerating the water that cools the fermentation vessel. Bacterial species cannot be used for single-cell protein production if they are pathogenic for plants, animals or humans.

Bacterial single-cell protein may be produced in conventional batch systems in which all of the nutrients are supplied to the fermentor initially; the cells are harvested when they have consumed the nutrients and stopped growing. However, in the more advanced production methods the nutrients are supplied continuously in the concentrations needed to support bacterial growth and the cells are harvested continuously once the population reaches the desired concentration.

The concentration of the carbon and energy source usually ranges from 2 to 10 per cent in batch processes. In the continuous process the supply of the carbon source is regulated so that the concentration in the growth medium does not exceed that required by the growing bacterial cells. This concentration will generally be lower than those used in batch processes.

Maintaining sterile conditions during single-cell protein production is very important because contaminating microorganisms grow very well in the culture medium. The incoming air, the nutrient medium and the fermentation equipment must be sterilized in all bacterial single-cell protein processes, and sterile conditions must be maintained throughout the production cycle.

A system for the continuous production of bacterial single-cell protein with methanol as the carbon and energy source shows features that are common to most production methods (Fig. 6.1). After the nutrients are sterilized they are introduced

Fig. 6.1 Schematic diagram of the process for the continuous production of bacterial single-cell protein. The inoculum containing the bacteria is introduced into the fermentation tank (F). The necessary nutrients (methanol, ammonia, phosphoric acid and mineral salts in this case) must be sterilized before they are added to prevent the introduction of bacteria that might contaminate the product. The fermentor contents must be supplied with air and cooled to dissipate the heat generated by the growing bacteria.

In continuous processes the nutrients are replenished as they are consumed to maintain the concentrations needed by the bacteria. The solution containing the bacteria is drawn off, treated to cause the bacteria to agglomerate or flocculate, and centrifuged. The liquid may then be recycled to the fermentor while the bacteria are spray-dried and ground to yield the final product.

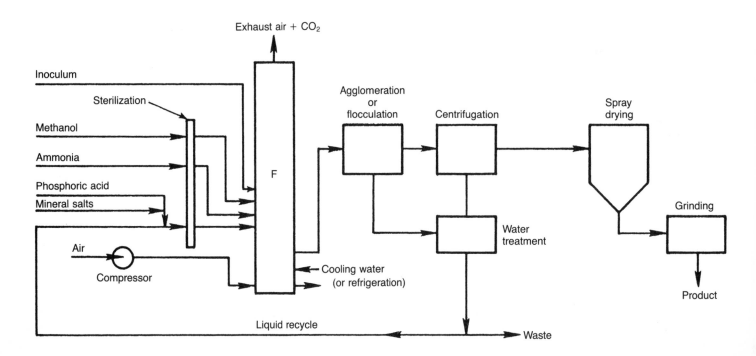

into a fermentation vessel and inoculated with the bacteria to be grown. The vessel, which is known as a 'bioreactor', must be supplied with sterile air and with cooling water to prevent the heat released during fermentation from building up and killing the cells. The cooling water is circulated in either the outer jacket of the fermentor or through internal cooling coils.

The vessels are also fitted with instruments that measure and control the pH and temperature of the contents and the concentration of dissolved oxygen. The exhaust air from the bioreactor contains carbon dioxide that may be separated and compressed for sale to industrial users of carbon dioxide.

After the bacteria are removed from the fermentation tank, they must be separated from the culture broth, which is usually done by adding chemicals that will cause the cells to clump and then centrifuging them. The separated cells are dried to yield a product that will be stable during shipment and storage. Finally, there must be equipment for grinding and packaging the cells and a system for treating and recycling the spent culture fluid.

Oxygen transfer to the cells in the fermentor is a critical factor in obtaining growth rates and yields that are economically satisfactory. A variety of fermentor designs can provide suitable aeration. The most commonly used are the baffled stirred-tank reactor (Fig. 6.2) and the air-lift fermentor (Fig. 6.3).

Although considerable research was conducted on the production of bacterial single-cell protein during the 1960s and early 1970s, Imperial Chemical Industries plc (ICI) in the United Kingdom developed the only process to reach a commercial scale of operation. In the ICI process the bacterium *Methylophilus methylotrophus*, which has a generation time of about 2 hours, is grown continuously with methanol as the substrate and additional nutrients including ammonia and the minerals phosphorus, calcium and potassium. The company developed for the process a unique air-lift fermentor with a capacity of 1500 cubic metres. The fermentor design minimizes the requirements for cooling the vessel and the problem of oxygen limitation.

In 1980 ICI commissioned a plant, with the capacity of producing 50 000 metric tons of single-cell protein every year, at Billingham, England. The plant has since been operated intermittently, with a production of 6000 metric tons per month. The bacteria grow on methanol as their energy source. Two metric tons of methanol yield about 1 metric ton of dry 'Pruteen' single-cell protein. The dried product, which contains about 72 per cent protein and 8 per cent moisture, has been sold as an animal feed supplement in Western European markets.

With soybean meal now costing just $190 per metric ton, however, 'Pruteen' is no longer competitive as an animal

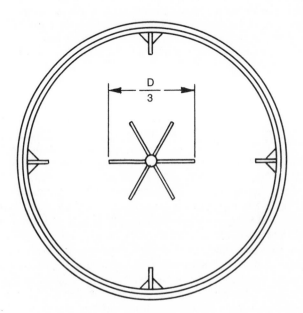

Fig. 6.2 Schematic diagram of a baffled stirred tank fermentor. The air introduced into the fermentor is dispersed by the propellor-like agitator. The baffles projecting from the side of the tank (shown in cross-sectional view) help to ensure that the contents of the tank are thoroughly mixed and oxygenated.

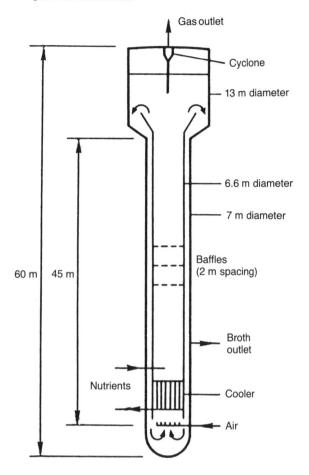

Gas outlet

Cyclone

13 m diameter

6.6 m diameter

7 m diameter

Baffles
(2 m spacing)

60 m   45 m

Broth
outlet

Nutrients

Cooler

Air

Fig. 6.3 Schematic diagram of the ICI pressure cycle fermentor. This fermentor has a capacity of 1500 cubic metres. Compressed air is introduced at the bottom of the fermentor and the contents are mixed by the rising air bubbles. The baffles help to keep the fermentor contents well oxygenated by breaking-up the bubbles, which expand as they rise against the hydraulic pressure gradient in the apparatus.

feedstuff and the plant is not being operated on a commercial scale at present. Nevertheless the development of the ICI process for making the bacterial single-cell protein exemplifies the application of modern chemical engineering to the field of biotechnology.

During the development of 'Pruteen' ICI scientists investigated the possibility of improving the conversion of methanol to single-cell protein by genetically modifying the ability of *M. methylotrophus* to use ammonia. They introduced into the bacteria a gene for an ammonia-assimilating enzyme that is more efficient than the endogenous bacterial enzyme. Although the new gene was stable in the bacteria and was expressed there, only a 3 to 5 per cent increase in single-cell protein yield was obtained with the genetically modified strain of bacteria.

Yeasts

Modern technology for producing yeast single-cell protein has largely developed since World War II. Today, yeast products for human or animal consumption are produced on a commercial scale in many countries (Table 6.5). In addition, baker's yeast, which is grown on molasses, is sold as a food flavouring and nutritive ingredient in addition to being used as a leavening agent.

Yeast can be grown on a number of substrates. These include carbohydrates, both of the complex type, such as starch, and of the simple type, such as the sugars glucose, sucrose and lactose. Alternatively, sugar-containing raw materials such as corn syrup, molasses and cheese whey can be used. Some yeasts are able to grow on straight-chain hydrocarbons, which are obtained from petroleum, or on ethanol or methanol.

In addition to a carbon source, a nitrogen source is required. The nitrogen can be provided by addition of ammonia or ammonium salts to the culture medium. A supplement of mineral nutrients is also required.

The requirements for production of yeast single-cell protein are similar to those previously described for production by bacteria. The yeast should have a generation time of about 2 to 3 hours. It should be pH- and temperature-tolerant and genetically stable, giving satisfactory yields from the substrate used, and not cause disease in plants, animals or humans.

The technology for producing yeast single-cell protein is also similar to that for making the bacterial products (Fig. 6.4). The baffled, stirred-tank fermentor is the most common type of vessel for yeast single-cell protein production, but air-lift fermentors are also used. As in the bacterial cultures, heat is released during yeast growth and the fermentor must be provided with a cooling system.

The yeast fermentations may be operated either in the batch or continuous mode, or by a third mode called 'fed-batch'. In fed-batch processes the substrate and other nutrients are added in an incremental manner to meet the growth requirements of the yeast while maintaining very low nutrient concentrations in the growth medium at any time. This method yields 3.5 to 4.5 per cent dry weight of product compared with the 1.0 to 1.5 dry weight of product yielded by batch cultivation. Cells grown by fed-batch processes are harvested as they are in the batch mode of production.

Although batch and fed-batch culture systems have been used in baker's yeast production for many years, only recently has the technology been available for monitoring and adjusting the pH and substrate concentrations to permit the continuous type of operation. Yeast cell concentrations of up to 16 per cent (dry weight) can be obtained in the continuous culture operations.

Yeasts have certain advantages over bacteria for production of single-cell protein. For one, the yeasts tolerate a more acid environment, in the range of 3.5 to 4.5 instead of the near-neutral pHs preferred by bacteria. Consequently, yeast processes can be operated in a clean but non-sterile mode at a pH of 4.0 to 4.5 because most bacterial contaminants will not grow

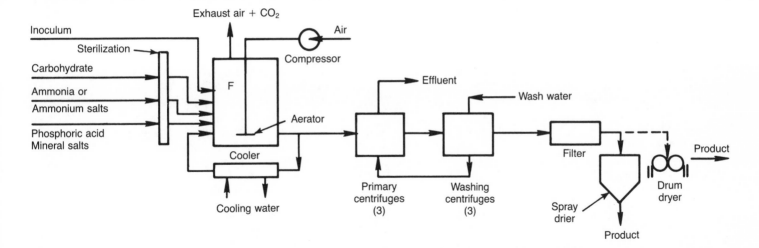

Exhaust air + CO_2

Inoculum

Sterilization

Carbohydrate

Ammonia or
Ammonium salts

Phosphoric acid
Mineral salts

Air

Compressor

F

Aerator

Cooler

Cooling water

Effluent

Wash water

Filter

Primary
centrifuges
(3)

Washing
centrifuges
(3)

Spray
drier

Drum
dryer

Product

Product

Fig. 6.4 Schematic diagram for yeast single-cell protein production from carbohydrate. The yeast-containing inoculum is added to the fermentor (F) together with the necessary nutrients – which include ammonia, phosphoric acid, mineral salts, and carbohydrate as the energy and carbon source. The tank contents are aerated with compressed air and cooled with water. The yeast cells are separated from the nutrient broth by centrifugation and then the cells are washed, filtered and dried in a spray or drum dryer.

Fig. 6.5 Fermentation tanks for producing food-grade torula yeast at the plant operated by the Pure Culture Products Division of Hercules, Inc., in Hutchinson, Minnesota. The plant has the capacity to make 7500 short tons of torula yeast per year. This yeast product is used as a dietary supplement for both animal and human consumption. (By kind permission of Pure Culture Products Division, Hercules, Inc.)

well in this degree of acidity. For another, yeast cell diameters are about 0.0005 centimetres as compared with 0.0001 centimetres for bacteria. Because of their larger size yeasts may be separated from the growth medium by centrifugation, without the need for a flocculation step.

Yeast single-cell protein production depends upon meeting the oxygen demand of the growing cultures. Yeasts grown on carbohydrates generally require about 1 kilogram of oxygen per kilogram dry weight of cells and when grown on hydrocarbons they need about twice that much. Air, which is sterilized by filtration, is supplied to the fermentor through a perforated screen or perforated pipes in the bottom of the vessel, or by a rotating aeration wheel or air-lift device similar to those used for culturing bacterial cells.

Yeast single-cell protein may be produced either under sterile, or clean but non-sterile, conditions. In a typical batch or fed-batch non-sterile operation in which carbohydrate is used as the carbon and energy source the medium is sterilized by passing it through a heat-exchanger and then charged into clean fermentors. Contamination control is based on having a pH of 4.0 to 5.0, supplying sterile air, and maintaining large populations of yeast cells in the fermentor to overwhelm any small numbers of contaminating microorganisms. In some continuous yeast fermentations that use hydrocarbons or ethanol as the substrate, completely sterile conditions may be needed to achieve the desired yields and product quality.

Candida utilis, which is known as torula yeast and used both as an animal feed supplement and for human consumption, is manufactured from a wide range of raw materials, among which are ethanol, the sulphite waste liquor from paper mills, normal paraffin hydrocarbons and cheese whey. The Pure Culture Products Division of Hercules, Inc., produces *C. utilis* in a plant in Hutchinson, Minnesota. The plant has a capacity of 7500 short tons per year (Fig. 6.5).

The plant is operated in a continuous, sterile manner with ethanol as the energy and carbon source. The yeast cells are drawn off continuously, washed, and spray-dried to yield a

Table 6.5. *Typical commercial-scale yeast single-cell protein processes*

Raw material	Yeast	Scale	Product	Organization
Cheese whey	*Kluyveromyces fragilis*	5000 tons/year	Food yeast, fermentation nutrient	Universal Foods Corporation, Juneau, Wisconsin
Ethanol	*Candida utilis*	10 000 tons/year	Torula yeast, food ingredient	Pure Culture Products, Hutchinson, Minnesota
N-Paraffins	*Candida guillienmondis*	20 000–40 000 tons/year	Food yeast	All-Union Research Institute of Protein Biosynthesis, USSR
Sulphite waste	*Candida utilis*	15 tons/day	Torula yeast	Rhinelander Paper Corporation, Rhinelander, Wisconsin

food-grade product that is further processed to produce flavouring ingredients for food. The typical yield is about 0.7 metric tons of yeast, on a dry weight basis, per metric ton of ethanol consumed. The protein content of the product is in the range of 50 to 55 per cent.

Commercial-scale plants in the United States and Europe also produce *C. utilis* from sulphite waste liquor. In a typical process the sulphite liquor, which contains a mixture of sugars, is treated with lime and by steam-stripping to remove sulphur dioxide, sulphites, and other sulphur compounds that inhibit the growth of the yeast. The operation is of the 'clean but non-sterile' type that was described previously. The product is recovered by centrifuging, washing and drying it.

Either a human food or an animal feed product can be made from sulphite liquor, depending on the process system and the product quality controls that are imposed. The sulphite waste liquor methods yield about 1 metric ton of dry yeast for every 2 tons of sugar in the liquor.

In the 1960s British Petroleum Company Ltd and others, notably Kanegefuchi Chemical Industry Company Ltd in Japan, developed a large-scale process for making *Candida* yeast single-cell protein with straight-chain paraffin hydrocarbons as the substrate. During the early 1960s British Petroleum constructed a plant with a capacity of 100 000 metric tons per year in Sardinia. The process used by the plant required paraffin hydrocarbons of 95.7 to 97 per cent purity, which were prepared by molecular sieving. An aerated-agitated fermentor was designed to operate continuously under sterile conditions.

Although the plant was completed, it was never operated on a commercial scale because of a dispute between British Petroleum and the Italian government over the product quality, especially the presence of hydrocarbon residues. At the present time the All-Union Research Institute of Protein Biosynthesis in Russia is operating the only large-scale facility for the production of yeast single-cell protein from hydrocarbons. The capacity of the operation is reported to be in the range of 20 000 to 40 000 metric tons per year. The product is used for animal feed.

The yields of yeast single-cell protein from paraffin hydrocar-bons are of the order of 0.9 metric tons of dry yeast per metric ton of hydrocarbon. The protein content of the *Candida* species used in the hydrocarbon processes is in the range of 60 to 65 per cent, which is somewhat higher than the protein content of *C. utilis*.

About 14 million kilograms of whey are produced every year in cheese manufacture in the United States. Cheese whey contains about 4 per cent lactose, which is readily used as a substrate by fragilis yeast (*Kluyveromyces fragilis*).

Universal Foods Corporation of Milwaukee, Wisconsin, operates a plant at Juneau, Wisconsin, for producing fragilis yeast from the cheese whey that is collected from nearby cheese plants. After inoculation of the whey-based growth medium the yeast grow to a steady-state concentration of 1 billion cells per millilitre in 8 to 12 hours. The process is then operated with a continuous feed of diluted whey and nutrients and withdrawal of the yeast product. Yields are about 0.45 to 0.55 kilograms of dry yeast per kilogram of lactose consumed.

Both animal feed and human food-grade products are manufactured by this process. The animal feed product contains about 45 per cent crude protein and 3 per cent moisture whereas the food-grade product contains about 50 per cent protein. Bel Fromagerie makes a yeast single-cell product by a similar process in a plant in Vendôme, France.

The Phillips Petroleum Company in Bartlesville, Oklahoma, has recently developed an innovative process for single-cell yeast production that gives much higher cell concentrations than older methods. In this process, which was patented in 1983, the novel design of a mechanically agitated fermentor achieves very high heat and oxygen transfer rates.

Cell concentrations reach 12 to 16 per cent of dry weight material when *C. utilis* is grown on ethanol, sucrose or molasses or when K. fragilis is grown on cheese whey. More conventional fed-batch processes give concentrations of around 4 per cent, as mentioned previously. Because of the high cell concentrations obtained by the Phillips method, the culture medium can be fed directly to the spray drier without first being concentrated. The result is a considerable saving in energy costs.

Moulds and higher fungi

Several of the higher fungi are used as human foods. The common commercial mushroom *Agaricus campestris* is one such example. These mushrooms are grown in manure or compost beds. The Japanese grow the shiitake mushroom *Cortinellus berkelyanus* on the logs of certain trees that have been inoculated with suspensions of the fungus spores. In China the padistraw mushroom *Volvaria volvacea* is cultured on moistened rice straw that is inoculated with the spores of this fungus.

Moreover, many fermented rice, fish and soybean foods are produced in Asia with the aid of moulds. Other moulds that have contributed to the human diet include *Trichosporon pullulans*, which was grown during World War I. This mould was cultivated in shallow trays holding a sugar-containing liquid medium. The mycelium, the filamentous network formed by the growing mould, is rich in fat and was harvested and used as a paste.

The present-day production of mould single-cell protein is achieved by methods similar to those used for the production of the yeast products. The simple sugars or the raw materials that contain them are suitable substrates for a variety of moulds. The carbohydrate concentrations in the growth media are typically around 10 per cent. Ammonia or an ammonium salt is the usual nitrogen source and supplemental mineral nutrients are generally included in the media.

The growth rates of moulds and higher fungi, which have generation times ranging from 4 to 16 hours, are usually slower than those of bacteria and yeasts. The moulds and higher fungi grow well at temperatures of 25 to 36 °C and pHs of 3.0 to 7.0. Most are cultivated at pHs below 5.0 to minimize contamination by bacteria, however.

Batch, fed-batch or continuous process can be used for the production of mould single-cell protein. In most cases batch processes in conventional aerated fermentors constitute the least costly alternative. The operations are conducted under sterile conditions if the product is to be a human food, whereas moulds for animal consumption can be produced under non-sterile, clean conditions. As in any other fermentation, cooling must be provided to control the heat released during the growth of the moulds.

Moulds and higher fungi, when cultured in aerated fermentors, may grow in either a filamentous or pellet form, depending on the species and the aeration conditions. This simplifies the recovery of the product because the filamentous mycelium or the pellets can be readily separated from the medium by screens, rotary vacuum filters or filter presses at low cost. However, mechanically agitated vessels are not suitable for growing the organisms because the mould filaments may concentrate around the impeller and not be uniformly dispersed throughout the growth medium. The use of fermentors in which the aeration also provides the agitation can avoid this problem.

A number of companies have developed processes for producing mould single-cell proteins (Table 6.6). In the United Kingdom Ranks Hovis McDougall Ltd is making a product called Mycoprotein, which is based on the mould *Fusarium graminearum*, for the human food market. Glucose serves as the energy source for the mould, which has a generation time of about 5.5 hours. The yield is about 0.5 kilograms dry weight of cells for every kilogram of sugar that is consumed.

The material is treated to reduce the content of ribonucleic acid because consumption by humans of more than 2 grams per day may cause kidney stones or gout. After filtration the filamentous mycelium cake can be flavoured and formed into a product that resembles chicken white meat and has a protein content of 45 per cent. The capacity of the Ranks Hovis McDougall plant is 50 to 100 tons per year, but at present the product is only being produced in small quantities for test-market studies in the United Kingdom.

Table 6.6. *Selected large-scale mould single-cell protein processes*

Raw material	Organism	Scale	Product	Organization
Glucose (food grade)	*Fusarium graminearum*	50–100 tons/year	Mycoprotein (human food)	Ranks Hovis McDougall, High Wycombe, UK
Cheese whey	*Penicillium cyclopium*	300 tons/year	Animal feed	Heurty, S.A., France
Coffee wastes	*Trichoderma harzianum*	40 000 litres	Animal feed	ICAITI, Guatemala and El Salvador
Sulphite waste liquor	*Paecilomyces varioti*	10 000 tons/year	Animal feed (Pekilo protein)	Tampela and Finnish Pulp and Paper Research Institute, Jämsänkoski, Finland
Pulp mill wastes	*Chaetomium cellulolyticum*	1 ton/day	Animal feed (Waterloo process)	Envirocon Ltd, Vancouver, BC, Canada; University of Waterloo, Ontario

During the 1960s several processes were developed for producing mushroom mycelium for use as a food flavouring ingredient. For example, the Special Products Division of Mid-American Dairymen, Inc., of Springfield, Missouri, grew morel mushroom mycelium in an aerated submerged culture, rather than on the solid substrates usually used to grow whole mushrooms. The growth medium, which contained glucose, corn-steep liquor and ammonium phosphate, was inoculated with the mushroom spores in a conventional aerated fermentor. After about 4 days the mycelium was harvested to yield a product for flavouring soups, sauces, gravies and other foods. However, the operation was discontinued when imported, dried mushrooms became available at lower cost.

In some operations moulds not only serve as a source of single-cell protein but also help to clean up the wastes from food processing and industrial plants. The Central American Research Institute for Industry (ICAITI) of Guatemala has investigated the use of the mould *Trichoderma harzianum* for treating the wastes from coffee processing in El Salvador while at the same time recovering a single-cell protein for supplementation of animal feeds. The system that the company tested was operated under non-sterile conditions. The content of microbial solids increased by 3.2 grams per litre in 24 hours and the product contained 56 per cent protein on a dry weight basis. This pilot plant has been operated as a demonstration unit.

In Finland the Finnish Pulp and Paper Research Institute and Tampela have collaborated in the development of the 'Pekilo' process, which used the fungus *Paecilomyces varioti* to reduce the polluting capabilities of spent sulphite waste liquor. The fermentation produces nearly 3 grams of fungal mycelium per litre of growth medium per hour. The single-cell protein is recovered by filtration and then washed and dried. The product, which can be used as an animal feed, has a protein content of 55 to 60 per cent. The Finnish plant had the capacity to produce 10 000 tons of the material per year, a figure which makes this operation the largest producer of mould single-cell protein. However, current economic conditions do not justify the continuing production of the Pekilo protein.

The value of agricultural, forestry or food-processing wastes as animal feeds can also be upgraded by 'solid substrate' fungal treatments. In these operations water is removed from the waste substrate to give a semi-solid sludge with a moisture content of 50 to 80 per cent and then commercial fertilizer is added to the material to provide nitrogen and phosphorous. The fermentors used for the solid substrate methods include simple aerated trays, windrows, aerated towers, rotating drums, scraped tubular vessels and stirred tanks.

Investigators at the University of Waterloo in Ontario, Canada, developed a method for using the fungus *Chaetomium cellulolyticum* to convert the cellulosic carbohydrates in agricultural, forestry or paper mill wastes to a single-cell protein product. Envirocon Ltd has constructed a pilot plant in Vancouver, British Columbia, that is based on the Waterloo process and has the capacity of producing 1 ton of single-cell protein per day.

Sterilized sludge from a pulp mill is incubated with *C. cellulolyticum* and appropriate nutrients in a 1400-litre fermentor for 24 hours. The product from this fermentor is then transferred to a second fermentor with a capacity of 14 000 litres. After 36 hours the fermentation reaches a stage at which continuous harvesting of the product can be performed. Alternatively, the fermentor can be run on a batch basis. Approximately 0.5 kilogram of single-cell protein, which contains 42 per cent protein, is obtained per kilogram of sludge.

Product quality and safety

The protein percentages cited above for the various single-cell protein products are crude values based on measurements of the total nitrogen contents of the materials. Although these values range from about 45 to approximately 60 per cent of the dry weight of the cells, a significant fraction is contributed by nucleic acids, which contain nitrogen and constitute 5 to 15 per cent of the dry weight of the cells. These substances have no nutritional value for non-ruminant animals, and, as previously mentioned, intakes by humans must be limited to 2 grams per day. Treatment of the cells with acid, alkali or enzymes can remove the nucleic acids.

In addition, single-cell protein may be deficient in certain of the essential amino acids that animals cannot synthesize for themselves. For example, studies of broiler chickens and pigs have shown the importance of supplementing single-cell protein products with methionine, which is the amino acid most likely to be present in concentrations insufficient for optimum nutrition. The contents of the amino acids arginine and lysine may also have to be adjusted in the rations fed to domestic animals. Broiler chickens grow most efficiently on feed that contains 7 to 15 per cent single-cell protein.

Currently in the United States, the regulations of the Food and Drug Administration permit the use in human food of the dried cells of baker's, torula and fragilis yeasts, and of a concentrate of protein that is extracted from baker's yeast. In the United Kingdom the Ministry of Agriculture, Fisheries and Food has allowed test-market studies of the chicken substitute, Mycoprotein, that is produced by Ranks Hovis MacDougall from *F. graminearum*. The Ministry also permits the use of the animal feed made by Imperial Chemical Industries from the dried cells of *M. methylotrophus*.

The Protein Advisory Group of the United Nations has developed guidelines for the production and evaluation of single-cell protein products. These guidelines give criteria for determining the nutritional quality and safety of a product as assessed by feeding studies in rats and in livestock such as chickens, pigs and veal calves. Government regulatory agencies may require additional testing of the products to determine whether they can cause cancer, birth defects or gene mutations.

The economics of single-cell protein production

The factors that influence the economic feasibility of single-cell protein production include: the capital costs of establishing the production facilities; the manufacturing costs of raw materials, energy, labour, maintenance, waste treatment and depreciation; and the location of the plant site in relation to the suppliers of the raw material and to the markets for the product.

In the mid-1970s the capital costs for producing feed-grade single-cell protein from methanol were in the range of $660 to $1000 per metric ton of annual capacity for plants that could produce 50 000 to 100 000 metric tons per year. Today, the capital costs would be nearly twice these values. Consequently it is not surprising that very few large-scale plants have been built since the mid-1970s. However, improvements and additions have been made to existing facilities.

The extent of the markets for single-cell protein products as animal feedstuffs will depend on the prices of the products and how efficiently they promote the growth of broiler chickens, laying hens, turkeys and pigs as compared with the performance of existing protein feedstuffs such as soybean and fish meal. The extensive livestock feeding studies that have been conducted by British Petroleum and Imperial Chemical Industries on their single-cell protein products exemplify the kinds of feeding performance results that are needed to satisfy potential users and government regulatory agencies of the value of these products for use as feeds for livestock, including chickens and pigs, which are intended for human consumption.

Flavour and texture, in addition to nutritional value, are important determinants of the acceptability of single-cell proteins as human foods. At present the major market for food-grade products is in such applications as flavouring or leavening of foodstuffs. For example, yeast protein derivatives have been used as food flavourings for many years. Torula yeast is added to processed meats as a flavouring agent and baker's yeast is, of course, used to make breads and other leavened products. In addition, any new single-cell protein product will have to satisfy the regulatory requirements of government agencies before it can be marketed for human or animal feeding.

The future

What is the future for single-cell protein products? The technological feasibility of the large-scale manufacturing of single-cell proteins has been amply demonstrated. A number of commercial operations are now being conducted to a limited extent in countries throughout the world. The introduction of new single-cell products will be limited more by economic, market and regulatory considerations than by technological constraints.

Single-cell proteins will most probably be used in human foods primarily as protein supplements or as ingredients that perform specific functions – acting as flavours or leavening agents, for example. Single-cell proteins for animal-feed applications will be most attractive in those areas where low-cost substrates such as waste carbohydrates are available and where conventional protein feedstuffs such as soybean and fish meal are in short supply.

Additional reading

Davis, P. (1984). *Single Cell Protein*. Academic Press, New York.

dePontanel, H. G. (ed.) (1972). *Proteins from Hydrocarbons*. Academic Press, New York.

Ferranti, M. P. and Fiechter, A. (eds.) (1983). *Production and Feeding of Single Cell Protein*. Applied Science Publishers, Barking, Essex.

Goldberg, I. (1985). *Single Cell Protein*. Springer-Verlag, New York.

Gutcho, S. (1973). *Proteins from Hydrocarbons*. Noyes Data Corporation, Park Ridge, New Jersey.

Klausner, A. (1986). Algaculture: food for thought. *Bio/Technology*, **4**, 947–53.

Litchfield, J. H. (1983). Single cell protein. *Science*, **219**, 740–6.

Mateles, R. I. and Tannenbaum, S. R. (eds.) (1968). *Single Cell Protein*. MIT Press, Cambridge, Massachusetts.

Reed, G. (ed.) (1982). *Prescott & Dunn's Industrial Microbiology*, 4th edn. AVI Publishing Co., Westport, Connecticut.

Rockwell, P. J. (1976). *Single Cell Protein from Cellulose and Hydrocarbons*. Noyes Data Corporation, Park Ridge, New Jersey.

Rose, A. H. (ed.) (1979). *Economic Microbiology*, vol. 4, *Microbial Biomass*. Academic Press, New York.

Senez, J.-C. (ed.) (1983). *International Symposium on Single Cell Proteins*. Technique et Documentation, Paris.

Shelef, G. and Soeder, C. J. (eds.) (1980). *Algae Biomass Production and Use*. Elsevier–North Holland, New York.

Solomons, G. L. (1983). Single cell protein. *CRC Critical Reviews in Biotechnology*, **1**, 21.

Tannenbaum, S. R. and Wang, D. I. C. (1975). *Single Cell Protein II*. MIT Press, Cambridge, Massachusetts.

7 Bacterial leaching and biomining

David Woods and Douglas E. Rawlings

Microorganisms have been forming and decomposing minerals in the earth's crust since geologically ancient times. Mining operations have long benefited from the activities of such naturally occurring microbes, especially from the ability of some bacteria to leach the metals from insoluble ores. As early as 1000 BC mine workers in the Mediterranean basin recovered the copper that was leached into mine drainage waters by bacteria, although they could not have been aware of the microbial involvement. The Romans in the first century and subsequently the Welsh in the sixteenth century and the Spanish who worked the Rio Tinto mine in the eighteenth century almost certainly used microbial leaching for recovering metals.

The contributions of bacteria to metal leaching were not recognized until relatively recently, however. The first reports that certain bacteria, which were then unidentified, are involved in the leaching of zinc and iron sulphides did not come until the 1920s. The essential role of bacteria in the leaching of mineral ores was largely neglected until 1947 when Arthur Colmer and M. E. Hinkle of West Virginia University in Morgantown identified a bacterium, now called *Thiobacillus ferrooxidans*, as the organism principally responsible for the leaching of metal sulphide ores.

Bacterial leaching is now being used successfully in many countries throughout the world to recover metals from a wide variety of ores. The principal metals recovered are copper and uranium, although cobalt, nickel, zinc, lead and gold may also be obtained. In 1983 copper production in the United States was second only to that in Chile, and more than 10 per cent of the US copper was recovered through bacterial leaching. In 1985 copper worth $350 million and uranium worth $20 million were recovered through microbial means in the United States alone. Gorham International, Inc., has projected that the annual production of the worldwide microbial metal industry will amount to $90 billion by the year 2000.

In the past few years bioleaching has been receiving increased attention because the technology has the potential to alleviate many of the problems being experienced by the mining industry. One major problem is the depletion of high-grade mineral deposits and the consequent need to mine at greater depths. In many instances it may be possible to use bacteria to leach the desired metal out of deep or low-grade deposits without removing them from the ground, thereby save the cost of bringing vast tonnages of ore and waste rock to the surface.

Moreover, many conventional procedures for processing metal ores, particularly those ores refractory to heat treatment, consume large quantities of energy. The bioleaching of ores and concentrates, which requires little energy, may provide an energy-saving alternative.

Bioleaching technology also has potential environmental benefits. A long-standing problem in many mining operations has been the uncontrolled release of metals and acids that occurs when the natural leaching bacteria act on the wastes in mine dumps and tailings dams. The controlled leaching of waste rock from conventional mining operations can result both in the recovery of valuable metals from the sites and in the protection of the environment from this source of pollution.

In addition, vast reserves of such fossil fuels as coal and oil cannot now be burned without unacceptable environmental pollution because of their high sulphur contents. The combustion of sulphur-containing fuels leads to the formation of sulphuric acid in the atmosphere and thus contributes to the production of 'acid rain' which is damaging plant and animal life especially in eastern Canada, the northeastern United States and some parts of western Europe. Bioleaching has the potential of removing the sulphur from fossil fuels.

Bioleaching bacteria

Several different microorganisms have been isolated from the sites where bioleaching takes place (Fig. 7.1). Those most important for the leaching process are *Thiobacillus ferrooxidans*, *T. thiooxidans*, *Leptospirillum ferrooxidans*, and a number of species belonging to the genus *Sulfolobus*. All of these organisms are adapted for living in the harsh conditions of their natural environment. They can thrive in acidic solutions, often work at high temperatures, and obtain the energy they need to grow by oxidizing inorganic substances.

Fig. 7.1 Scanning electron micrographs of microorganisms associated with ore particles. The ore particles in (*a*) and (*b*) are covered with rod-shaped bacteria. Spiral-shaped bacteria are attached to the particles in (*c*), and in (*d*) the particles carry both bacteria and the long filamentous projections of a fungal mycelium. (By kind permission of Alex Hartley, University of Cape Town, South Africa.)

T. ferrooxidans, which is the best studied of the bioleaching bacteria, is a small, rod-shaped bacterium that grows best in highly acid solutions; its optimum pH is in the range of 1.5 to 2.5 (Pure neutral water has a pH of 7.) This bacterium is able to derive its energy from the oxidation of ferrous iron ($Fe2+$) to ferric iron ($Fe3+$) and from the oxidation of reduced forms of sulphur to sulphuric acid. Oxygen is the preferred acceptor of the electrons removed during the oxidation reactions, but in the absence of oxygen the organism can use ferric iron as an alternative electron acceptor for the oxidation of reduced sulphur.

T. ferrooxidans has remarkably modest nutritional requirements. All strains are obligate autotrophs, which means that they use the carbon dioxide from the atmosphere as the sole source of carbon for synthesizing their organic constituents. In fact, they cannot grow on organic carbon sources. Moreover, several *T. ferrooxidans* strains are able to fix nitrogen, converting the molecular nitrogen of the atmosphere to ammonia and the other nitrogen-containing nutrients they need. Aeration of a sample of iron pyrite (FeS) in acidified water is sufficient to ensure the growth of most strains at the expense of the pyrite.

Although *T. ferrooxidans* is considered the most important organism in the leaching of metals from ores, it apparently does not work by itself. It is often found growing in close association with other bacteria, including *T. thiooxidans*, which oxidizes sulphur, *T. acidophilus* and *Acidiphilium cryptum*. Mixed cultures of the bacteria are often more efficient at ore decomposition than is *T. ferrooxidans* alone.

The bioleaching of metals from sulphur ores can occur at temperatures ranging from 20 to 80°C or even higher. *T. ferrooxidans* grows best in the temperature range 20 to 30°C. At a temperature of about 50°C the TH strains, which are so called because they are moderately thermophilic (heat-loving) and resemble the *Thiobacillus* bacteria in some respects, can be isolated. At still higher temperatures, from 60 to 80°C, members of the genus *Sulfolobus* leach metals from a variety of ores. These organisms are not used in commercial leaching operations at present, but may eventually prove valuable as they are especially efficient at attacking mineral ores that resist leaching by other bacteria.

Leaching reactions

Leaching reactions generally involve the conversion of insoluble metal ores, which are often sulphides, to soluble compounds from which the desired metal can be more readily isolated (Fig. 7.2). The leaching bacteria may achieve this conversion directly by oxidizing the metal sulphides to produce ferric iron, sulphuric acid and metal sulphates, the identity of which will depend on the type of ore. Some examples of the leaching reactions that result from direct bacterial attack are:

$$4FeS_2 \text{ (pyrite)} + 15O_2 + 2H_2O \rightarrow 2Fe_2(SO_4)_3 + 2H_2SO_4 \quad (1)$$

$$4CuFeS_2 \text{ (chalcopyrite)} + 17O_2 + 2H_2SO_4 \rightarrow$$
$$4CuSO_4 + 2Fe_2(SO_4)_3 + 2H_2O \quad (2)$$

$$2FeAsS \text{ (arsenopyrite)} + 7O_2 + 2H_2O \rightarrow 2FeAsO_4 + 2H_2SO_4 \quad (3)$$

$$CuS \text{ (covellite)} + 2O_2 \rightarrow CuSO_4 \quad (4)$$

Alternatively, bacterial leaching of minerals may be indirect. As shown in reactions (5) to (8), the ferric iron and sulphuric acid generated by the direct oxidation of metal sulphides are themselves capable of oxidizing certain ores to form oxides and sulphates that are soluble in acid solutions. When iron is present, both direct bacterial action and indirect chemical attack are likely to contribute to the leaching of metal ores.

$$2FeS_2 + 2Fe_2(SO_4)_3 \rightarrow 6FeSO_4 + 4S \quad (5)$$

$$CuS + Fe_2(SO_4)_3 \rightarrow CuSO_4 + 2FeSO_4 + S \quad (6)$$

$$UO_2 + Fe_2(SO_4)_3 \rightarrow UO_2SO_4 + 2Fe_2SO_4 \quad (7)$$

$$UO_3 + H_2SO_4 \rightarrow UO_2SO_4 + H_2O \quad (8)$$

Leaching operations

Bioleaching can be applied to the recovery of valuable metals from low-grade mineral deposits, vast tonnages of which are available. According to one estimate, more than 33 billion kilograms of copper are located in mine dumps in the western United States. A single dump, location at Bingham Canyon, Utah, contains 4 billion tons of low-grade waste material with a copper content of less than 0.5 per cent.

Mine dumps are designed to take advantage of the natural lay of the land. They are often built in natural valleys so that the water used to irrigate the dumps flows on its own to a natural collection site or dam. From there, the 'pregnant' liquor containing the soluble metals that have been leached from the ore is pumped through pipes to the recovery plant.

At its simplest, dump leaching entails watering the waste ore in a dump site in cycles of 7 to 10 days interspersed with rest periods of equal length. The rate of metal recovery from the dump may be greatly increased by using instead of water a leaching solution that has an acid pH of 1.7 to 2.4 and contains 2 to 4 grams per litre of ferric iron.

At the recovery plant, copper is most frequently removed from the pregnant leaching liquor by precipitating the metal with metallic iron to produce a cement copper, which is further refined by smelting. The copper can also be recovered by electrically depositing it on suitable electrodes or by extracting it with a solvent.

After the minerals have been removed, the leaching solution can be regenerated and recycled to the dump. Regeneration is

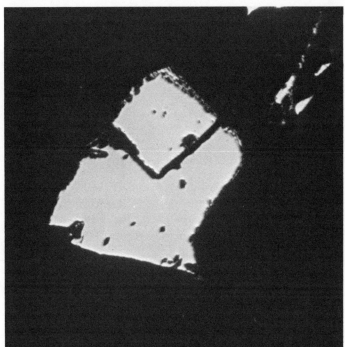

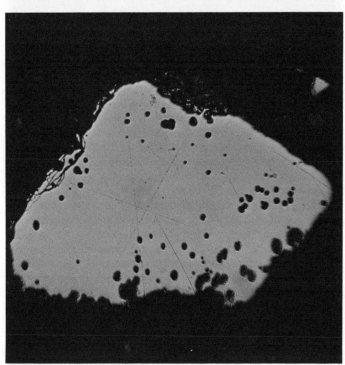

Fig. 7.2 Pyrite particles with deeply penetrating pores that were formed during bioleaching. The particles are about 80 micrometres across. (By kind permission of M. Southwood, Council for Mineral Technology, South Africa.)

achieved by aerating the spent solution and allowing iron-oxidizing bacteria such as *T. ferrooxidans* to convert the ferrous iron produced during chemical leaching back to ferric iron and sulphuric acid (reactions 9 and 10):

$$4FeSO_4 + O_2 + H_2SO_4 \rightarrow 2Fe_2(SO_4)_3 + H_2O \qquad (9)$$

$$S_8 + 12O_2 + 8H_2O \rightarrow 8H_2SO_4 \qquad (10)$$

However, addition of some sulphuric acid may be necessary to maintain a desirable pH. Because dump leaching is slow, a relatively constant amount of copper may be released from a dump for a number of years.

Controlled leaching of minerals from dump sites may help minimize what would otherwise be a serious source of environmental pollution. The leaching of minerals from mine dumps occurs naturally, resulting in the contamination of surrounding areas with heavy metals and acids. Deliberate leaching, in which the run-off fluids are collected should prevent much of this pollution.

Bioleaching of higher-grade ores is handled in a somewhat different fashion from the leaching of the lower-grade wastes. The higher-grade materials are heaped on an impermeable base to ensure that the leach solutions are not lost (Figs. 7.3, 7.4 and 7.5). Efficient irrigation and liquor-collection facilities are built into the heaps. Consequently, they operate with a greatly reduced leach-cycle time. The leaching operation can be completed in months, compared with the years it may take to leach a waste dump.

Vat leaching, in which high-grade ores or concentrates are treated in aerated tanks, is even more efficient than the heap methods and can be completed in a matter of days. Vat leaching has been successfully used to decompose gold-containing arsenopyrite ores before gold recovery.

Ore deposits can also be leached *in situ*, without removing them from underground. This avoids the cost of bringing ore to the surface and may allow the economic mining of low-grade deposits. After the ore body has been fractured by suitably placed explosive charges, the leaching solution is pumped in. The pregnant leaching liquors may then be collected in recovery wells and pumped to the surface for metal extraction. The leaching solutions are usually generated on the surface by using bacteria to oxidize pyrite or other appropriate ores.

Fig. 7.3 Diagram of a heap-leaching operation for copper recovery. After the leaching solution is sprayed on the ore, the fluid that drains through the material is collected and sent to the precipitation plant where copper is removed. The spent leaching solution can then be regenerated by bacterial oxidation and recycled through the dump.

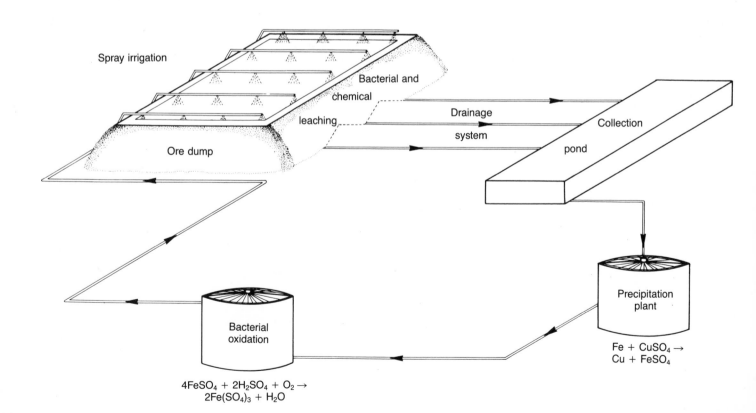

Another application of *in situ* leaching is the recovery of metals from the halos of low-grade ore that remain after a mine has been worked out. At the Stanrock mine in Canada, simply hosing the roof, walls and floors of working and worked-out stopes with a leaching solution at intervals of three months produced an acid wash water that contained sufficient concentrations of soluble uranium to make recovery of the metal economically feasible. A uranium production of 7.5 tons per month was obtained in this way.

Genetic improvement of leaching bacteria

Although non-biological methods of mining and metal processing are likely to dominate in the immediate future, the economic competitiveness of bioleaching for recovering large quantities of copper and uranium from low-grade ores has been amply demonstrated. Investigators are now turning their attention to the improvement of the bacterial strains used for leaching – another area of current research that may benefit from the advent of recombinant DNA technology. Genetic manipulation of the bacteria may ultimately permit the development of new strains with better growth rates, increased leaching efficiencies, and improved resistance to metals that have toxic effects on the organisms.

Before any of this can be accomplished, however, a great deal more will need to be learned about the molecular genetics of *T. ferrooxidans* and the other bacteria that contribute to bioleaching. Compared with what is known about the well-studied *Escherichia coli* and *Bacillus subtilis*, the bacterial species that have so far been the mainstays of recombinant DNA research, knowledge of the members of the *Thiobacillus* genus is relatively sparse.

T. ferrooxidans is a prime target for genetic improvement because of its central role in bioleaching, although alteration of the other bacteria that contribute to the process may also be necessary. Development of a system for altering the genetic constitution of *T. ferrooxidans*, or any bacterial species for that matter, requires several steps. First, suitable characteristics must be identified that can serve as markers for the identification and selection of bacterial strains that have been transformed by the acquisition of new genetic material. Second, vector DNA molecules that can be used to introduce new genes into the bacteria will have to be identified and characterized. Third, a means of introducing the vectors into recipient *T. ferrooxidans* strains must be developed. And finally, the genes coding for the desired characteristics must be isolated and inserted into the vector molecules for transfer into the bacteria being manipulated.

All of these steps have been well worked out for *E. coli*, but comparable research on *T. ferrooxidans* is still in the beginning stages. Techniques that have proved effective with *E. coli* may well prove unsuitable for the bioleaching bacterium. Whereas *T. ferrooxidans* thrives in acid pHs and is an autotroph that uses only carbon dioxide as a carbon source, *E. coli* requires a nearly neutral pH and is a heterotroph that needs at least simple organic compounds to live. Researchers have more or less had to start from scratch in devising methods for the genetic manipulation of *T. ferrooxidans*.

Selectable markers for *T. ferrooxidans*

No method for transferring new genes into cells is 100 per cent efficient and a means of selecting those cells that have acquired the transferred genetic material is essential. Transformed cells of bacteria such as *E. coli* are readily identified by selection for a newly acquired property, which might be resistance to poisoning by a heavy metal or an antibiotic, or an alteration in the nutritional requirements of the cells.

Metal ion tolerance is a particularly attractive marker for genetic studies of *T. ferrooxidans* because increased metal resistance has the potential for conferring an industrially significant characteristic on the bacterium. Although *T. ferrooxidans* is naturally resistant to high concentrations of a wide range of heavy metals, most strains are sensitive to mercury, arsenic, silver, uranium and molybdenum. Exposure to these metals could reduce the effectiveness of the bacteria. For example, the leaching of gold-bearing arsenopyrite ores or concentrates is hindered by the release of toxic arsenic compounds.

Genes coding for resistance to arsenic and mercury have been identified in other bacteria. If such genes could be introduced

Fig. 7.4 Primary leaching pad at the copper mine of the Sociedad Minera Pudahuel in Santiago, Chile. The leaching pads are 20 metres wide by 40 metres long and have two travelling bridge cranes for loading and unloading the ore. The pads, which contain a 2.4-metre thick layer of crushed ore, are of concrete covered with Hypalon 'Water Saver' and are equipped with drainage systems to collect the fluids that percolate through the ore. The leaching solution is called raffinate and is recycled from the solvent extraction plant, which removes most of the copper leached in the previous cycle. Raffinate contains about 8 grams per litre of sulphuric acid and 0.3 to 0.5 grams per litre of residual copper. During the 18-day primary leaching cycle it is applied at a rate of 0.8 litres per minute per square metre. Primary leaching extracts 90 per cent of the copper oxide and 45 per cent of the copper sulphides. The leached ore is then removed from the pads and transported by truck to the secondary leach area, which is about 1.5 kilometres away. (By kind permission of E. M. Domic, Sociedad Minera Pudahuel Ltd, Santiago, Chile.)

Fig. 7.5 Secondary leaching pads at the Sociedad Minera Pudahuel copper mine. The pads, which are 6 metres high, are leached for 120 days during which they are sprayed with the leaching solution at a rate of 0.05 litres per minute per square metre. The secondary leach recovers about 50 per cent of the remaining copper. (By kind permission of E. M. Domic, Sociedad Minera Pudahuel Ltd, Santiago, Chile.)

Fig. 7.4

Fig. 7.5

into *T. ferrooxidans*, they could not only serve as selective markers for transformation of the organism but might also improve its leaching capabilities.

Antibiotic resistance genes have often been used as selective markers but their application to *T. ferrooxidans* is complicated by the unusual growth requirements of the organisms. The low pH and high concentration of metal ions such as iron in *T. ferrooxidans* growth media render many antibiotics unstable. The situation is further complicated by the slow growth of the bacterium, which requires that the antibiotics remain active over a long time. Consequently, *T. ferrooxidans* may appear to be resistant to an antibiotic only because the bactericidal agent is inactivated in the low pH and high iron content of the growth medium.

Although the organism is sensitive to several antibiotics, including chloramphenicol and ampicillin, it appeared to be resistant to several others, including tetracycline and streptomycin, when grown under the normal conditions. However, the apparent resistance turned out to be caused by the inactivation of the antibiotics, possibly by the iron in the medium. By developing a medium for growing *T. ferrooxidans* in the absence of iron and at a pH of 4.0 instead of the usual 1.5 to 2.5 investigators were able to show that the organism is sensitive to tetracycline. Appropriate alteration of growth conditions may make it easier to determine whether *T. ferrooxidans* is truly resistant to an antibiotic.

Gene mutations that make the growth of a bacterium dependent on the addition to the culture medium of some amino acid, vitamin or other nutrient that is not required by the wild-type organism have proved extremely useful over the years for the study of molecular genetics. Such mutant strains can serve as the basis of a selection system for genetic manipulation because the reintroduction of a good copy of the mutated gene relieves the dependency of the bacteria on the nutrient, thereby allowing them to grow in its absence. So far at least, *T. ferrooxidans* does not appear to be amenable to this selection strategy. The organism does not require any organic compounds for its growth and treatment with mutagenic agents such as ultraviolet light and the chemical *N*-methyl-*N*'-nitro-*N*-nitrosoguanidine (NTG) has not yet produced any mutant with such a requirement.

Recently, John Cox of the Martek Corporation in Columbia, Maryland, demonstrated that NTG treatment produces *T. ferrooxidans* mutants that lack rusticyanin, a copper-containing protein that participates in the electron transfer reactions of the bacterium. The result shows that conventional mutagenesis with the chemical works and should stimulate researchers to isolate other important mutants.

Vector construction

In addition to attempting to find selectable markers for *T. ferrooxidans*, researchers are also beginning to devise vectors

for carrying new genes into that organism. Plasmids, which are small circular pieces of DNA that replicate independently of the bacterial chromosome, have proved to be very useful vectors for gene transfer into *E. coli* cells. A vector must be able to replicate in the industrial organism that is to be manipulated. Whether any of the *E. coli* vectors will work in *T. ferrooxidans*, which has a very different physiology, is currently an open question. However, plasmids have been isolated from a number of *T. ferrooxidans* strains that could serve as the basis for constructing vectors for that organism.

Another requirement for a vector is the ability to shuttle between the industrial host and one of the well-characterized hosts, such as *E. coli* or *Bacillus subtilis*, in which genes are usually cloned. This means that the vector origin of replication, the nucleotide segment which contains the necessary signals for initiating reproduction of the DNA, must either be functional in both hosts or the vector must contain an origin of replication for each species. Douglas Rawlings and David Woods of the University of Cape Town, South Africa, have made some progress towards developing a suitable shuttle vector for *T. ferrooxidans*.

They separately cloned four plasmids from the bioleaching bacterium in an *E. coli* plasmid vector. The recombinant plasmids, which also carry genes for resistance to chloramphenicol and tetracycline, were able to replicate in *E. coli* from an origin of replication located on the *T. ferrooxidans* DNA. The origin of replication of one *T. ferrooxidans* plasmid also worked in the bacterium *Pseudomonas aeruginosa*. Because of the broad host range of this sequence, the plasmid containing it has the potential of being a very useful shuttle vector for future genetic manipulation experiments involving *T. ferrooxidans*.

Although the *T. ferrooxidans* plasmids reproduce in *E. coli*, the genes they carry do not seem to be expressed there. The reason for this lack of expression is currently unclear. Conceivably, the protein-synthesizing machinery of *E. coli* fails to recognize the gene control signals of the other bacterium. According to Rawlings, Woods, and their colleague Eugenia Barros, the *T. ferrooxidans* plasmid genes are expressed in a cell-free *E. coli* extract, however.

Gene transfer systems for *T. ferrooxidans*

The three methods by which genes are transferred into bacterial cells are transduction, transformation and conjugation. Transduction depends on the use of a bacterial virus to deliver DNA into the cells it infects. Because no viruses have been discovered for *T. ferrooxidans*, it is not possible at present to use this technique.

Transformation involves the simple uptake of naked DNA by cells. However, this technique is applicable to a rather restricted range of natural bacteria and the conditions required for DNA uptake may vary substantially from bacterium to bacterium. Finding the critical conditions for transformation often requires serendipity. Extensive attempts to transform cells of various *T. ferrooxidans* strains have been unsuccessful to date, but the efforts are continuing.

In conjugation, DNA is transferred directly from one bacterial cell to another. Conjugation is mediated by large plasmids that contain more than 30 kilobases of DNA and possess transfer (*tra*) genes. Some plasmids that carry the *tra* genes can transfer only themselves. However, researchers have discovered a group of plasmids with broad host ranges that can not only move themselves but can also mobilize the transfer of other plasmids that are not capable of moving between bacterial cells on their own. The existence of the broad-host-range plasmids means that conjugation is a broadly applicable method for transferring plasmids between bacteria.

Because self-transmissible plasmids have yet to be isolated from *T. ferrooxidans*, Rawlings and Woods are investigating whether cloned recombinant *T. ferrooxidans* plasmids can be mobilized to move betweeen bacterial cells by other plasmids. So far they have shown that certain of the broad-host-range plasmids can cause the recombinant *T. ferrooxidans* plasmids to move at high frequencies between cells of different *E. coli* strains. However, they have not been able to demonstrate the movement of the recombinant plasmids from *E. coli* to *T. ferrooxidans*. This is not surprising in view of the very different growth requirements of the two bacterial species. Conjugation requires energy and it has not been possible to devise a medium in which both bacteria can meet their energy needs at the same time.

An approach to solving this problem is to use a third bacterial species, one which can grow both with *E. coli* and with *T. ferrooxidans*, as an intermediate. The idea is that the plasmid would be transferred in two stages, first from *E. coli* to the intermediate and then from the intermediate to *T. ferrooxidans*. Certain Thiobacillus species, including *T. novellus* and *T. intermedius*, may be appropriate intermediates. They can grow, as *E. coli* does, at neutral pH in a organic medium and they can also tolerate the conditions of low pH and an inorganic medium that are required by *T. ferrooxidans*.

The first stage of plasmid transfer, from *E. coli* to *T. novellus*, has been achieved. The second stage, transfer to *T. ferrooxidans* from *T. novellus*, is under investigation. The experiments take months to complete because *T. ferrooxidans* grows very slowly. Meanwhile, the ability of the *T. ferrooxidans* plasmids to be mobilized implies that the bacterium does contain self-transmissible plasmids and that a search for them might be worth while.

Cloning and expression of *T. ferrooxidans* genes

The cloning of *T. ferrooxidans* genes has a number of potential advantages. For one, it would permit analysis of the genes' structures and functions, thereby leading to a better understand-ing of the molecular biology of the organism. For another, it might provide genes that could be used for the genetic improvement of *T. ferrooxidans*.

Those genes currently available for this purpose are largely derived from unrelated species such as *E. coli*. At present, very little is known about how gene expression is controlled in *T. ferrooxidans*, nor have investigators determined whether *E. coli* genes can be expressed in the leaching bacterium. Use of cloned *T. ferrooxidans* genes would obviate any expression problems that might be encountered with genes from other species.

Although *T. ferrooxidans* is extremely efficient in scavenging nitrogen in the form of ammonia, the scarcity of nitrogen in the leaching liquors may limit the ore-leaching efficiency of the organism. The nitrogen-fixing and assimilating reactions of the bacterium are thus possible targets for genetic improvement.

The enzyme glutamine synthetase catalyses one of the main reactions by which ammonia is assimilated:

$$\text{L-glutamate} + \text{NH}_4 + \text{ATP} \rightarrow \text{L-glutamine} + \text{ADP} + \text{P}_i$$

Recently, Woods and Rawlings cloned in *E. coli* the *T. ferrooxidans glnA* gene, which codes for glutamine synthetase.

The *T. ferrooxidans* enzyme was made in *E. coli*, with the gene expression proceeding from a *T. ferrooxidans* promoter. (A promoter is one of the regulatory sequences needed for initiation of gene expression.) Whether this promoter is the normal one for expression of the *T. ferrooxidans glnA* gene is unknown, but the expression of the gene in *E. coli* is the first evidence that a promoter from an obligate autotroph is functional in a heterotroph.

The isolation and characterization of enzymes from *T. ferrooxidans* is hampered by the slow growth of the bacteria in inorganic culture media, which results in low cell yields. The cloning and expression of *T. ferrooxidans* genes in *E. coli* overcomes these problems and allows the production of sufficient quantities of enzyme for characterization. The *T. ferrooxidans* glutamine synthetase resembles the enzyme of other species in at least some ways. The typical glutamine synthetase protein consists of 12 identical subunits, each having a molecular weight of about 60 000. The protein encoded by the *T. ferrooxidans glnA* also has a molecular weight of approximately 60 000. Moreover, the *T. ferrooxidans* and *E. coli* enzymes are antigenically related, which suggests that they may have simi¹ ˙ amino acid sequences, even though the nucleotide sequences of the corresponding genes appear to be unrelated. This may happen because the genetic code is degenerate; that is, some amino acids are specified by more than one nucleotide codon. When the nucleotide sequence of the *T. ferrooxidans glnA* gene is determined, it can be compared with that of other *glnA* genes, to see whether the leaching bacterium displays a different pattern of codon usage.

The ability to fix nitrogen is important for any organism that inhabits environments lacking in that nutrient. (See Chapter 9 for a discussion of nitrogen fixation.) Mary Mackintosh of the Microbiological Research Establishment at Porton Down, England, showed in 1978 that at least one strain of *T. ferrooxidans* is able to incorporate atmospheric nitrogen into its cellular material. However, studies have not demonstrated nitrogen fixation by the bacteria in heap-leaching operations.

Nevertheless, the evidence indicates that the capability is widespread in *T. ferrooxidans* strains. According to Woods, Rawlings, and Inge-Martine Pretorius of Cape Town, the DNA from five strains contains the *nifH*, *nifD* and *nifK* genes, which are three of the structural genes needed for synthesizing the nitrogen-fixing machinery. Moreover, the genes are arranged on the DNA in the same order as they are in *Klebsiella pneumoniae*, a bacterium that is one of the prototypes for nitrogen fixation studies.

Evidence also suggests that *T. ferrooxidans* contains the *ntrA*, *ntrB* and *ntrC* genes, three genes that have central roles in regulating nitrogen fixation as well as several other aspects of bacterial nitrogen metabolism. Studies are under way to determine whether the organization and control of the *T. ferrooxidans* *ntr* and *nif* genes resemble those of analogous genes in other bacterial species.

The physiology of *T. ferrooxidans* displays an interesting contradiction with regard to its energy requirements and nitrogen-fixing ability. When the organism obtains its energy by oxidation of ferrous iron, it must use oxygen as the final acceptor of the electrons removed from the metal ion. The *nif* proteins are, however, rapidly inactivated by oxygen and nitrogen-fixing systems can only function when protected from it. Mackintosh demonstrated that *T. ferrooxidans* is nonetheless able to fix atmospheric nitrogen by using ferrous iron as an energy source. Oxygen had to be supplied in limited concentrations that were sufficient to allow enough iron oxidation to support the nitrogen fixation, but insufficient to inactivate the nitrogen-fixing enzymes. Further work is needed to elucidate the mechanism by which *T. ferrooxidans* resolves its apparently conflicting requirements for the presence of oxygen for energy generation and the absence of oxygen for nitrogen fixation.

Studies on the molecular genetics of other members of the genus *Thiobacillus* are under way. Although these investigations are at an even earlier stage of development than those of *T. ferrooxidans*, the basics of a genetic system for several of the bacteria exist. As mentioned previously, researchers have demonstrated the conjugational transfer of plasmids between *E. coli* and *Thiobacillus* species, including *T. novellus*. In addition, Saul Yankofsky and his colleagues at the University of Tel Aviv, Israel, were able to return certain mutants of *T. thioparus* to normal by transforming them with DNA from the wild-type bacterium. Finally, Anne Summers and her associates at the University of Georgia in Athens have used transposons to create mutants of *T. novellus* and *T. versutus*. (Transposons are segments of DNA that can insert themselves into the cellular genome. If they interrupt a gene, they can destroy its function.)

Conclusions

Considerable progress in the development of a system for the genetic manipulation of *T. ferrooxidans* has been made over the past few years. Plasmids have been isolated from the organism that may serve as vectors for carrying new genes into the bacterial cells. In addition, a *T. ferrooxidans* gene has been shown to be expressed in *E. coli*, an indication that the gene control sequences of one species may be recognized by the other. Still lacking, however, is a means of introducing the vector DNA into *T. ferrooxidans* cells, although some steps towards developing such a method have been taken.

The application of recombinant DNA and associated genetic techniques to *T. ferrooxidans* and related bacteria should result in a better understanding of the molecular biology of the organisms. For example, the generation of new mutants would permit investigations into the genetics and regulation of iron and sulphur oxidation. An examination of the ability of *T. ferrooxidans* to use ferric iron, instead of oxygen, as an electron acceptor when oxidizing reduced sulphur compounds could have practical consequences. An enhancement of this ability may improve the performance of *T. ferrooxidans* in leaching operations in which oxygen is in short supply.

An advantage of recombinant DNA technology is that it can be used to introduce multiple copies of a gene into cells. Such gene amplification may permit the identification of the enzymes that act as metabolic 'bottlenecks'. This knowledge would provide new insights into how to improve the growth rate and leaching performance of the bacteria. Another possible application of recombinant DNA technology that is of particular practical importance is improvement of the tolerance of *T. ferrooxidans* to certain heavy metals and other inhibitory ions encountered in the leaching environment.

An important consideration in the development of genetically modified strains of *T. ferrooxidans* is the organism's ubiquitous distribution in the environment. A particular ore body tends to be associated with a natural strain that is especially well-adapted to that ore. A genetically altered strain from another source may not be able to outcompete the indigenous bacteria. It may therefore be necessary to genetically manipulate the local isolate in order to produce a strain suited to a given ore body.

Additional reading

Brierley, C. L. (1982). Microbiological mining. *Scientific American*, **247** (August), 42.

Brierley, C. L. (1978). Bacterial leaching. *Critical Reviews in Microbiology*, **6**, 207.

Curtin, M. E. (1983). Microbial mining and metal recovery: corporations take the long and cautious path. *Bio/Technology*, **1**, 229.

Monticello, D. J. and Finnerty, W. R. (1985). Microbial desulfurization of fossil fuels. *Annual Review of Microbiology*, **39**, 371.

Kelly, D. P., Norris, P. R. and Brierley, C. L. (1979). Microbiological methods for the extraction and recovery of metals. In *Microbial Technology: Current State and Future Prospects*, ed. A. T. Bull, D. C. Ellwood and C. Rattledge, p. 263. Cambridge University Press, Cambridge, England.

Woods, D. and Rawlings, D. (1985). Molecular genetic studies on the *Thiobacillus* and the development of improved biomining bacteria. *Bioessays*, **2**, 9.

Bacteria and the environment

8 Tamar Barkay, Deb Chatterjee, Stephen Cuskey, Ronald Walter, Fred Genthren and Al W. Bourquin

As the past several chapters have demonstrated, microorganisms have remarkable synthetic and degradative capabilities that are becoming increasingly important to biotechnology. Some of the applications currently under development will require the release of microbes into the environment. These applications include using bacteria or viruses to control insect pests. In addition, bacteria can degrade wastes and toxic chemicals and therefore have potential for cleaning up chemical spills and other forms of pollution. Bacteria are even finding application in mining, because certain of the organisms can extract valuable minerals from low-grade ores that could not otherwise be profitably mined.

Many of the microorganisms that are to be used in these applications are being genetically altered by recombinant DNA methods. For example, new chemical-degrading capacities can be bestowed on bacteria by introducing into them the genes for the enzymes that catalyse the desired reactions. Any deliberate release of an organism into a new environment raises questions about whether environmental damage might ensue, but the proposed releases of genetically altered microbes will require especially careful evaluation.

Natural habitats already contain their own indigenous populations of microorganisms, including bacteria, yeast and fungi. Moreover, these microbes play a central role in the carbon cycle, in which they break down dead plant and animal matter and release carbon dioxide and other nutrients that can be used by plants to synthesize new organic materials. Soil microorganisms are also essential contributors to the nitrogen cycle. Among other things they can 'fix' nitrogen, taking inert nitrogen gas from the atmosphere and converting it to forms needed by plants.

Releasing genetically altered organisms might disturb this delicate web of nature or possibly lead to the emergence of organisms with increased infectivity for insects or pathogenicity for animals or plants, including crop plants. A necessary aspect of the proposed releases thus includes a careful assessment of the impact that the microbes might have on the environment and human health.

Agents for microbial pest control

In the past several decades the use of synthetic organic chemicals as broad-sprectrum insecticides has saved millions of lives that might otherwise have been lost to insect-borne diseases, such as malaria. The chemicals have also contributed to the substantial increases in farm yields that have occurred during this time. However, these improvements in the quality of life have come at a high price that has only become evident within the past two decades.

The toxic residues of the synthetic insecticides have become environmental contaminants with deleterious effects on wildlife, including birds and such beneficial insects as bees, and could conceivably have adverse effects on humans as well. Moreover, the continued application of some of the insecticides has fostered the emergence of resistance pest variants that are no longer killed by the chemicals. For example, malaria-carrying mosquitoes in many parts of the world have become resistant to the insecticide DDT (dichlorodiphenyltrichloroethane) which for years was the agent of choice for controlling this pest.

Because of these problems with toxic residues and resistant pests, several insecticides have either been banned from production in the United States or can only be used in certain emergency situations after exhaustive governmental review. The latter is true for persistent chlorinated pesticides, such as heptachlor, which is used to treat for termites under building foundations.

For these and other reasons, academic, governmental and industrial scientists have for some time been actively exploring how pests might be controlled with natural or biological agents. This can be done in any of several ways, among which are the release of sterile male insects to mate non-productively with females of the pest species, and the application of pheromones to disrupt insect feeding or reproductive behaviour.

Another area of investigation that appears to have a great deal of promise for biological pest control is the use of microorganisms that have the innate ability to produce toxins that specifically kill certain insects. Microbial control of insects is attractive because the microbes generally have very narrow host ranges. Consequently, they affect only certain insect species at a given location without causing the widespread destruction of beneficial, as well as harmful, insects. Moreover, in the cases studied thus far, the development of resistance to the insect pathogens occurs less frequently than the development of resistance to the synthetic chemical pesticides. Finally, the build-up of toxic residues in the environment does not occur.

Bacillus thuringiensis

More than 1500 biological control agents or products may have the potential for commercial development. Two of the most extensively studied and field-tested of these are the bacterium *Bacillus thuringiensis* and the viruses of the Baculoviridae family.

In the 20 years in which *B. thuringiensis* has been used for biological pest control, some 30 different varieties have been shown to possess insecticidal activity against more than 100 species of insects in the biological orders Lepidoptera and Diptera. The latter order includes almost all filter-feeding mosquitoes and the blackfly. The use of *B. thuringiensis* to control mosquitoes is becoming standard practice, and firms in the United States, China, Israel, India and Nigeria produce large quantities of the organism commercially.

Application of the microbe is relatively simple because it forms spores that can be dispersed in a water mixture and applied over the location to be treated. The spores, which have a crystalline appearance when examined under the microscope, contain protein toxins. When the spore crystals are eaten by insect larvae, they release a toxin, called delta-endotoxin, that damages the membranes of the cells lining the larval gut. The toxin acts very quickly – mosquito larvae can be killed within minutes of ingesting the spore crystals.

Many field trials have been performed with *B. thuringiensis* and they show that it causes an excellent reduction in the numbers of mosquito larvae. However, repeated applications are generally required to sustain the larvicidal activity. This problem may be solved by using recombinant DNA techniques to move *B. thuringiensis* toxin genes into other microorganisms that have a better ability to persist in what may be hostile environments, such as the lakes where mosquitoes breed. The delta-endotoxin gene has been cloned in the bacterium *Escherichia coli*. In addition, Terry Graham and Lidia Watrud from the Monsanto Corporation in St. Louis, Missouri, have cloned it in the bacterium *Pseudomonas fluorescens*, which lives around plant roots.

The US Environmental Protection Agency (EPA) is currently reviewing applications to conduct field trials with these genetically engineered microorganisms. *B. thuringiensis* is widely accepted as the safest mosquito larvicide that has been developed so far. Whether genetically engineered microbial pesticides will be equally safe and effective is a matter of intense study.

Baculovirus

A number of viruses which are also capable of infecting and causing the death of specific insect pests are either being used or are under consideration for use as biological control agents. Among these are the viruses of the Baculoviridae group, which are effective for controlling such pests as the cotton bollworm, the Douglas fir tussock moth and the gypsy moth.

The baculoviruses, like *B. thuringiensis*, infect only certain insects and are very active as insecticides. In addition, the viruses can be applied easily and they have a long shelf-life.

The viral particles are usually applied in a water-based mixture that is sprayed on the foliage in the area to be protected. When insects ingest the virus-containing foliage the viral particles infect the gut tissue. The virus reproduces there and then spreads to other tissues, thereby causing a systemic infection and the death of the insect. The insect carcass, which is laden with mature viral particles, can spread the virus to additional susceptible hosts.

Studies with *B. thuringiensis*, the baculoviruses and other biological agents have shown that they can successfully control many insect pests. The economic feasibility of biological control might be improved when – or if – genetic engineering is used to produce microbial and viral agents that have enhanced resistance to the physical stresses that they will encounter in the field. These stresses include exposure to dryness, heat, cold and ultraviolet light from the sun. In addition, genetic manipulations that broaden the host range of microbial control agents would make them more attractive from a commercial point of view.

Current investigations are also concerned with producing the control agents more economically and satisfying governmental regulations regarding the potential impact of the agents' release into the environment. Intensive studies to assess the risks of releasing genetically engineered microorganisms in the field are already under way. The goal is to preserve the benefits already obtained with the broad-spectrum organic pesticides, while reducing or eliminating the deposition of toxic residues and other potential environmental problems.

Agricultural uses of genetically altered organisms

Recombinant DNA technology shows great promise for other agricultural applications in addition to the production of genetically engineered bacterial and viral pesticides. One recent example involves the use of a genetically engineered microbe to protect crop plants against frost damage.

The common bacterium *Pseudomonas syringae* resides on the surface of many plant species and contains a protein in its cell membrane that causes water to form ice crystals at the relatively high temperature of 0 to 2 °C thereby producing ice damage to the plant. In the absence of such ice-nucleating agents plants do not show frost damage until temperatures of about −6 to −8 °C are reached.

Using genetic engineering techniques, Steven Lindow and his colleagues at the University of California at Berkeley, deleted the gene coding for the ice-nucleating protein from the *P. syringae* chromosome. The expectation is that spraying plants or seeds with the genetically engineered ('ice-minus') bacterium will allow it to colonize the plant surfaces before the natural strains do, thereby rendering the plants more frost resistant.

Natural isolates of *P. syringae* that have spontaneously lost the gene for the ice-nucleating protein are also available, as are ice-minus mutants that have been produced in the laboratory by conventional methods without recombinant DNA technology. These bacteria might be used in the same way as the genetically engineered versions. However, because the exact changes that have occurred are known only for the bacteria that have been produced by recombinant DNA technology, these strains are considered to have the least risk associated with their environmental release.

The Berkeley group, and also scientists from the EPA, will field-test a genetically engineered ice-minus strain of *P. syringae* at a site in northern California under strictly controlled experimental conditions. The trial is aimed at assessing the bacterium's efficacy in making plants more frost-resistant, its potential environmental effects, and the procedures for monitoring the organism after release.

The application of recombinant DNA technology is not limited to bacteria and other microbes, but can also be used to modify higher organisms. A particularly exciting area in agricultural research involves the genetic manipulation of crop plants (see Chapter 11). By introducing new genes into plants, researchers are beginning to develop improved strains that are resistant to pathogenic bacteria, viruses and herbicides.

Carbon-cycling by microorganisms

Without the recycling activities carried out by bacteria, yeast and fungi, such biologically important elements as carbon, nitrogen and sulphur could become permanently locked in organic material. Imagine what would happen, for example, if the leaves that fall from the trees every autumn were not degraded by microorganisms, but left to lie for ever on the forest floor.

Microbes are extremely versatile in the reactions that they can perform to break down organic compounds. Enzymatic pathways that have evolved over millions of years allow microorganisms to use a wide range of compounds to provide energy and the building blocks needed for synthesizing cellular constituents. Studies of these enzymatic pathways show that, in general, compounds with chemical structures similar to those of cellular building blocks, including amino acids and sugars, are degraded by shorter pathways than those needed for breaking down more complex compounds, which often require several steps for complete degradation.

Sources of synthetic compounds in nature

For the last several decades, microorganisms have been exposed to thousands of novel organic compounds that have been made by man. These compounds, which challenge even the versatile degradative abilities of the microbes, are released into the environment in agricultural, industrial and domestic wastes. They may be spilled accidentally or dumped deliberately. The chemicals fall into different categories, including fungicides, pesticides, herbicides, plasticizers, solvents, detergents, flame retardants and coolants – and their number is ever increasing.

Some of these synthetic organic compounds are degraded by the microbial community and do not pose any environmental threat. They are broken down because they are structurally similar to natural compounds and consequently can be acted upon by pre-existing microbial enzymes. However, other groups of synthetic compounds are much less susceptible to microbial attack and persist in nature for long periods of time. These agents thus have the potential to cause serious pollution problems.

Even compounds that are readily broken down in some environmental conditions may last for a long time in other situations. For example, the insecticide parathion usually disappears from the soil within a month, but in appropriate conditions may persist for 16 years or more after the last application. Such prolonged persistence may reflect the absence from the soil of the necessary degradative microbes as a result of some biochemical or environmental condition that hinders their growth.

Microbial metabolism of halogenated compounds

Synthetic compounds that contain halogens (the elements bromine, chlorine, fluorine or iodine) are the cause of toxic pollution at numerous hazardous-waste sites. Microorganisms are unable to break down many of these compounds, which are widely used. DDT belongs to this category, and although its use has been greatly curtailed since 1975 it still persists in the soil in many areas. Polychlorinated biphenyls (PCBs), which serve as an insulating material in electric equipment, and the dry-cleaning solvent carbon tetrachloride, are also common halogenated organic chemicals.

The presence of halo-organic compounds is not totally the fault of man, however. More than 200 halogenated natural products have been identified. Seventy-five per cent of them contain chlorine and many originate in the salt water of the oceans. The existence of these natural halo-organic chemicals is an indication that microorganisms may have enzymes for degrading the compounds – and that these capabilities might be put to work in ridding the environment of contamination by man-made chemicals.

In general, two types of microbial degradation exist. In one, compounds are broken down to support the growth of an organism and to provide essential nutrients such as carbon, nitrogen or sulphur. In the second type of metabolism, which is commonly known as 'co-metabolism', the compounds neither support the organism's growth nor serve as nutrient sources. Co-metabolism is generally the fortuituous result of the lack of absolute substrate specificity on the part of microbial enzymes. Certain unnatural compounds, such as halo-organic chemicals, may resemble the natural enzyme substrate sufficiently to undergo the same enzymatic reaction. Compounds that do not resemble any natural substrates will be refractory to microbial degradation unless they are altered by light, water or other environmental factors in a way that makes them susceptible to microbial enzymes.

Improving microbial degradation by genetic engineering

Although researchers are isolating microorganisms that can break down halogenated organic chemicals, the microbes are often limited in what they can do. For example, a *Pseudomonas* strain has been identified that can degrade 3-chlorobenzoic acid, but not the closely related compounds 4-chloro- and 3,5-dichlorobenzoic acids.

Another *Pseudomonas* strain is capable of degrading 2,4-D (2,4-dichlorophenoxyacetic acid) but cannot act on 2,4,5-T (2,4,5-trichlorophenoxyacetic acid) which differs from 2,4-D only in having a third chlorine on the molecule. Both of these compounds have been widely used as herbicides, and 2,4,5-T has persisted in the environment.

The application of the new genetic methods to constructing novel microbial strains that have improved capacities for degrading various synthetic compounds is currently an active area of research. However, before such strains can be constructed the researcher must answer the question 'Why are the toxic chemicals not being degraded normally?' This requires a detailed knowledge of the pathways for degrading structurally related compounds, including an understanding of the specificities of the enzymes involved and their regulation.

Microbes may not be capable of accomplishing the complete degradation of a particular compound for a number of reasons. The organisms may not have all the enzymes needed, or the enzymes may be present but too slow-acting. Such problems could be solved by introducing into the organism new genes that code for enzymes with the desired specificities. Alternatively mutations could be introduced into the appropriate gene of the microbe to change the enzyme specificity to allow recognition of the desired substrate. If the problem is the slowness of the degradative pathway, the solution could be to increase the amount of enzymes produced by cloning the corresponding genes in a plasmid that would be present in multiple copies in the cell.

Another possible explanation for the lack of degradation of a chemical is its failure to enter the microbial cell. This might be overcome by introducing genetic mutations into an existing transport system to widen its specificity range to include the chemical in question. Finally, the synthesis of the degradative enzymes may be closely regulated so that they are normally turned off and are not turned on by the chemical in question. In that event the solution would be to select for regulatory mutants that produce the desired enzymes without the need for specific activation of the genes. Methods are available for performing all of these manipulations mentioned above.

Walter Reineke and Hans Knackmuss of the Universität-Gesamthochschule in Wuppertal, West Germany, have shown that it is possible to construct total pathways for chemical degradation by combining the genetic capacities of two bacterial strains. The genes required for degrading synthetic chemicals are normally carried on plasmids, rather than in the chromosomal DNA. For example, the genes that allow the previously described strain of *Pseudomonas* to degrade 3-chlorobenzoic acid, but not 4-chloro- or 3,5-dichlorobenzoic acids, are located on a plasmid.

Plasmids can readily be transferred between species of *Pseudomonas*. Reineke and Knackmuss introduced into the *Pseudomonas* strain that has the capacity to degrade 3-chlorobenzoic acid a second plasmid, called the TOL plasmid, from another *Pseudomonas* species. The TOL plasmid carries genes coding for enzymes that allow the conversion of 3-chloro-, 4-chloro- and 3,5-dichlorobenzoic acids to the corresponding chlorinated catechols. The resulting bacterium can completely degrade both of the monochlorinated benzoic acids (Fig. 8.1).

Growing this strain in the presence of 3,5-dichlorobenzoic acid then allowed the isolation of a mutant that could degrade all three of the chlorinated benzoic acids. These manipulations did not require recombinant DNA techniques, incidentally, but showed that simple genetic procedures can be exploited to develop novel bacterial strains through forced evolution.

Using a different procedure, Ananda Chakrabarty and his colleagues at the University of Chicago Medical Center in Chicago isolated a strain of *Pseudomonas cepacia* that can break down 2,4,5-T. The DNA structures of some of the naturally occurring plasmids that participate in the degradation of aromatic compounds are very similar, a finding which suggests that the plasmids are evolutionarily related. The current assumption is that plasmids with new genetic functions may have evolved from a common ancestor in ways that are poorly understood but probably involved recombinations between the plasmid genes, gene deletions and mutations. Consequently, adding new plasmids to those already in the indigenous bacteria of contaminated soils may provide an adequate gene pool for the evolution of new genes that can degrade chemical contaminants.

Chakrabarty and his colleagues obtained the *P. cepacia* strain with the ability to break down 2,4,5-T by mixing bacterial cultures from areas contaminated with the herbicide with plasmid-carrying laboratory strains of the bacterium, and then selecting for 2,4,5-T-degrading bacteria. The strain that they isolated can decontaminate soil containing as much as 20 000 parts per million of the 2,4,5-T and restore plant growth.

The bacterium works by releasing the chlorine from the 2,4,5-T molecule. Moreover, it appears that the dehalogenating activity is non-specific because the bacterial strain can also release the halogens from a wide variety of other phenols that contain chlorine, bromine or fluorine.

The resistance of any compound to degradation depends in large part on the halogens or other substituent groups that it contains. Removal of these groups from the compounds is thus

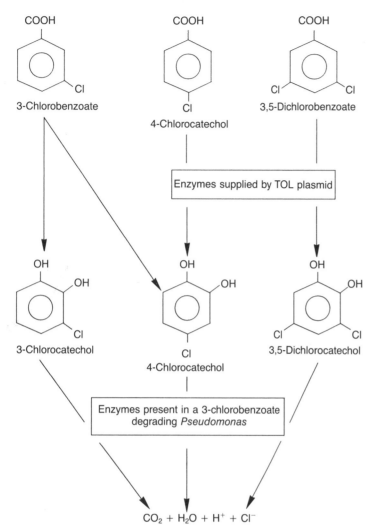

Fig. 8.1 Hybrid pathway for the complete degradation of chlorobenzoic acids that was constructed by plasmid transfer. The original *Pseudomonas* strain could completely degrade 3-chlorobenzoic acid to carbon dioxide, water, and the hydrogen and chloride ions. However, it could not degrade 4-chloro- and 3,5-dichlorobenzoic acids because it could not convert them to the corresponding catechols, a reaction that is an intermediate step in the complete degradation. Introduction of the TOL plasmid into the bacterium permits the catechol conversion of the two benzoic acids, thereby enabling their complete breakdown.

important because it generates compounds that can be broken down by pre-existing enzyme systems. The next challenge is to clone the gene or genes needed for the dehalogenating activity in vectors that can be introduced into bacteria to produce even more effective strains for degrading halo-organic chemicals.

Although bacteria have a great deal of potential for degrading contaminants in the environment, the reactions carried out by the organisms do not always produce less toxic substances. Sometimes the products are more dangerous than the parent compounds. For example, DDT is converted by bacteria in some soil and aquatic environments to DDD (dichlorodiphenyl-dichloroethane), which is a more potent insecticide than the original compound and also persists longer in the environment and is more toxic to higher animals.

The use of bacteria in the mining industry

The possibilities of tailoring microbes to perform beneficial functions extends surprisingly far beyond the scope of pharmaceutical and agricultural applications. Bacteria are being used commercially in the copper and uranium mining industries. More than 10 per cent of the copper produced in the United States is mined by bacteria (see also Chapter 7).

Bacterial mining depends on the ability of the microbes to oxidize the insoluble metal compounds of the ores to produce soluble compounds, but bacteria can also be used in a more passive way to recover metals from solution. Bacterial cell walls contain negatively charged molecules that make them excellent traps for binding positively charged metal ions. This 'biosorption' occurs even with dead cells and is reversible, which means that the metals can readily be removed from the organisms.

Biosorption can serve a dual purpose. It can be used both to recover or concentrate valuable metals from industrial wastes and to clean up such wastes by removing toxic metals.

Bacteria are not the only organisms that can be used in this way. The cell walls of fungi contain chitin, a structural polysaccharide that is also an effective binder of metals. Bacterial and fungal processes have already been developed for treating the wastes from nuclear power plants and for recovering precious metals.

The environmental impact of genetically engineered microorganisms

The future use of genetically engineered microorganisms in the applications just described depends on whether they can safely be released into the environment. This is a controversial and often bitterly debated issue. While some previous releases of organisms into new environments have proved to be beneficial, or at least harmless, others have been deleterious.

The establishment of a novel organism in an ecosystem depends on the physical and chemical conditions existing there and on the organism's interactions with indigenous species. Whether a genetically engineered organism will have an advantage or disadvantage in a particular ecological niche will therefore depend both on its characteristics and on those of the receiving ecosystem.

When a foreign organism does become established in a new ecosystem it may or may not produce deleterious effects, but it will almost certainly cause changes of some type. They may be so inconsequential as to be undetectable, or they may be catastrophic. For example, the introduction of the gypsy moth (*Lymanthria dispar*) into the United States has had devastating effects on plant foliage because the insect has no natural predators in North America. Frances Sharples of Oak Ridge National Laboratory in Oak Ridge, Tennessee, has reviewed numerous additional examples of environmental damage caused by releases of novel organisms.

A novel microorganism can affect an environment either directly or indirectly. Direct effects include pathogenicity or toxicity to individual species or to communities of species. Genetic manipulation of many microorganisms is aimed precisely at increasing their pathogenic or toxic effects to pest species in the environment. Such microorganisms, although designed to have a limited host range, may also pose a hazard to non-target organisms.

Indirect effects result from changes in the structure or function of the indigenous community. A new organism can alter community structures in any of several ways. It might, for example, act through predation or by outcompeting indigenous species for essential nutrients. Daniel Simberloff of Florida State University in Tallahassee has reviewed 854 cases in which organisms were introduced into new environments, and analysed the effects of the newcomers on the composition of the indigenous community. Although the effects of the introductions varied, some environments suffered a loss in numbers of species. Predation caused the species losses in 50 of these cases and competition in another 51.

Microbial activities can alter the acidity and oxygen content of soil and water, the availability of nutrients for plants, and the chemical form of metals. Because the productivity of an ecosystem depends on the decomposition of organic matter and nutrient cycling by microbes, any disruption of microbial communities as a result of the introduction of a novel microorganism could alter the entire ecosystem.

The ultimate fate of natural and manipulated genetic material in the environment depends on the survival, establishment and growth of the hosts that harbour the genetic material. The survival, establishment and growth of the host organisms in turn depend on their genetic constitutions and on the physical and chemical characteristics of the environment, as well as on their interactions with the other organisms present there.

The physical factors that help to determine whether a new organism can become established include temperature, atmospheric pressure, sunlight, the surfaces of plants and sediments, and the spatial relations that contribute to the protection, nutrient accumulation or chemical requirements of the cells. The chemical factors include the availability of carbon sources, inorganic nutrients, growth factors and water; the ion composition of the soil and water; the environmental pH; gas composition; and the presence of toxic substances.

Because microbial communities are such an intricate part of their biological, chemical and physical environment, assessing the potential hazards of releasing a new microorganism requires an experimental system that reflects the complexity of the mic-

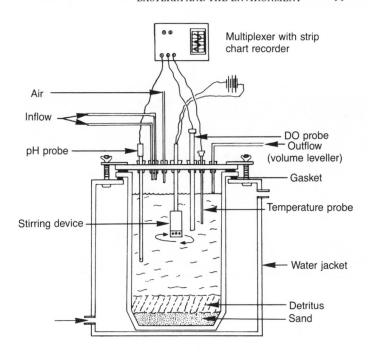

Fig. 8.2 A microcosm used to study the response of aquatic microbial communities to environmental perturbations. The tank containing the microcosm is surrounded by a water jacket that can be used to cool or warm the contents. It is fitted with intake tubes for air and for water and other materials and with an outflow tube that is used to maintain the appropriate fluid level. Probes measure the pH, temperature and dissolved oxygen (DO) concentration of the water in the microcosm. A chart recorder keeps a continuous record of these values, which give information about the responses of the microorganisms living in the microcosm.

robial environment. 'Microcosms' which are laboratory-operated enclosures containing a piece of the environment, are often used for this purpose (Fig. 8.2).

A microcosm can be constructed either by bringing in a natural sample or by reconstituting it in the laboratory. However a microcosm is constructed, its utility depends on how closely it resembles the situation in the field.

Detecting microorganisms in environmental samples

The risk assessment of microbial releases into the environment requires that the organism in question be detected and counted. This is not an easy task because the cells must be detected in complex uncharacterized environmental samples, such as soils and aquatic sediments. Microorganisms in the environment may become undetectable by conventional methods, yet still remain viable and active. Despite these inherent difficulties, microbial ecologists have developed methods for enumerating microbes in environmental samples that meet the analytical and operative criteria required for any monitoring procedure. These criteria include *sensitivity*, which allows a very few genetically engineered microorganisms to be differentiated from a complex background of indigenous microbes. The methods must also be *reproducible* enough to provide measurements that can be repeated under standardized conditions, regardless of any variability in the origin and nature of the sample. They have to be *universal* to allow application to samples from diverse ecosystems, such as leaf surfaces or freshwater reservoirs. They have to be *specific* for the genetically manipulated microorganism or its products. And finally, the methods must be *operationally feasible*; that is, they have to be cost-effective and capable of being performed with the level of professional expertise that exists in most environmental quality laboratories.

One of the oldest methods for detecting a particular microbial species in a complex sample involves the use of selective culture media. This technique is based on the inclusion in the medium used for growing microorganisms of a compound that either inhibits the growth of all species except the one of interest or else specifically reacts with the organism of interest or its products to give an easily observable characteristic (Fig. 8.3).

A wide variety of such selective media are available. If the organism to be detected is resistant to an antibiotic, for example, then it will be able to grow in the presence of that antibiotic while other organisms will not. A requirement for some unusual nutrient can also be used to select for a particular microbial species.

Other methods depend on probes that can specifically pick out the microorganism being analysed. The widely used probes include antibodies that react with some component of the microbial cell. The antibodies are usually labelled with a fluorescent dye so that they can easily be detected (Fig. 8.4).

Recent advances in hybridoma technology allow the production of monoclonal antibodies to essentially any purified component of the microbial cell wall, thereby greatly increasing the range and specificity of this methodology. Fluorescent monoclonal antibodies are sufficiently specific to permit researchers to distinguish different species of nitrogen-fixing *Rhizobium* bacteria, which are closely related and otherwise difficult to differentiate. The antibodies have also proved reliable for detecting microorganisms in their 'non-culturable' life stages.

Cloned genes are another type of highly effective probe. Certain treatments can cause the two strands of the DNA double helix to separate. The single strands can re-form the base pairs that hold the double helix together only with other DNA strands from very similar genes. Consequently a cloned gene strand, which has been combined with a dye, radioactive isotope, or some other easily detectable label, can be used to detect comparable genes in the DNA of an organism.

DNA probes were originally developed as a method in recombinant DNA technology for detecting specific gene clones. Because the DNAs of no two species are exactly alike, the probes can also be used to detect specific organisms in environmental samples and pathogens in water systems and food products.

This method can be adapted for detecting non-culturable organisms by analysing microbial DNA that has been directly isolated from environmental samples. It can be used to estimate the abundance of a genetically manipulated organism in a sample because the foreign gene contained by such an organism could be readily detected in the total DNA with an appropriate probe.

Detecting specific microorganisms by their DNA compositions can be complicated because bacteria can exchange genetic material with one another. However, the genes coding for the RNA of the ribosomes are not transferred in this way. The ribosomal RNA genes and the corresponding RNA sequences are specific for each organism and can be used for its unequivocal identification. They can be viewed as the personal monograms of bacterial species. Analysis of the ribosomal RNA has been applied to determining the relatedness of bacteria and for characterizing natural microbial communities.

Gene transfer among bacteria in the environment

Foreign genes may become established in the environment even if the genetically engineered microorganisms that originally carried the genes fail to survive. This is because foreign genes may be transferred to indigenous microorganisms. The potential risk posed by these gene transfers may be substantial because the genes may be more stably maintained by microbes that are already established in an ecological niche than by the genetically engineered 'interlopers'.

Genes may be transferred by conjugation as a result of direct contact between a donor and recipient cell; by transduction, which results when the gene becomes incorporated into the genome of a bacterial virus that infects another bacterial species; by transformation, in which the naked DNA is transported through the recipient cell wall and becomes a part of the new host's genome; and by the fusion of two bacterial cells.

The extent of the occurrence and significance of these transfer methods in microbial communities in the natural environment has not yet been fully evaluated. Although the available information suggests that the genetic exchanges occur at low frequencies, several questions must be answered to assess the risk that may be involved in the release of genetically engineered organisms that may transfer foreign genes to indigenous species: Does genetic exchange occur and at what frequencies? What are the environmental conditions that affect the transfers? How do such transfers affect the structure and function of the indigenous community? Do the locations of the foreign genes in the genomes of engineered organisms affect their transfer? Can there be transfer from dead cells? These questions are just beginning to be addressed, but methods that were originally developed by molecular geneticists are now helping microbial ecologists to perform the necessary studies.

The environmental effects of genetically engineered microbes may not be limited to the area of their release because microorganisms have the potential to spread. Released organisms can

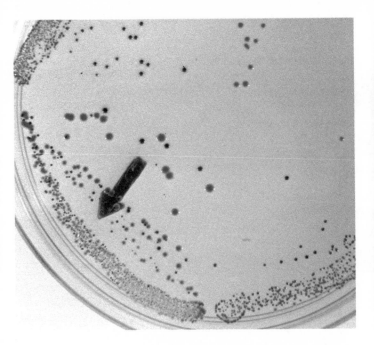

Fig. 8.3 Use of a dye to identify the presence of a particular kind of microorganism. *Rhodopseudomonas capsulata* bacteria are grown in culture on a selective medium containing 5-bromo-4-chloro-3-indolyl-β-D-galactoside as a chromogenic indicator. The blue colonies indicate clones of the organism in which genes needed for lactose breakdown have been cloned. Bacteria containing the cloned genes produce the enzyme β-galactosidase, which removes the galactose from the indolyl group. Release of the indolyl group inside the cells causes them to turn blue.

Fig. 8.4 Detection of a particular variant of the cholera-causing bacterium *Vibrio cholerae* in an environmental sample. The *V. cholerae* cells are stained with a monoclonal antibody, which has been labelled with a fluorescent dye. The stained cells fluoresce green, while bacteria that do not have the antigen recognized by the antibody are stained orange in colour. (By kind permission of Phillis Brayton and Rita Colwell, University of Maryland, College Park.)

reach new sites by way of transport by winds, run-off waters and migratory animals. An organism designed to function and be contained in one environment may, with an as yet unknown likelihood, become uncontrollable in a new site. The transport of released organisms is therefore a crucial issue for risk assessment.

A great deal is already known about the spread of human pathogens as a result of irrigation with treated waste waters and about the global transmission of infections by some plant pathogens, especially fungi that spread by spores. The data bases, methods and mathematical models that were developed to study these phenomena could be used to follow the transport of genetically engineered organisms.

Concerns regarding the establishment of engineered organisms, the stability of their genetic material, their spread in the environment, and their potential for perturbing the ecosystem define the issues of biotechnology risk assessment. As studies are conducted and information pertinent to risk evaluation becomes available, the criteria for determining that biotechnological products are safe will be established. This scientific regulatory procedure is essential for realizing the industrial and therapeutic potential of biotechnology.

Additional reading

Curtin, M. E. (1983). Microbial mining and metal recovery: corporations take the long and cautious path. *Bio/Technology*, **1**, 229.

Fox, G. E. and others (1980). The phylogeny of prokaryotes. *Science*, **209**, 457.

Halvorson, H. O., Pramer, D. and Dogul, M. (eds.) (1985). *Engineered Organisms in the Environment: Scientific Issues*. Proceedings of a cross-disciplinary symposium held in Philadelphia, 10–13 June 1985. American Society for Microbiology, Washington, DC.

Jacobson, M. (1975). Introduction. In *Insecticides of the Future*, ed. M. Jacobson. Marcel Dekker, New York.

Miller, L. K., Lingg, A. J. and Bulla, L. A. Jr (1984). Bacterial, viral, and fungal insecticides. In *Biotechnology and Biological Frontiers*, ed. P. H. Abelson, pp. 214–29. American Association for the Advancement of Science, Washington, DC.

Pritchard, P. H. and Bourquin, A. W. (1984). The use of microcosms for evaluation of interactions between pollutants and microorganisms. In *Advances in Microbial Ecology*, vol. 7, ed. K. C. Marshall, pp. 133–215. Plenum Press, New York.

Reineke, W. and Knackmuss, H. J. (1980). Hybrid pathway of chlorobenzoate metabolism in *Pseudomonas* sp. B13 derivatives. *Applied and Environmental Microbiology*, **142**, 467.

Sharples, F. E. (1983). Spread of organisms with novel genotypes: thoughts from an ecological perspective. *Recombinant DNA Technology Bulletin*, **6**, 43.

Teich, A. H., Levin, M. A. and Pace, J. H. (eds.) (1985). *Biotechnology and the Environment: Risk and Regulation*. American Association for the Advancement of Science, Washington, DC.

9 Biological nitrogen fixation

by Andrew W. B. Johnston

Water, water, everywhere
Nor any drop to drink
Samuel Taylor Coleridge, 'The Rime of the Ancient Mariner'

The problem of nitrogen

Though the words above refer to the compound water, they could equally apply to the element nitrogen, N_2. In this form, nitrogen is literally all around us – it constitutes nearly 80 per cent of the atmosphere we breathe – but it is inaccessible to us and to all animals, plants and fungi, and to virtually all bacteria. Nonetheless, nitrogen in an organic form is a major component of all living things. Proteins, nucleic acids, vitamins, and numerous other 'vital' molecules all contain it. How then is the vast reservoir of inert nitrogen gas made available for the assembly of the repertoire of nitrogenous organic molecules?

The answer is biological nitrogen fixation. A relatively small number of bacterial species have the special ability to reduce or 'fix' atmospheric N_2 to form ammonia, a product that can be used by plants and other microbes as a building block for the synthesis of amino acids and thence other nitrogenous compounds. On a global scale the amounts of nitrogen fixed by these bacteria are impressive. Estimates are in the region of 200 million tons each year. At a more parochial level, a clear demonstration of the ecological and agronomic importance of nitrogen fixation can be seen in the longest-running agricultural experiment in the world.

In 1843, John Bennet Lawes and Henry Gilbert instigated the so-called Broadbalk Experiment at the Rothamsted Experimental Station in England. Strips of land have been sown with wheat every year since the start of the experiment. Each strip has been subjected to a particular fertilizer regime over the entire period and the yield of grain has been measured annually. The results from three strips, each of which received a different nitrogen fertilizer treatment, provide some food for thought.

One of these plots was treated continuously with nitrogen fertilizer in amounts comparable to those normally used in Western European agriculture. The other two have never been given nitrogen fertilizer. The two unfertilized plots differ in that one has been weeded whereas no attempt has been made to remove the weeds from the other.

Not surprisingly, the wheat yields of the plot that has been treated with nitrogen fertilizer have been high – equivalent to approximately 6 tons per hectare per year – compared with the yields of the unfertilized plots. At first glance the results for these two plots look paradoxical. The one with the weeds has an annual yield of 4 tons of wheat per hectare, about double

the yield of the plot without weeds. However, these are not just any weeds. They are vetches and clovers, members of one plant family, the Leguminosae.

The legume plants can thrive under conditions of nitrogen limitation because they form an efficient symbiosis with the nitrogen-fixing bacteria of the genus *Rhizobium* (Fig. 9.1). The bacteria provide fixed nitrogen to support the growth both of themselves and of the legumes. Then, when the plants die, the soil is also enriched with sufficient fixed nitrogen to increase the wheat yield of the unweeded plot by about 2 tons per hectare over that of the weeded plot. The experiment clearly demonstrates that in the field biological nitrogen fixation can substitute, at least in part, for commercial nitrogen fertilizer, especially when a nitrogen-fixing soil microbe associates with a higher green plant.

Nevertheless, even the unfertilized, weed-free plot yields 2 tons of wheat per hectare. The sources of nitrogen to support this growth are threefold: the formation of nitrous oxides and other pollutants by lightning and industrial plants; the slow chemical release of nitrogenous compounds from the soil; and biological nitrogen fixation by soil bacteria that do not form symbiotic relationships with leguminous plants.

A major long-term goal of research programmes in biological nitrogen fixation is reducing the current reliance on chemical fertilizers for crop plant growth by improving the efficiencies and ranges of the nitrogen-fixing organisms. Around 40 million tons of nitrogen fertilizer are manufactured each year, nearly all by the Haber–Bosch process in which gaseous nitrogen and hydrogen are passed over a catalyst at high temperature and pressure to form ammonia.

Because one of the feedstocks for this reaction is hydrogen gas, ammonia synthesis is energy-expensive. Although energy costs do not currently appear as economically threatening as they did during the energy crisis of the early 1970s, predicting how the vagaries of politics and economics will affect fuel costs in the medium and long term future is difficult. In addition, large areas of the agricultural world are, even now, too poor to buy enough commercial nitrogen fertilizer to give optimum crop yields.

Shifting from chemically to biologically fixed nitrogen is also desirable from an environmental point of view. Much of the chemical fertilizer that is applied to crops is wasted by leaching

Fig. 9.1 Demonstration of the benefits of *Rhizobium* bacteria to the growth of *Phaseolus* beans. The three plants were grown in compost with no added nitrogen fertilizer. The plant on the left had no *Rhizobium* added; the centre one was inoculated with a poor nitrogen-fixing strain of *R. phaseoli*; and the plant on the right with a selected high-performance strain of the bacterium.

Table 9.1. *Selected nitrogen-fixing bacteria*

Species	Bacterial group	Comments
Klebsiella pneumoniae	Gram-negative	Model system for *nif* genetics
Azotobacter vinelandii	Gram-negative	Fixes nitrogen in air; contains a protein that protects nitrogenase from oxygen damage
Rhizobium species	Gram-negative	Fixes nitrogen in legume root nodules
Rhodospirillum, Rhodopseudomonas	Gram-negative	Purple-green photosynthetic bacteria
Frankia	Gram-positive actinomycete	Fixes nitrogen in root nodules of various woody trees and shrubs
Clostridium	Gram-positive	Obligate anaerobe
Anabaena	Filamentous blue-green alga	Fixes nitrogen in specialised cells called heterocysts; some species associate with higher plants such as cycads and *Azolla*
Methanococcus	Archaebacterium	
Azospirillum	Gram-negative	Associated with roots of grasses

from the soil. At worst, this can generate unacceptably high concentrations of nitrate in drinking water and can lead to the eutrophication of water systems with a consequent overgrowth of algae and other plant-life and a decline in fish and shellfish populations.

The range of nitrogen-fixing organisms

In any list of nitrogen-fixing bacteria (Table 9.1), a number of features are apparent. First, all are bacteria: simple non-nucleated prokaryotes. Despite many searches, no nitrogen-fixers have yet been found among the nucleated eukaryotes. As the Broadbalk Experiment illustrates, legume weeds, which have indirect access to biologically fixed nitrogen, have a strong selective advantage over other weeds in nitrogen-poor soils. If nitrogen-fixing capabilities confer a selective advantage on an organism, why then are there no nitrogen-fixing yeasts, for example? And why have nitroplasts not evolved, as chloroplasts and mitochondria apparently have, from bacteria that have taken up permanent residence in cells and established the most intimate of symbioses? The answers to these questions are unknown. It may simply be that the ability to fix nitrogen has evolved relatively recently and that insufficient time has elapsed for a eukaryotic nitrogen-fixer to evolve.

Second, the members of the nitrogen-fixing club are taxonomically both wide-ranging and sporadic in occurrence. Representatives have been found in groups as diverse as the Cyanobacteria (the blue-green algae), the Archaebacteria, and in both Gram-positive and Gram-negative bacteria (two major bacterial subgroupings that are distinguished on the basis of their ability to take up the Gram stain that was devised by Danish physician Cristian Gram). Yet within each of these groups only occasional strains or species can fix nitrogen. Lateral evolution, in which the nitrogen fixation genes – *nif* genes as they are called – have been transferred from one bacterial type to another, may be the cause of this taxonomic distribution.

Third, many of the bacteria do not fix nitrogen by themselves, but work in a symbiotic interaction with higher plants. The reason for this comes back to the question of energy. The triple bond that joins the two nitrogen atoms in a molecule of gaseous nitrogen is a tough one to crack. Just as there is a high energy cost for the chemical production of ammonia, there is a high energy burden for the nitrogen-fixing bacteria. When the bacteria associate with carbon-fixing green plants, the result is a nice nutritional trade-off. The plant gets its fixed nitrogen and the bacterium acquires the fixed carbon it needs for energy.

The diversity of such symbioses is great. In addition to the interaction between legumes and various species of the *Rhizobium* genus, there is, for example, the association between a nitrogen-fixing blue-green alga and the water fern *Azolla* (Fig. 9.2), which has been used for centuries as a source of nitrogen fertilizer in rice paddies. The symbiotic interactions are not even confined to those between bacteria and plants. Termites, which devour wood, contain in their guts populations of nitrogen-fixing bacteria that help the insects overcome the nitrogen deficiencies of their staple diet.

Fig. 9.2   A single frond of the water-fern *Azolla caroliniana*. Each leaf of the fern contains a small cavity for holding the nitrogen-fixing blue-green alga *Anaebaena azollae*. The fern is used to fertilize rice paddies in Southeast Asia. The magnification of the image on the slide is about × 1. (By kind permission of Steven Dunbar, Batelle-Kettering Research Laboratory, Yellow Springs, Ohio.)

Biochemistry of nitrogenase

The special ability of nitrogen fixing bacteria to reduce N_2 to ammonia depends on the possession of an enzyme system called the 'nitrogenase complex'. The complex appears to be very similar in all the nitrogen-fixers studied to date and the information gleaned from one is almost certainly applicable to others.

Current knowledge indicates that the nitrogenase complex is composed of six proteins and contains two different enzyme activities, one called simply nitrogenase and the other called nitrogenase reductase (Fig. 9.3). The nitrogenase component of the complex contains four subunits, two copies of each of two different proteins. Its structure also includes a cofactor, iron–molybdenum cofactor, which contains the metals iron and molybdenum. The structure of the cofactor is unknown despite many years of study.

The precise details of how nitrogenase works are not completely clear. The N_2 almost certainly binds to the cofactor, after which it is reduced to ammonia by the addition of electrons and hydrogen ions. The hydrogen ions are obtained from water through a number of steps, the exact nature of which is still a matter of some debate.

The reduction of the N_2 is energy-expensive; 20 to 30 molecules of adenosine triphosphate (ATP), the cell's energy currency, are required to support the reduction of one molecule of nitrogen to ammonia. Moreover, the nitrogenase reaction is inherently wasteful in that is also reduces hydrogen ions to molecular hydrogen, H_2, which is given off as a gas.

Nitrogenase reductase has a molecular weight of 60 000 and consists of two identical protein subunits that have a characteristic brown colour because they contain clusters of iron and sulphur (Fig. 9.4). As its name suggests, the enzyme reduces nitrogenase, thereby replenishing the electrons used to reduce the N_2. The reductase acquires the electrons it transfers from other proteins, the exact identities of which vary in the different nitrogen-fixing bacteria.

The oxygen problem

One more very important point about nitrogenase is that it is poisoned by oxygen. Exposed to the atmosphere, the enzyme irreversibly loses half its activity in 30 seconds – a problem

Fig. 9.3   The *nif* gene cluster of *Klebsiella pneumoniae*. The seventeen *nif* genes are arranged in eight transcriptional units, the dimensions and orientations of which are indicated by the arrow. Where known definitely, the roles of the individual *nif* genes are shown. *nifH* specifies the polypeptide of nitrogenase reductase (Kp2) and *nifD* and *nifK* specify the two polypeptides of nitrogenase (Kp1). *nifQ* is likely to be involved in the uptake of molybdenum and *nifS* and *nifU* may be involved in the processing of nitrogenase. *nifA* and *nifL* participate in the regulation of the other genes of the complex. Fe–Mo cofactor, iron–molybdenum cofactor. (By kind permission of Ray Dixon, Unit of Nitrogen Fixation, University of Sussex.)

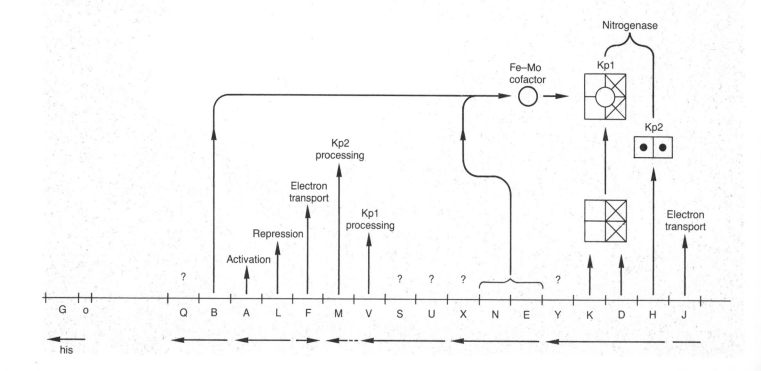

The blue-green algae, which can both carry out photosynthesis and fix nitrogen, would seem to be playing with fire because oxygen is actively liberated during photosynthesis. They have at least two ways of dealing with this potential hazard, however. In some blue-green algae, there is a temporal separation of nitrogen fixation, which occurs at night, and photosynthesis, which takes place during the day. Various filamentous blue-green algae, such as *Anabaena*, have developed a more sophisticated solution. The cells of the filaments are generally photosynthetic, but under conditions that favour nitrogen fixation some of them differentiate to produce morphologically distinct non-photosynthetic cells that fix nitrogen (Fig. 9.5).

Genetics of nitrogen fixation

An increasing number of nitrogen-fixing bacteria are being studied with the techniques of genetics and molecular biology, but the original experimental organism for such studies and the one that has been analysed in by far the greatest detail is *Klebsiella pneumoniae*. This bacterium is non-symbiotic and grows well in culture. Moreover, it is subject to the same genetic trickery that was established for *Escherichia coli*, to which it is related.

What has been learned about the *nif* genes of *K. pneumoniae* not only provides an intellectual framework for the study of the biochemical genetics of nitrogen fixation, but has also provided tools that have greatly facilitated the analysis of nitrogen fixation in other bacteria that are less amenable to genetic analysis. Much of the work on identifying and isolating the *nif* genes of *K. pneumoniae* was done at the Unit of Nitrogen Fixation of the University of Sussex, England, which is headed by John Postgate, and in the laboratories of Winston Brill at the University of Wisconsin (Madison) and Frederick Ausubel of Harvard University in Cambridge, Massachusetts.

A relatively small region of the chromosome of *K. pneumoniae*, when transferred to *E. coli*, allows the recipient cells to fix nitrogen, a result which shows that the *nif* genes of *K. pneumoniae* are clustered on the chromosomal DNA. Although the reduction of N_2 to ammonia may appear to be a straightforward reaction, *K. pneumoniae* devotes no less than 17 genes, which occupy about 22 kilobases of DNA in the *nif* gene cluster, to the reaction (Fig. 9.3).

The *nifH* gene specifies the nitrogenase reductase protein and the *nifD* and *nifK* genes encode the two protein components of nitrogenase. Five genes (*nifB, Q, V, N* and *E*) are involved in some as yet unspecified way in the synthesis of the iron–molybdenum cofactor and two genes (*nifF* and *J*) determine polypeptides needed for electron transfer to nitrogenase reductase. Three genes (*nifM, S,* and *V*) are required for the maturation of the complete, functional nitrogenase complex and two, *nifA* and *nifL*, have been shown to regulate the expression of all the other *nif* genes. Finally, the functions of *nifX* and *nifY* are as yet unknown.

Fig. 9.4 Purification of nitrogenase. Nitrogenase obtained from cultures of *K. pneumoniae* was purified by column chromatography. The nitrogenase proteins (the dark area) are brown because they contain iron. (By kind permission of John Postgate, Unit of Nitrogen Fixation, University of Sussex.)

that seems to be common to the nitrogenases of all nitrogen-fixers. The different organisms have found varying ways, some mundane but some ingenious, of solving the oxygen problem.

One strategy, used by bacteria of the *Clostridium* genus, is to live in an oxygen-free environment; for these bacteria, the risks of oxygen-damage never arise. Other nitrogen-fixers, such as the bacterium *Klebsiella pneumoniae*, can live either with or without oxygen, but only fix nitrogen when they are growing anaerobically.

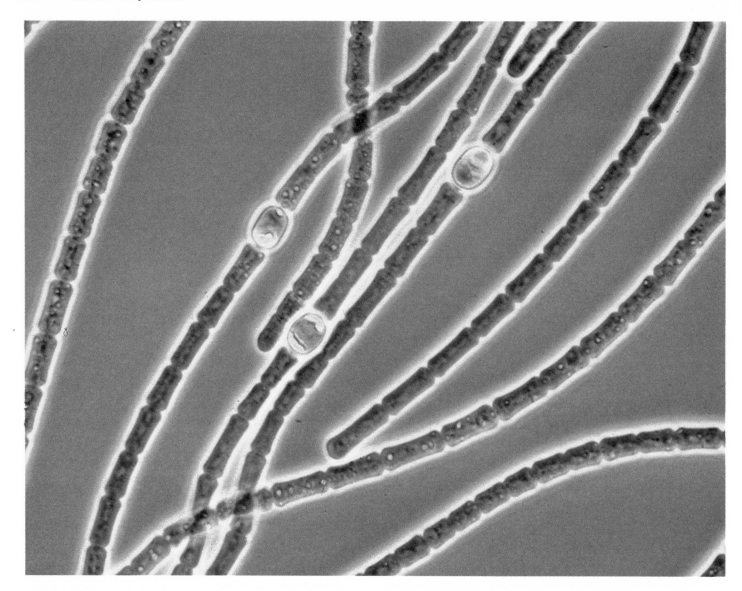

Fig. 9.5 Filaments of the blue-green alga *Anabaena*. The rectangular cells carry out photosynthesis, but do not fix nitrogen. The nitrogen-fixing enzymes would be poisoned by the oxygen released during photosynthesis. The larger, oval cells are the 'heterocysts', which do fix nitrogen. The heavy walls of the heterocysts prevent oxygen from coming in and inactivating the nitrogen-fixing enzymes. (Photographs by Sue Barns. By courtesy of Norman R. Pace and S. Barns, Indiana University in Bloomington, Indiana.)

Several of the *nif* genes of other nitrogen-fixing organisms have proved to be very similar in structure to those of *K. pneumoniae*, although in the other nitrogen-fixers the genes are usually scattered about the genome instead of being tightly clustered as they are in *K. pneumoniae*.

Regulation of *nif* gene expression

If nitrogen-fixing bacteria have a suitable source of fixed nitrogen, such as ammonia, glutamate or asparagine, the transcription of the *nif* genes is shut down so that the organisms do not waste the energy needed for synthesizing the proteins nor the ATP required for driving the reduction reaction. The genes are also not expressed when the cells are exposed to air. This, too, makes good biological sense. What is the point of synthesizing nitrogenase if it is to be strangled at birth by the toxic effects of oxygen?

Again, the studies of *K. pneumoniae* have provided the model for analysis of *nif* gene regulation in other nitrogen-fixers (Fig. 9.6). The studies show that *nif* gene regulation is very complex, involving both local control by genes within the *nif* complex

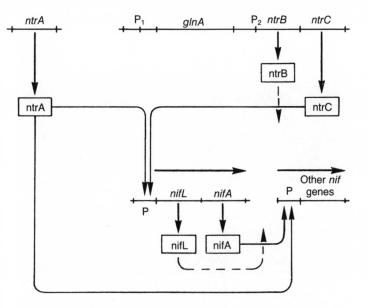

Fig. 9.6 The regulation of *nif* gene expression in *Klebsiella pneumoniae* and the link between *nif* gene transcription and the *ntrA*, *ntrB* and *ntrC* genes. As indicated in the text, *nifA* activates the promoters of the other *nif* genes but, in the presence of oxygen and ammonia, *nifL* prevents this activation (dashed line). The *nifA*-mediated activation also requires the *ntrA* gene product. Similarly, *ntrC*, in concert with *ntrA*, activates the *nifL,A* promoter, but in high ammonia *ntrB* blocks this activation. The *ntrB* and *ntrC* genes can be transcribed from either of two promoters, one of which (P_1) is also involved in transcription of *glnA*, the structural gene for glutamine synthetase. (By kind permission of Ray Dixon, Unit of Nitrogen Fixation, University of Sussex.)

Fig. 9.7 The promoter sequences upstream from the *nif* genes and from *E. coli* genes. The start of transcription is shown as +1 (bp is the abbreviation for base pairs). The *E. coli* promoters (*a*) have conserved sequences at −35 and −10 relative to the start of transcription, whereas the *nif* promoters (*b*) are different both in the identities and in the spacing of the conserved sequences.

and more global control by regulatory genes located elsewhere in the genome. To start with, expression of the *nif* genes, like that of all others, requires an RNA polymerase enzyme to transcribe the DNA into messenger RNA.

Promoters are the control regions at the beginning of genes to which the RNA polymerase must bind during initiation of transcription. The nucleotide sequences of the *nif* gene promoters are very different from those of certain, well-studied *E. coli* genes (Fig. 9.7). This suggests that the RNA polymerase that recognizes the *nif* gene promoters is different from the enzyme that binds to the promoters of the other genes. In fact, this suggestion has recently been confirmed.

Mutations in a gene called *ntrA* (*ntr* stands for nitrogen regulation), which is not a member of the *nif* gene complex, completely abolish nitrogen fixation. Boris Magasanik of the Massachusetts Institute of Technology in Cambridge, Massachusetts, and S. Kustu of the University of California at Davis have established that the *ntrA* gene codes for a protein, called a sigma factor, that confers on RNA polymerase the ability to recognize the *nif* gene promoter. Without this protein the *nif* genes, including the regulators *nifA* and *L*, could not be transcribed and nitrogen fixation would not occur. Other sigma factors help RNA polymerase to recognize other gene promoters.

Just having the correct RNA polymerase is not sufficient for *nif* gene transcription to occur, however. A gene from within the complex, *nifA*, makes a protein that is also necessary for inducing *nif* gene transcription. Expression of the *nifA* gene is itself turned on or off in response to environmental conditions.

The *ntrA* gene and three additional genes – *glnA*, *ntrB* and *ntrC* – are part of a global system that regulates many aspects of nitrogen metabolism in bacteria. In the absence of nitrogen sources such as ammonia and glutamate, the *ntrC* product switches on genes that allow bacteria to use other nitrogen-containing compounds that would normally be less favoured sources of the element. Among the genes switched on by the *ntrC* product is *nifA*, which then activates the *nif* gene complex. The product of the *ntrB* gene inhibits this *ntrC*-mediated acti-

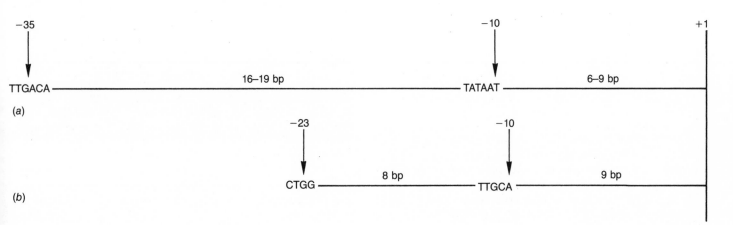

vation of the *nif* genes when ammonia concentrations are high. Finally, the *nifL* gene also contributes to the lack of expression of the *nif* genes in the presence of a nitrogen source such as ammonia or of oxygen. When the *nifL* product is exposed to ammonia or oxygen it apparently prevents the *nifA* protein from inducing the transcription of the other *nif* genes.

Symbiotic nitrogen fixation

The most important nitrogen-fixing bacteria, both agriculturally and ecologically, are those that interact with plants in symbioses that may be either simple or complex. In one simple interaction, the bacterium *Azospirillum* lives around the root surfaces of grasses, although there are questions about whether the plants receive any significant contribution of nitrogen from this association.

No such question arises about the more complex interactions between bacteria of the *Rhizobium* genus and legumes or between the bacterium *Frankia* and a variety of trees and shrubs, including alder. Legumes can flourish in nitrogen-poor soils because of their nitrogen-fixing symbionts. The importance of this family of plants in agriculture is enormous. Many of the world's seed crops are legumes – including soybeans, peas, beans, peanuts, lentils and chickpeas – and to these must be added the forage legumes, such as clover and alfalfa, which contribute to animal nutrition. None of the hosts for *Frankia* is used in agriculture, but they are frequently found as pioneer species that can help to reclaim derelict land.

Understanding these symbiotic interactions requires analysis not just of the *nif* genes, but also of the special plant and bacterial genes that allow them to engage in such complex interactions. At first sight, members of the genus *Rhizobium* appear to be ordinary rod-shaped bacteria (Fig. 9.8). Most strains cannot be coaxed to fix nitrogen when they are grown by themselves in culture. Yet these bacteria have the unique capacity to recognize and invade particular legumes and induce in the host plant a coordinated response that includes organized cell division and the synthesis of an array of proteins.

Normally the infection site is at the tip of a growing root-hair, which curls, branches or corkscrews in response to the invading bacteria (Fig. 9.9). The bacteria enter through the infection thread, a tube made by the plant that grows down through the root-hair, and multiply there. The presence of the infection thread, possibly combined with signals from the *Rhizobium* on the root surface, induces the division of cells in the root, thereby forming an incipient nodule. As the nodule develops, the initial infection thread continues to grow and branch, proceeding through and between the root cells of the host plant. The bacteria within the infection thread are pinched off, surrounded by a plant-specified membrane, and then released into the cytoplasm of the nodule cells. (Fig. 9.10).

The liberated *Rhizobium* cells of the nitrogen-fixing nodule are known as 'bacteroids'. It is in this form that the bacteria

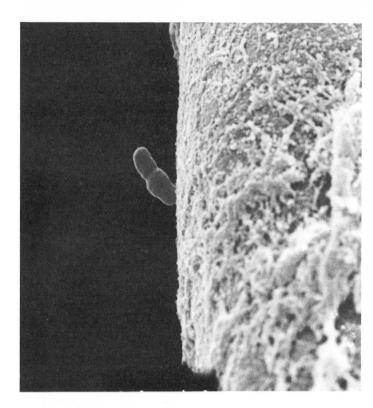

Fig. 9.8   *Rhizobium* bacterium on a root-hair. The scanning electron micrograph shows a rod-shaped bacterium adhering to a pea root-hair. (By courtesy of Jeremy Burgess, John Innes Institute, Norwich.)

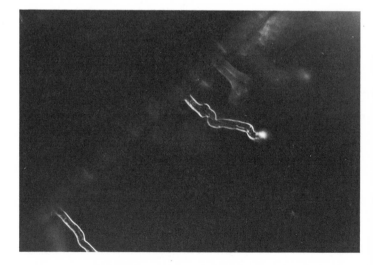

Fig. 9.9   A curling pea root-hair. After infection by a *Rhizobium* bacterium, a root hair becomes deformed. The infection thread, which carries the bacteria into the plant, can be seen in the centre of the hair. (By kind permission of Brian Wells, John Innes Institute, Norwich.)

The *Rhizobium*–legume symbioses often display host-range specificity; particular legume species are nodulated only by certain species of the bacteria (Table 9.2). For some bacteria,

Table 9.2. *Selected* Rhizobium *species and their hosts*

Bacterial species	Host plant	Comments
Rhizobium leguminosarum	Peas, *Vicia* (Broad bean) lentils	Species are very closely related to each other
R. trifolii	Clover	
R. phaseoli	*Phaseolus* bean	
R. loti	Lotus	Lotus is also nodulated by *Bradyrhizobium*
R. lupinii	Lupin	
R. meliloti	Alfalfa	
R. sesbania	Sesbania	Induces stem and root nodules on *Sesbania*. Also fixes nitrogen in free-living culture
R. fraedii	Soybean	Induces non-fixing nodules on most soybean cultures
Bradyrhizobium japonicum	Soybean	Some strains fix nitrogen in free-living culture
B. 'cowpea miscellany'	Nodulates several tropical legumes including cowpea and also the non-legume *Parasponium*	Has the ability to nodulate a non-legume host

including *R. leguminosarum*, *R. trifolii* and *R. phaseoli*, which nodulate peas, clover and *Phaseolus* beans respectively, the bacterial host range is apparently the only characteristic that distinguishes the different species.

Other *Rhizobium* species have more basic biochemical differences, over and above their variations in host-range specificity. The most striking illustration concerns the distinction between the fast- and slow-growing species. On the basis of biochemical characteristics and DNA homologies, these two groups have very few similarities other than their ability to nodulate legumes. For this reason, the slow-growing species have recently been classified as members of a new genus, the *Bradyrhizobium* – literally meaning the slow *Rhizobium*. Despite these differences, a nitrogen-fixing nodule can be much the same, irrespective of whether it is induced by a fast- or slow-growing species.

Not surprisingly, the morphological differentiation occurring during nodule development is reflected at the biochemical level. One obvious example in *Rhizobium* is the turning on of the *nif* genes. But a comparison of all the proteins of the free-living

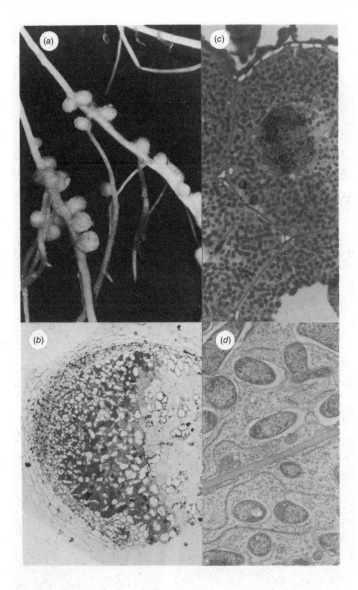

Fig. 9.10 The organization of pea nodules induced and occupied by *Rhizobium leguminosarum*. (*a*) The macroscopic appearance of the nodules on pea roots. (*b*) A micrograph that shows a longitudinal section through a pea nodule. (*c*) Individual pea cells crammed with nitrogen-fixing bacteroids. (*d*) Bacteroids surrounded by the peribacteroid membrane. (By kind permission of Nicholas Brewin, John Innes Institute, Norwich.)

normally turn on their *nif* genes and excrete the resulting ammonia to the host plant, which assimilates the ammonia by condensing it with glutamic acid to form glutamine. The glutamine in turn is used to disseminate the fixed nitrogen to the rest of the plant.

bacteria with those of the bacteroids reveals that there is quite a different protein population in the two states. Many genes that are off in the free bacterial cells are on in the bacteroids, and vice versa.

The biochemical constitution of the proteins made by the plant nodule also shows major changes compared with that in uninfected roots. At least 50 new proteins, called nodulins, have been detected specifically in the nodule and the total of the nodule-specific proteins may be many times that number. The functions of only a few nodulins are known. Some, such as the enzymes glutamine synthetase and uricase, are needed for ammonia assimilation. The most abundant nodulin, leghaemoglobin, transports oxygen to the nitrogen-fixing bacteroids.

Leghaemoglobin causes the characteristic pink coloration of nitrogen-fixing nodules (Figs. 9.11 and 9.12). It is present exclusively in the plant-cell cytoplasm, but it is, in a sense, a microcosm of the symbiosis. The host genome specifies the protein portion of the leghaemoglobin molecule, whereas the bacteroid almost certainly makes the haem portion. Although leghaemoglobin is widely thought to protect nitrogenase from oxygen damage, its primary function is delivering oxygen to the bacteroids, which require so much oxygen for energy during nitrogen fixation that their immediate environment is very low in oxygen.

The leghaemoglobin genes of the soybean are remarkably similar to the mammalian haemoglobin genes. Moreoover, the root nodules formed on non-leguminous plants by *Frankia* also contain leghaemoglobin. These findings raise questions about the evolutionary history of the molecule. Was it originally present in all plants, but then lost from the majority of them? Did it evolve independently in the different plant groups? And if so, was this evolution by conventional means or does leghaemoglobin represent one of nature's transgenic creations? That is, was a haemaglobin gene somehow transferred from an early mammal to a plant? These questions remain unanswered, as do many others about legumes' responses to infection by Rhizobium bacteria.

Genetic analysis of *Rhizobium* bacteria

Rhizobium bacteria are much simpler and more amenable to genetic analysis than their legume hosts. Not surprisingly then, progress in identifying the bacterial genes that are needed for symbiotic nitrogen fixation and nodulation has been much faster than progress in identifying the plant genes that contribute to these activities. Moreover, isolation of the *nif* genes of *K. pneumoniae* has greatly aided in identifying the corresponding genes in the *Rhizobium* bacteria. The nucleotide sequences of the *nif* genes of the different bacteria are so similar that the *K. pneumoniae* genes can be used as probes to 'fish out' the *nif* genes of other species.

Using this approach, investigators including Frank Cannon and his colleagues in the Sussex group have shown that the *nif*

Fig. 9.11 Nitrogen-fixing and non-nitrogen-fixing nodules on a pea root. The pea was innoculated with two strains of *R. leguminosarum*. One was able to fix nitrogen and induced pink nodules that contain the leghaemoglobin. The other was defective in nitrogen fixation and, consequently, the nodule is very pale because it contains very small quantities of leghaemoglobin.

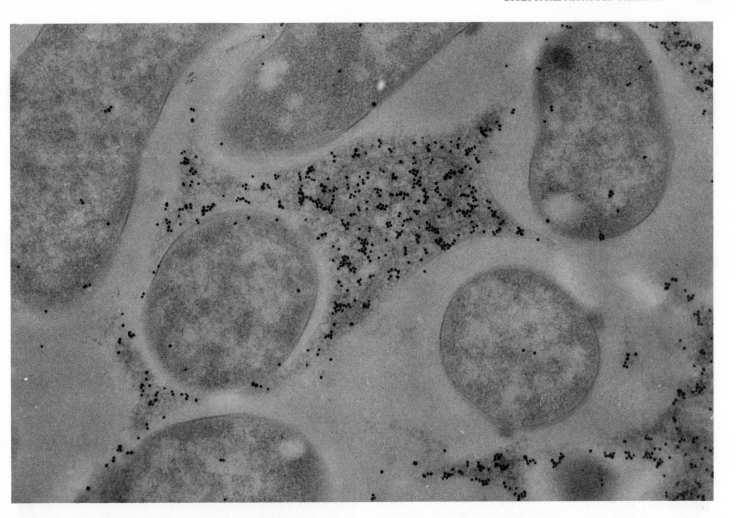

Fig. 9.12 Localization of leghaemoglobin in the plant cell cytoplasm. A nitrogen-fixing pea nodule was sliced in very thin sections that were bathed in an antibody specific for leghaemoglobin. The antibody was then labelled with gold particles, which show up as dark dots. The dots are found exclusively in the plant cytoplasm, not in the bacteroids or the space surrounding them. (By kind permission of Brian Wells and John Robertson, John Innes Institute, Norwich.)

genes of the fast-growing *Rhizobium* bacteria are located on large plasmids, called symbiotic plasmids. (Fig. 9.13). The same plasmids also contain the *nod* genes, which are needed for the bacteria to induce nodule formation by the host plant. In *Bradyrhizobium* bacteria the *nif* and *nod* genes are on the bacterial chromosome.

The most interesting current question concerning *nif* genes in *Rhizobium* is why they are expressed only in the nodule and not in bacteria that are in the free-living state. The regulatory circuitry that affects *nif* gene expression in *Rhizobium* species is clearly very similar to that in *K. pneumoniae*. The species *R. leguminosarum*, *R. meliloti* and *B. japonicum* have been shown by workers in Ausubel's laboratory and in those of Andrew Johnston at the John Innes Institute in Norwich, England, and of Hanke Hennecke at the Eidgenossische Technische Hochschule in Zurich, Switzerland, to have genes similar in sequence to the regulatory *nifA* gene of *K. pneumoniae*. In addition, the *Rhizobium* and *Klebsiella nif* promoters have similar sequences. In fact, transcription from the *Rhizobium nif* promoters can be activated by the *nifA* gene product of *K. pneumoniae*.

These similarities notwithstanding, simply lowering the concentrations of oxygen and ammonia in the growth medium of *K. pneumoniae* immediately turns on nitrogen fixation in that organism, but has no effect on nitrogen fixation by most *Rhizobium* bacteria. The special components in the root nodules that coax the *Rhizobium* bacteria to transcribe their *nif* genes represent yet another unknown of nitrogen fixation research.

Although the identification of many of the *Rhizobium nif*

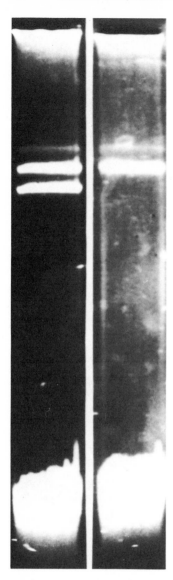

Fig. 9.13 Visualization of a symbiotic plasmid of *R. leguminosarum.* The plasmid DNA was extracted from cells of the wild-type bacterium (left track) and of a non-nodulating mutant (right track). Separation of the different plasmids by electrophoresis on an agarose gel showed that the wild-type bacteria have three plasmids, while the mutant has lost the smallest one, the symbiotic plasmid. (By kind permission of Dulal Borthakur, John Innes Institute, Norwich.)

Fig. 9.14 Appearance of a *R. leguminosarum* mutant that is defective in exopolysaccharide. The top dish shows the wild-type strain and the lower one contains a mutant that no longer makes exopolysaccharide nor nodulates peas.

genes has been greatly facilitated by using the information available from *K. pneumoniae*, that non-symbiotic nitrogen-fixer has been of no help when it comes to analysing the genes needed for nodulation. The work has to be done, from scratch, with *Rhizobium*.

The surface of *Rhizobium* bacteria, which must at some point make contact with the root of the host plant, might be expected to be an important determinant of nodulation and specificity. To a certain extent this has been borne out by genetic evidence. Some mutants, which fail to make an extracellular polymer called exopolysaccharide, are also unable to nodulate their normal host plants (Fig. 9.14). The story is not straightforward, however. Other mutant strains that are defective in exopolysaccharide can nodulate but cannot fix nitrogen. Still others seem

unimpaired in either nodulation or nitrogen fixation. These variations may reflect the as yet unknown subtleties of the interaction between the exopolysaccharide and the host plant.

As mentioned earlier the same large plasmids that carry the *nif* genes of the fast-growing *Rhizobium* species also contain a cluster of *nod* genes that participate in nodule formation. Several investigators have shown that transfer of DNA containing the *nod* genes of *R. leguminosarum*, which nodulates peas, to other *Rhizobium* bacteria that would ordinarily nodulate clover or beans, confers on the recipient bacteria the ability to form normal nodules on peas.

This does not mean that the plasmid *nod* cluster is all that is required for *Rhizobium* bacteria to nodulate, however. Bacteria of the genus *Agrobacterium* are closely related to the *Rhizobium* bacteria, although the *Agrobacterium* species do not have nodulating or nitrogen-fixing abilities. Introduction of the *R. leguminosarum nod* genes into *Agrobacterium* cells does not give the bacteria the ability to form normal nodules on pea roots, although they induce 'pseudo-nodules' that are devoid of bacteroids. The results show that *R. leguminosarum* must contain additional genes that are required for normal nodule development but are absent from *Agrobacterium* species.

Analysis of the rhizobial *nod* gene cluster has revealed some clues concerning the genes' functions and regulation. The research groups that have contributed to this work include those of Johnston, Adam Kondorosi of the Biological Research Centre of the Hungarian Academy of Sciences in Szeged, Sharon Long of Stanford University in Palo Alto, California, and Barry Rolfe of the Australian National University in Canberra.

Eight genes – which are which are designated *nodA* to *nodF*, *nodI* and *nodJ* – have been identified in the cluster (Fig 9.15 and Table 9.3). Mutations in *nodD*, *A*, *B* and *C* abolish nodulation, whereas mutations in the remaining four genes only delay the onset of nodule development and reduce the numbers formed. Despite the differences in the host-range specificities of the various *Rhizobium* species, their corresponding *nod* genes are similar in sequence, location and function. This similarity is an advantage because deduction of the role of a given *nod* gene in one species may be extrapolated to provide information on its function in others.

The *nodA*, *B*, *C*, *I* and *J* genes are transcribed as a single unit and appear to specify proteins that are associated with the bacterial membrane. The *nodI* gene encodes a protein that may be involved in membrane transport, although the identity of

Table 9.3. *Properties of nodulation genes in* Rhizobium

Gene	Size of gene product (kilodaltons)	Effects of mutation[a]	Comment on role of gene product
nodA	18	Nod⁻Rhc⁻	Membrane-bound
nodB	23	Nod⁻Rhc⁻	—
nodC	46	Nod⁻Rhc⁻	Membrane-bound
nodI	34	Nod delay	Transport protein?
nodJ	27	Nod delay	Membrane-bound?
nodD	34	Nod⁻Rhc⁻	Regulatory
nodF	10	Nod delay	Similar to acyl carrier protein
nodE	48	Nod delay	—

[a]Nod⁻, fails to nodulate; Rhc⁻, fails to curl root-hairs.

the substance that it might carry across the membrane is unknown. The *nodE* and *F* genes are also transcribed together. The *nodF* gene is similar to acyl carrier protein, a protein that participates in the synthesis of fatty acids. Whether the *nodF* gene is also involved in fat synthesis remains to be discovered. The *nodE* gene may help to prevent nodulation of legumes other than the normal host for a particular *Rhizobium* species.

Regulation of *nod* gene expression

The last of these *nod* genes, *nodD*, has been shown to be regulatory, controlling the transcription both of itself and of the other *nod* genes in the cluster. When rhizobial bacteria are maintained in a minimal culture medium the *nodD* gene is expressed at high levels while the remaining *nod* genes are not transcribed. A dramatic difference is found, however, when the bacterial cells are exposed to exudate from pea, clover or alfalfa roots. Transcription of all the *nod* genes, except *nodD*, then increases some seventy-fold.

This induction is dependent on the presence of an active *nodD* gene, a result which shows that the gene has a regulatory role. Some substance, present in the root exudates, acts through the *nodD* gene to cause the active transcription of the other *nod* genes. Just upstream from the beginning of both the *nodA*, *B*, *C*, *I*, *J*, and *nodF*, *E* regulatory units is a short, conserved DNA segment that may be involved in their regulation. The *nodD* gene, at least in *R. leguminosarum*, is itself subject to negative regulation by its product, high concentrations of which repress transcription of the gene.

Analysis of the root exudate of pea, alfalfa and clover plants has identified several low molecular weight compounds that are capable of activating transcription of the *nod* genes. The molecules are compounds of the flavone or flavanone type (Fig. 9.16). Commercially available compounds of these classes have been found to have inducing activity for the *nod* genes.

Although a great deal has been learned about the location

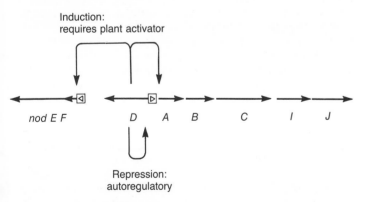

Induction: requires plant activator

nod E F   D A B C I J

Repression: autoregulatory

Fig. 9.15 Representation of the *nod* genes of *R. leguminosarum* and the way in which they are regulated. The dimensions and direction of transcription of the eight *nod* genes are indicated. The regulatory regions are shown as triangles in squares. The *nodD* gene regulates its own transcription in minimal medium, but activation of the other *nod* genes requires factors (flavones or flavanones) from peas, as well as the *nodD* gene product.

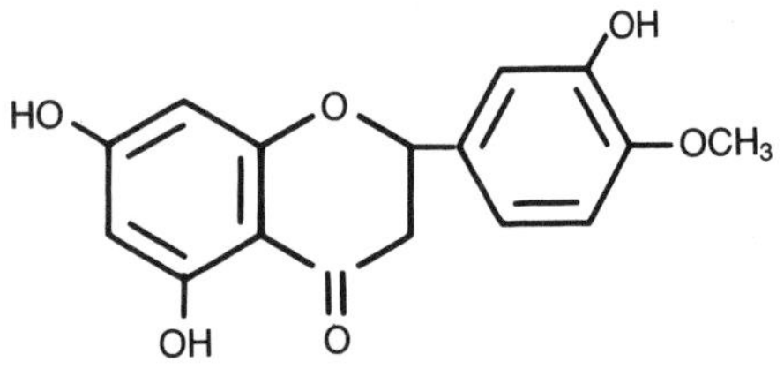

5,7,3′-Trihydroxy-4′-methoxyflavanone
(hesperetin)

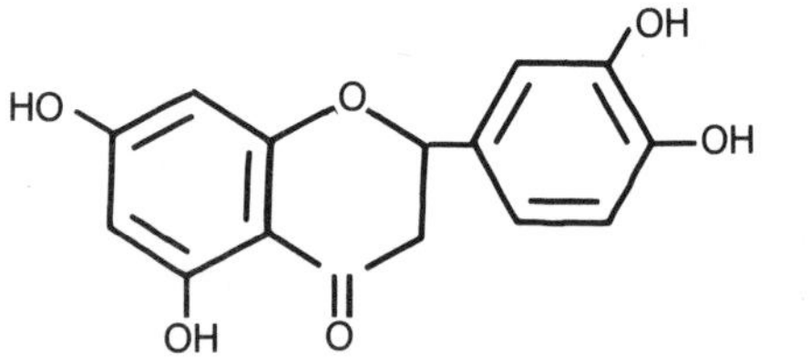

5,7,3′,4′-Tetrahydroxyflavanone
(eriodictyol)

Fig. 9.16 The chemical structure of two flavanones that activate *nod* gene transcription.

and structure of the genes that are involved in nodulation and nitrogen fixation, much remains to be done to translate the information gleaned at the DNA level into a true biochemical and biological understanding of infection by the rhizobial bacteria. Molecular analysis of the *Rhizobium*–legume symbiosis may have wider implications, quite apart from the importance of the interaction itself.

The rhizobial bacteria and other organisms that engage in complex interactions with plants, whether these be symbiotic or pathogenic, can be viewed as eliciting specific biochemical and morphological responses in the host. If the precise mechanisms can be established by which *Rhizobium* bacteria induce the curling of root-hairs, the formation of infection threads and the division of root cells the information might lead to a better understanding of how the growth and differentiation of plant cells and organs is generally determined.

Cost-free nitrogen fertilizer: how real a prospect?

The study of biological nitrogen fixation spans an array of scientific disciplines ranging from inorganic chemistry through enzymology, microbial physiology, molecular genetics and plant – microbe interactions to agronomy and ecology. Where in these

various areas of endeavour can we look for the developments that will, in the future, provide sources of fixed nitrogen for plant nutrition that are economically, socially and environmentally more acceptable than the current heavy reliance on commercial fertilizers? Which prognostications might be considered 'flights of fancy' and which have real prospects of being accomplished?

The chemistry and biochemistry of nitrogen fixation are currently receiving less attention than perhaps they deserve in comparison to the more fashionable areas of molecular biology and genetics. In nitrogenase, nature has provided the world with an enzyme that reduces nitrogen at atmospheric pressure and at ambient temperature. Elucidating the structures of the catalyst, namely the iron–molybdenum cofactor of nitrogenase, and of the intermediates between the substrate N_2 and the final product ammonia should be high on the list of research priorities.

Genetics can be a very powerful tool in such work. Mutants that block the synthesis of the iron–molybdenum cofactor at different stages are available. They could be used to illuminate how the cofactor is made and assist in the description of its structure. Such studies might lead to catalysts for ammonia production that do not require high pressures and temperatures to reduce nitrogen and, perhaps even more important, that obtain the hydrogen for the reduction reaction not from energy-expensive hydrogen gas but from water, which is the probable hydrogen source in the enzymatic reaction.

Two other drawbacks of nitrogenase are its extreme oxygen sensitivity and its wasteful tendency to liberate gaseous hydrogen during nitrogen reduction. Given the ubiquity of these properties in all nitrogenases studied so far, these characteristics may be absolutely associated with the action of the enzyme. But are they? The ratio of the number of electrons directed towards the formation of hydrogen to the number of electrons used for the reduction of nitrogen is not a constant. It is affected by the relative abundance of the substrates for the two reactions and the energy status of the cell.

Attempts to isolate mutant forms of nitrogenase that are less sensitive to oxygen have been lacking. In addition, there has been no serious effort aimed at obtaining *nif* gene mutants that specify nitrogenase proteins that liberate less hydrogen than wild-type strains. Although some strains of *Rhizobium* bacteria have 'hydrogen uptake' systems that can effectively recycle the hydrogen and use the energy it represents for fixing nitrogen, inoculation with these strains has so far produced only ambiguous effects on plant growth. Whether nitrogenase can be improved with respect to the hydrogen and oxygen problems cannot be predicted – but so far this approach has not received much attention.

Another strategy for improving biological nitrogen fixation involves manipulating and exploiting the non-symbiotic nitrogen-fixing microbes such as the blue-green alga *Anabaena*. All of these organisms tenaciously retain the ammonia they have fixed, assimilating it into amino acids and proteins as soon as it is made. However, *Anabaena* mutants that liberate some or all of the fixed nitrogen have been isolated. If such mutants could be used in turn to produce new strains with increased rates of nitrogen fixation, such selected blue-green algae might be incorporated into irrigation water with beneficial results. This strategy would be most appropriate in those parts of the world with high light incidence and would be particularly attractive in regions with subsistence levels of agriculture where any ammonia input would be better than that which currently prevails – that is, none at all.

A second group of non-symbiotic nitrogen-fixing bacteria that might be amenable to exploitation are those that live on the root surfaces of plants. In the mid-1970's there was intense interest in one such organism, *Azospirillum*, because of results suggesting that nitrogen fixation by these bacteria on the roots of grasses and of corn and other cereals benefited the hosts. Claims that corn could get as much biologically fixed nitrogen as soybeans were even heard. Subsequent research in the greenhouse and in the field led to a rapid diminution of the optimism, and now research on *Azospirillum* has subsided to a fairly low level. It is possible, however, that in the current climate of disillusion with the organism the baby has been thrown out with the bathwater.

This bacterium, after all, fixes nitrogen and associates with the roots of some of the world's major crop plants. It should be possible to obtain *Azospirillum* mutants that fix more nitrogen than the wild-type strains, are less prone to repressing expression of the *nif* genes in the presence of fixed nitrogen, and are more capable of liberating some of the nitrogen they fix instead of assimilating it immediately.

There is no guarantee that such mutants would confer any benefits on a cereal crop. They might, for example, be so enfeebled that they are unable to compete with the resident soil bacteria. Nevertheless, attempts to use molecular genetics to construct strains of root-dwelling bacteria with improved nitrogen-fixing capabilities have at least a chance of paying off. Such an approach need not be confined to selecting improved variants of natural nitrogen-fixing bacteria. There is no reason why the *nif* genes from *K. pneumoniae* could not be transferred into other bacteria that colonize roots in large numbers.

The rhizobial bacteria have already been recognized as the major contributors to the nitrogen economy of many of the world's crop plants. What are the prospects for exploiting biotechnology to enhance biological nitrogen fixation in legumes? Some recent experiments by Donald Phillips of the University of California at Davis, though hardly high-tech, have shown that conventional plant breeding may have much to offer. By selecting explicitly for alfalfa plants that are superior when grown on fixed nitrogen and that also grow well when inoculated with *Rhizobium* bacteria, Phillips and his colleagues rapidly obtained plants with increased yield, protein content and nitrogen-fixing capacity. The results indicate that, at least for this legume, the limitations on the amount of nitrogen fixed depend not so much on the bacterium as on the host plant.

Nevertheless, rhizobial bacteria may be absent from the soil or, if present, inefficient at fixing nitrogen. In such cases the legume seeds are inoculated with the bacteria by incorporating them into a seed coating. Several commercial organizations manufacture such inoculants. In fact, all the soybean nodules in North America are occupied by inoculated *Rhizobium* strains because the soybean and its nitrogen-fixing symbiont, being natives of eastern Asia, are normally absent from the North American continent.

To a large extent use of inoculated legume seeds has been successful, particularly in areas where there are no indigenous *Rhizobium* strains that can nodulate the legume in question. However, serious problems have been encountered when such indigenous bacteria are present. In general, the resident strains appear to be better adapted to nodulate under the particular field conditions. It can prove very difficult in practice to get the foreign inoculant strain to occupy the nodules. Competitiveness for nodulation is therefore a highly desirable trait for any *Rhizobium* that is to be used for inoculation.

Because the genetic and biochemical basis of competitiveness are not known, it is not now possible to state how genetic manipulation could be directed to making a strain more competitive. This might be done empirically, however, if two rhizobial strains were available: one that fixes a great deal of nitrogen but is not competitive for nodulation and another for which the reverse is true. Simply transferring DNA from the competitive strain to the other might create a hybrid that is both competitive and an efficient nitrogen-fixer. This could be done even if the precise nature of the transferred gene or genes is unknown.

A different approach to the problem of competitiveness could exploit genetic variation in the legume host, as well as in the bacterium. For example, a primitive line of peas, designated variety Afghanistan, fails to be nodulated by most strains of *R. leguminosarum*, the normal pea symbiont. The plant's resistance to nodulation is determined by a single recessive gene. However, a particular *R. leguminosarum* strain, which was isolated from Turkish soil, nodulates the Afghanistan variety pea, as well as the normal commercial strains. The extended host-range of the Turkish bacterium is due to the presence of a single gene. When this gene is transferred into conventional strains of *R. leguminosarum* it confers on them the ability to nodulate the Afghanistan pea variety.

These results suggest that it is at least theoretically possible to bypass the competition problem by genetically manipulating the legume plant and the nodulating *Rhizobium* bacterium. The first step would be the introduction of the nodulation resistance gene of the Afghanistan pea into a commercial pea strain. The second step would be transfer of the host-range genes of the Turkish strain of *R. leguminosarum* into an efficient commercial strain of the bacterium, which could then be used to inoculate the modified plant. The presence of the resistance gene in the crop plant should prevent nodulation by the resident soil population of *R. leguminosarum*, and thus the nodules should be occupied exclusively by the inoculating bacterial strain with its

118 ANDREW W. B. JOHNSTON

extended host range. This approach may be generally applicable to other legume and rhizobial species.

The most commonly conceived notion concerning the future application of research on biological nitrogen fixation is perhaps the most spectacular – the development of nitrogen-fixing non-legume crop plants. Two general scenarios for the construction of these marvellous organisms can be envisioned. In the first, the roots of potatoes, corn or tobacco plants are festooned with root nodules that are occupied by nitrogen-fixing *Rhizobium* bacteria. In the second, the *nif* genes are stitched directly into the genome of the host plant.

The probability of engineering non-legume plants that can form nitrogen-fixing nodules is very low. Very little is known about the plant's contributions to the development of the symbiotic interaction with *Rhizobium* bacteria, but, as mentioned previously, scores and possibly hundreds of plant genes are activated specifically in the nodule. For all the power of molecular biology and the growing ability to introduce new genes into plants (see Chapter 11), there seems to be little prospect of getting all these nodule-specific genes into non-legumes and having them expressed in a controlled, coordinated fashion.

In contrast, there appears to be no technical reason why *nif* genes, possibly from *K. pneumoniae*, could not be introduced into plants and expressed there. Although the job would be time-consuming, the *nif* gene transcriptional units could be hooked up to regulatory sequences that allow their expression in plants. These constructs could then be put into a vector, such as the Ti plasmid of *Agrobacterium tumefaciens*, that can be used to introduce the genes into the genome of an amenable plant. That is the easy part, however.

The regulation of the *nif* genes would have to be fine-tuned so that their various products are made in the correct ratios. But more than that, the toxic effect of oxygen on nitrogenase constitutes a major problem that must be addressed. The genes would have to be expressed in a part of the plant that has a low oxygen concentration. Such tissues include the central portions of large storage organs (potato tubers for example), the insides of germinating seeds and the inner regions of roots. It would thus be desirable to have the *nif* genes expressed from regulatory sequences that are active in, and preferably only in, these anaerobic plant tissues.

But even if this can be accomplished, there are other imponderables. Will the complex nitrogenase enzyme be assembled correctly in the foreign environment? Can the ammonia that is fixed be assimilated locally? If not, it is likely to be leaked to the outside. Lastly, even if nitrogen fixation is accomplished, will the energy demand be so great as to reduce the plant yields to unacceptably low levels? These and other questions must be answered before a field of nitrogen-fixing potatoes can be imagined. The exciting thing is that plant molecular biology has developed so rapidly during the past few years that the experimental protocols for addressing the questions are, or soon will be, available.

Additional reading

Dixon, R. A. (1984). The genetic complexity of nitrogen fixation. *Journal of General Microbiology*, **130**, 2745–2755.

Downie, J. A. and Johnston, A. W. B. (1986). Nodulation of legumes by *Rhizobium*: the recognised route? *Cell*, in press.

Halverson, Z. J. and Stacey, G. (1986). Signal exchange in plant–microbe interactions. *Microbiological Reviews*, **50**, 193–225.

Long, S. R. (1984). Genetics of *Rhizobium* nodulation. In *Plant–Microbe Interactions* vol. 1, *Molecular and Genetic Perspectives*, ed. T. Kesuge and E. W. Nester. Macmillan, New York.

Nadler, D. P. S. and Nadler, K. (1900) Legume–*Rhizobium* symbiosis: host's point of view. In *Genes Involved in Microbe–Plant Interactions*, ed. D. P. S. Verma and T. H. Hohn. pp. 58–94. Springer-Verlag, New York.

Sprent, J. I. (1986) Benefits of *Rhizobium* to agriculture. *Trends in Biotechnology*, **4**, 124–9.

10 Plant cell and tissue culture

Edward C. Cocking

Plant cell and tissue culture is fundamental to most aspects of plant biotechnology. A wide range of plant applications depend on the ability to grow plant tissues and cells in simple nutrient solutions of known composition. These applications include plant propagation; germplasm maintenance and storage, which is crucial for retaining the gene pools of plants that are not under active cultivation; the production of commercially useful chemicals; and plant genetic engineering.

The origins of plant cell culture methods date back to the early 1900s when Gottleib Haberlandt showed that it is possible to maintain certain types of plant cells in a healthy condition in culture. Although the cells did not divide, Haberlandt's work set the direction for the research that was to come in the future.

The inability to get cultured plant cells to divide was to persist for many years, however. The methodology did not enter the modern era until the 1950s. In the middle of that decade Folke Skoog of the Univeristy of Wisconsin in Madison discovered the cytokinins, a group of plant hormones that have a variety of effects, including stimulation of cell division. In addition, Robert Gautheret of the Pierre-et-Marie Curie University, Paris, France, found that auxins, another group of plant hormones, stimulate the division of callus cells. Callus is an undifferentiated mass of cells that forms on plant wounds and is also produced by plant tissues in culture (Fig. 10.1).

With the addition of cytokinins and auxins to the growth medium, which also contains salts and sugar, investigators became able to grow plant cells in culture, not just to maintain them there. They also learned that, in many cases, whole plants can be regenerated from callus or other types of plant cells that divide in culture. The ability to regenerate plants is a prerequisite for many of the biotechnological applications of plants that are currently under exploration.

In a review published in 1958, Albert Riker and Albert Hilderbrandt of the University of Wisconsin, Madison, commented that plant cell culture methods were opening the way for numerous fundamental and applied investigations of plants. The procedures used so successfully to study the molecular genetics of microorganisms could begin to be applied to higher plant tissues. Although Riker and Hildebrandt wrote their review before the advent of biotechnology as we know it today and even before the term biotechnology was coined, their prediction has been amply borne out.

The area of plant biotechnology is now burgeoning. Various plant cell and tissue culture methods have resulted, or in the near future are likely to result, in the production of useful plants or products. The methods are being used to generate new plant varieties and to produce valuable compounds that would be difficult to obtain by chemical synthesis. The ability to culture plant cells is also central to the use of genetic engineering techniques for introducing new genes into plant species.

The use of cell culture for plant propagation

Plant tissue culture was intially introduced to facilitate the clonal propagation of horticultural species. In clonal propagation plants are reproduced asexually so that the new individuals are all identical to the original plant; that is, they are all members of the same clone. Home gardeners are well acquainted with such common examples of clonal propagation as the generation of new African violet or philodendron plants from cuttings. This mode of reproduction is often advantageous in horticulture because many plants, especially hybrids, either cannot reproduce sexually or, if they do, lose the desirable characteristics that have been bred into them.

For example, a given stand of palm-oil trees usually contains one or two that are superior to the others and are therefore called 'elite'. If the elite trees are bred sexually, their superior characteristics are lost. Laurie Jones of Unilever Corporation in Bedford, England, has recently been able to clone elite palm-oil trees by using tissue culture methods, thereby opening the way to reproducing the trees without losing their superior characteristics.

The production of 'synthetic seeds' is another means of reproducing plants that are not amenable to sexual reproduction. The somatic cells of some plants can be propagated in culture and then induced to form embryos. (The somatic cells of an organism are all the cells other than the germ cells). Synthetic seeds are produced by encapsulating the embryos in a protective covering that allows them to be handled more or less like conventional seeds.

Fig. 10.1 Photograph of a culture of callus tissue from the wild tomato. Plants are beginning to regenerate from the callus cells.

Because the plant embryos are grown in suspension culture, the method is suitable for large-scale production of synthetic seeds. The culture can readily be carried out in spin-filter bioreactors, which allow the easy separation of the embryos from the culture fluids. There are numerous advantages to this approach, including environmental control and the ability to produce essentially unlimited numbers of embryos. But there are drawbacks that still need to be eliminated. These include the potentially high costs of the culture medium and the limited number of species that will form embryos in culture.

Plant pathologists are also finding the cell and tissue culture methods to be valuable. For example, these researchers can use the methods to produce the uniform populations of cells they need for studying the effects of plant pathogens. Among other things, some forms of cultured cells can be induced to divide synchronously – a situation that facilitates studies of the molecular biology of virus infection and replication.

Another application of the cell culture methods is in the production of pathogen-free plant strains. Cultured plant cells can be sterilized to remove bacteria or other pathogens that sometimes establish persistent infections of whole plants. Pathogen-free strains of forage grasses and legumes have been achieved in this way. In addition, 'cryopreservation', in which cultured plant cells are frozen for storage, is a means of keeping unusual plant variants for long periods.

The generation of somaclonal variation in plants.

Although cell culture techniques are being used to propagate plants clonally and thereby maintain desired characteristics in the progeny, the culture procedures often elicit genetic variability in the plants produced from the cultured cells. There is now as much interest in this variability, which may provide a way of producing desirable new characteristics in established varieties of crop species, as there is in maintaining stability through clonal propagation.

The variability that occurs in plants that have been regenerated from cultured cells or tissues came as something of a surprise. All the somatic cells of an individual plant should have the same genetic composition, and the plants regenerated from those cells were expected to be identical. Instead they often show a great deal of diversity in their characteristics. This somaclonal variation, as it is called, was recognized as ubiquitous only as recently as 1981. Although the causes of somaclonal variation have not yet been fully elucidated, it is likely to arise as a consquence of DNA transposition.

The genetic material was considered to be extremely stable until some 40 years ago when Barbara McClintock of Cold Spring Harbor Laboratory on Long Island, New York, discovered 'transposable elements' in maize. The elements, which may occur in all species, are segments of DNA that can move about the genome, sometimes causing mutations by inserting in and disrupting genes. The mutations can also be reversed if the elements leave the disrupted genes. The conditions used for plant tissue culture apparently stimulate the movements of transposable elements, thereby leading to the high frequency of somaclonal variation in plants derived from the cultured cells.

The production of somaclonal variation in lettuce and tomato plants illustrates how cell and tissue culture can be used for this purpose. Christine Browne, John Lucas and Brian Power performed the experiments on lettuce at the Univeristy of Nottingham, England, For the work, the investigators first produced callus from each of three lettuce varieties by culturing the cotyledons (the first leaves produced by the germinating seed) or the more mature leaves. Browne, Lucas and Power then regenerated whole plants from the callus tissue by means of a rapid, single-step procedure that the researchers developed for the lettuce.

The next step was to grow the regenerated plants under both glasshouse and field conditions and to assess the plants' offspring for variation in a number of characteristics, including morphology, seed weight, seedling vigour, and reaction to two lettuce pathogens, the downy mildew fungus and lettuce mosaic virus.

The regenerated plants of all three lettuce variants exhibited somaclonal variation in these characteristics, with reduced vigour, albinism and changes in the content of the plant pigment chlorophyll among the changes observed. The plants also varied with regard to their reactions to downy mildew and the lettuce mosaic virus: some were more susceptible and others less susceptible to the pathogens. One plant line displayed an interesting combination of increased yield and chlorophyll content combined with early flowering.

These traits are not novel, but the ease with which they can be recovered after tissue culture contrasts with the low frequency of variation encountered in conventional breeding experiments. The frequency of the variation that is induced by cell culture is greater than that obtainable by subjecting plants to irradiation or chemical treatments that cause gene mutations. Moreover, the regeneration step seems to have a 'cleansing' effect that helps to eliminate deleterious changes.

Variation, however it is induced, is more likely to be harmful to the plant than to be beneficial, but plants apparently do not regenerate as readily from cultured cells that have undergone harmful changes as they do from cells with beneficial variations. In any event, when plants are produced by conventional breeding methods, much larger numbers would need to be screened to detect variation equivalent to that arising during the culture procedures.

Fig. 10.2 Micrograph of protoplasts that were isolated from the leaves of a wild tomato species. Protoplasts are prepared by enzymatically dissolving away the rigid, cellulose-containing walls of plant cells. The small green granules within these protoplasts are the chloroplasts, which contain the green pigment chlorophyll and carry out the reactions of photosynthesis.

Protoplasts of different plant species may be fused to form somatic cell hybrids. This is a new way of obtaining hybrids of plant species that cannot be crossed by standard breeding methods, provided that whole plants can be regenerated from the hybrid protoplasts. In addition, plant genetic engineering can be achieved by introducing individual cloned genes into protoplasts. The success of this application also depends on the ability to regenerate whole plants from the protoplasts.

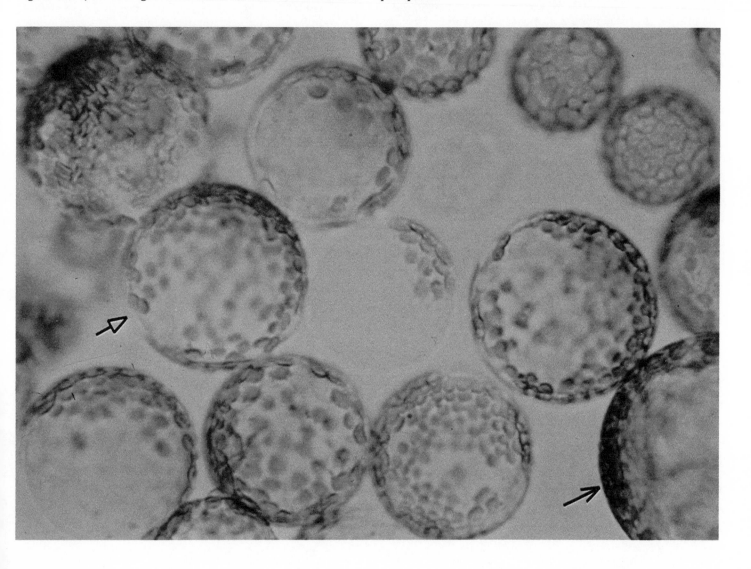

The lettuce plants that display somaclonal variation may have the potential for improving the crop species. For example, strains that have traits such as the increased resistance to downy mildew and lettuce mosaic virus and increased yield can be used in conventional breeding programmes that aim to produce better lettuce varieties. The early flowering trait could also prove useful in this regard by shortening generation times. These illustrations highlight the fact that the novel tissue culture procedures provide an addition to conventional breeding methods; they rarely serve as alternative technologies.

The lettuce work of the Nottingham group also illustrates the fact that simple, rapid procedures are always preferred and can sometimes have distinct advantages. Previous studies of somaclonal variation in lettuce have used protoplasts (Fig. 10.2), plant cells from which the rigid cell walls have been removed, rather than callus tissue. However, the experiments of Brown, Lucas and Power showed that regeneration of lettuce plants from callus is superior to regeneration from protoplasts. The protoplast procedures are time-consuming compared with regeneration from callus by the new method developed by the Nottingham group. Moreover, fewer lettuce varieties can be regenerated from protoplasts than from callus cells.

Because the frequency and types of variation observed in the callus-derived regenerants are similar to those found in the protoplast-derived regenerants, the use of protoplasts appears to be unnecessary. Moreover, all the lettuce plants regenerated from callus were diploid; that is, they had their normal complement of chromosomes. In contrast, a high proportion of the protoplast-derived regenerants were tetraploid, having twice as many chromosomes as they should, and showed a reduced fertility. The chromosome number normally doubles when cells are preparing to divide and the protoplasts may have become tetraploid because two daughter cells failed to form after the doubling. Alternatively, the protoplasts may have fused with one another.

The experimental approach to the generation of somaclonal variation in the tomato, which ranks among the world's top 30 crops (Fig. 10.3), has largely involved the use of young, fully expanded leaves taken from plants that have been grown in the glasshouse. Explants from the leaves are cultured on a medium known to permit regeneration of such explants. There has been little work on somaclonal variation in tomato plants regenerated from callus or from protoplasts. Most varieties are not responsive to protoplast regeneration techniques.

David Evans, Roderick Sharp, and their colleagues at DNA Plant Technology, Inc. (DNAP), in Cinnaminson, New Jersey, have carried out the most comprehensive study on cultured tomato leaf explants. As is usually the case with plants regenerated from cultured tissue, the tomato plants obtained from the leaf explants showed variations in chromosome number.

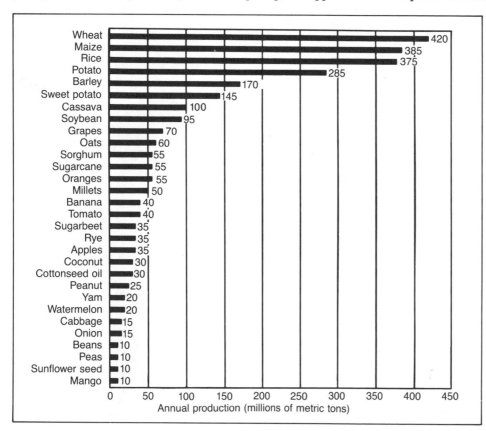

Fig. 10.3 Bar graph showing the world's top 30 crops (excluding grass). The cereals wheat, maize and rice lead the way. Potato is next on the list with an annual production of 285 million metric tons per year. The annual production of 40 million metric tons of tomatoes makes this crop sixteenth on the list.

The regenerated tomato plants also showed a high degree of somaclonal variation. Especially interesting was the discovery of single-gene mutations in plants that were regenerated from leaf explants from several varieties. The frequency of the mutations, which occurred spontaneously under the culture conditions, was as high as 1 in every 20 regenerated plants. Mutations in the nuclear genes included dominant, semi-dominant and recessive types. Mutations may also have occurred in the genes in the chloroplasts.

In addition, there was some evidence that single-gene mutants of a type not previously found after exposure of plants to conventional mutagenic treatments with chemicals or radiation could be recovered in the regenerated tomato plants. Among these were mutations resulting in tomatoes with a higher solids content, a feature of considerable importance in tomato processing. This illustrates the advantage of using tissue culture methods to produce minute changes in a plant variety in which other changes are not desired. Such somaclonal variation could be regarded as the mildest form of genetic manipulation.

Because somaclonal variation is basically simple and easy to achieve, it is not surprising that it has now assumed an important role in the biotechnological applications of plant cell and tissue culture. Prominent among these, in addition to the efforts to improve lettuce and tomato, are applications aimed at improving the vegetatively propagated crops, which often defy classic breeding and genetic approaches.

Few world crops rival the potato as a target for improvement through somaclonal variation. This crop plant is usually propagated vegetatively and some of the best cultivars are sexually impotent.

The development of better strains of commerical potatoes is a major concern of the food processing industry. For example, in the United Kingdom the potato variety called 'Record' is grown under contract exclusively for the production of potato crisps, but 'Record' is low yielding, susceptible to certain diseases, and accumulates reducing sugars, which causes the crisps to become dark and unacceptable when stored.

Somaclonal variation is currently being assessed in plants regenerated from 'Record' and also from a range of other potato varieties of interest to the food industry. Taken together, the studies on somaclonal variation in several plant species are challenging the traditional view that somatic mutations are rare and of little consequence in plant population dynamics.

The limitations of plant cell culture methods

The successful applications of plant cell culture methods in plant propagation and the generation of somaclonal variation does not mean that all the problems have been solved or that the researchers can now do everything that they would like to do with the technology. Thomas Orton of DNAP has highlighted the naivety of the simplistic view that multicellular organisms, such as plants, could be made to behave like microorganisms. If this were possible, it would enable researchers to induce mutations in plant cells, select for desired traits and characterize the new mutants, all in culture, and usher in a new era of understanding of the structure, function and developmental coordination of plant genes.

Plant cells do not behave like microorganisms in culture, however. Bacterial cells, for example, separate when they divide, forming equivalent individuals. But plant cells do not do this; the daughter cells remain attached and form a cell mass. Those inside the mass will be in an environment that is distinctly different from that of those on the outside and will not be exposed to the same selection conditions. Nutrients and other constituents of the culture medium will be less accessible to the inner cells, for example.

Difficulties such as these have permeated the large number of studies concerned with selecting better crop plants from cultured cells, so that there is now a better appreciation of the limitations of the technology. In addition to the propensity of plant cells to form masses in culture, these limitations include the inability to regenerate many varieties of crop plants from cultured cells. Plant regeneration is still more of an art than a science. Even when one variety of a given species regenerates, others may not. Finding the correct conditions for regeneration can be very time-consuming.

With regard to selecting agronomically useful mutants from plant cell cultures, Roy Chaleff of E. I. Dupont de Nemours and Company in Wilmington, Delaware, has made the pertinent observation that only traits that are expressed at the cellular level can be readily identified. Therefore, although some single-gene traits such as tolerances for heavy metals, salt and herbicides may be selected for in culture, other desirable characteristics, such as high yield and time to flowering are controlled by many genes and are not at present amenable to this type of approach.

Moreover, even if cells with increased resistance to salt, heavy metals or herbicides can be selected in culture, this does not necessarily mean that the whole plants regenerated from those cells will display the same tolerances. A plant's ability to withstand high salt concentrations, for example, may be more dependent on the capacity of its roots to prevent salt uptake than on the salt resistance of its individual cells.

Most agriculturally significant traits are as yet too poorly understood to permit the design of effective strategies for selecting improved plants in culture systems. Nevertheless, the basic studies that are being performed with cultured plant cells and tissues should result in a better understanding of those traits. Studies of mutants selected from tissue culture systems are likely to improve the knowledge of plant development and of the genetic and physiological basis of a range of plant responses. This will help lay a better foundation for the various genetic manipulations that are currently providing the main impetus for new developments in plant biotechnology.

Hybrid production by protoplast fusion

In addition to using plant cell and tissue culture for embryogenesis and other types of clonal plant propagation and as a source of somaclonal variation, the technology is finding increasing application for producing hybrid plants by protoplast fusion. Protoplasts must be used for these experiments because cells that retain their walls would not be able to fuse.

The basic technology of protoplast isolation and fusion is now well established. The cell walls are usually removed by digestion with a mixture of enzymes that break down the cell wall constituents. Protoplast fusion can be induced either chemically with materials such as polyethylene glycol or by applying an electric field.

The remaining challenge is to apply these technologies adequately to crop improvements. Current limitations on the work include difficulties in regenerating plants from cultured protoplasts and in selecting the hybrids that result from the fusions. Overcoming or bypassing these problems is central to further biotechnological developments.

The challenges are best appreciated by considering some of the anticipated applications of protoplast fusion. For example, the importance of improving rice, the world's most important cereal crop, has frequently been highlighted. Enhanced disease resistance and salt tolerance might be achieved by fusing the protoplasts of cultivated rice strains with those of wild species of rice. In addition, protoplast fusion could enable the direct transfer of cytoplasmic male sterility between strains. This trait, which is apparently encoded in the mitochondrial genome, prevents a plant from producing pollen and therefore from pollinating itself. Prevention of self-pollination is considered desirable by plant breeders who want to produce hybrids by cross-pollination with other plants. Otherwise the breeders have to undertake the tedious task of removing the pollen-producing organs from the plants by hand.

But before any of these ideas can be implemented there is the need to be able to regenerate whole rice plants reproducibly from protoplasts. Many attempts have been made to achieve this but the results have for the most part been equivocal. Because rice protoplasts that have been isolated directly from the plant do not undergo any reproducible, sustained reproduction, workers have used protoplasts isolated from cell suspension cultures. But these preparations may contain contaminating cells with intact walls that might themselves be capable of undergoing sustained division and ultimately regenerating into whole plants. This means that any regenerated plants might have come from the intact cells, which would not be capable of fusing, and not from the protoplasts.

Recently, however, John Thompson, Ruslan Abdullah and Edward Cocking of the University of Nottingham, England, have developed improved cultural procedures that also enable the identification of sustained division by individual protoplasts. Using these procedures the investigators have unequivocally demonstrated the efficient regeneration of whole rice plants from protoplasts by somatic embryogenesis (Fig. 10.4).

Fig. 10.4   A row of rice plants that have been regenerated from protoplasts and are growing in the glasshouse at the University of Nottingham. (By kind permission of John Thompson and Ruslan Abdullah, University of Nottingham.)

The techniques require that the rice cells first be induced to grow in suspension culture for several months. After this, protoplasts are produced and are themselves grown in culture, this time in laboratory dishes. The dividing protoplasts form colonies that develop somatic embryos and it is from these embryos that the whole plants can be regenerated. Similar procedures are likely to be applicable to a wide range of rice varieties, according to Cocking and his colleagues.

Other investigators have cultured individual protoplasts of *Brassica napus* (oil seed rape) in a microdroplet system that uses a synthetic medium. The investigators obtained protoplast survival rates of more than 70 per cent and division frequencies of up to 65 per cent. The work should increase the knowledge of the physiology of different cell types and, if it can be extended to a range of other species, should greatly improve the ability to culture protoplast fusion products.

Rice and the other cereals belong to the monocotyledonous subclass of plants, which are so called because they have only one cotyledon. Regeneration of monocot plants from protoplasts has generally proved much more difficult than regeneration of dicotyledonous plants. Legumes, for example, are dicots and for these plants most of the cell and tissue culture problems have been largely resolved, with the exception of those associated with the regeneration of plants from soybean protoplasts.

Legumes such as white clover (*Trifolium repens*) and alfalfa (*Medicago sativa*) can cause a potentially fatal condition called 'bloat' in cattle and other farm animals that eat them. Bloat-free clover and alfalfa strains might be achieved by fusing *T. repens* protoplasts with those of *T. arvense*, a clover species that accumulates in its leaves condensed tannins that act as antibloat factors. Alternatively, *T. repens* or *M. sativa* protoplasts might be fused with protoplasts of sainfoin (*Onobrychis viciifolia*), which possesses tannins in all of its vegetative parts, or of *Lotus corniculatus*, another bloat-safe forage legume.

Readily applicable methods of separating the hybrids from the parental protoplasts in the culture are needed before this and other potential applications of protoplast fusion can be realized, however. Hybrids can be isolated if each partner in the fusion bears a different selectable mutation. In that event only the hybrids will contain both mutations, and will therefore be distinguishable from the parental cells. Mutants suitable for such complementation selection are generally unavailable in the legumes and cereals.

Although mechanical isolation of fused protoplasts has been attempted, obtaining sufficient numbers for adequate assessment is not usually possible. The recent introduction of fluorescence-activated cell sorting to separate the hybrids from the other protoplasts is likely to revolutionize the rate at which assessments of hydrid production by protoplast fusion can be carried out.

Brassica improvement is similarly hindered by lack of hybrid selection methods. Most of the cell and tissue culture problems associated with plant regeneration from *B. napus* protoplasts have been resolved. The remaining challenge is the development of adequate selection methods. One goal of protoplast fusion with *Brassica* is the production of strains containing cytoplasmic male sterility factors, and species that are suitable for transfer of those factors must be identified.

Although hybrid protoplasts that are formed by fusing different mutants have been isolated by genetic complementation such selection methods require the development of new mutants, or of markers based on differential growth responses, for each combination of species. An alternative approach involves combining in one species an auxotrophic mutation, which imposes a new nutritional requirement on the plant, with a dominant mutation that confers resistance to an antibiotic.

Cocking, D. Pental and their Nottingham colleagues have recently produced such a double mutant of tobacco (*Nicotiana tabacum*) by crossing streptomycin-resistant tobacco plants with tobacco plants that have a mutation that makes them deficient in the enzyme nitrate reductase. Plants that lack this enzyme can no longer grow with nitrate as their sole nitrogen source, but must be provided with ammonia or some other alternative source of the essential element.

The Nottingham workers produced the double mutant tobacco plants both by standard sexual breeding methods and by fusing appropriate protoplasts. Streptomycin resistance is encoded in the chloroplast genome, whereas the nitrate reductase gene is located on a chromosome in the nucleus. Protoplast fusion was achieved in such a way as to produce hybrids in which the cytoplasm, which contains the chloroplasts, came from the strepotomycin-resistant plant and the nuclei came from the plant with the mutant nitrate reductase gene. The resulting hybrids can be identified because they are resistant to the antibiotic streptomycin and require a non-nitrate source of nitrogen.

Protoplasts of the tobacco double mutant can be used for somatic hybridization with protoplasts from a range of other plant species. Crosses can be achieved that would be difficult or impossible or obtain by standard breeding methods. For example, when protoplasts of the petunia (*Petunia hybrida*) are used as the fusion partner, hybrids can be produced that have the *P. hybrida* nuclear genome and the *N. tabacum* chloroplast genome. The hybrids can be identified because they are streptomycin resistant and have regained the ability to grow on nitrate as a result of the petunia component having provided the nitrate reductase enzyme.

When attempting to produce hybrid plants by protoplast fusion, it is ideal if plants can be regenerated from the protoplasts of both parents. However, earlier studies in the petunia have indicated that hybrid plant regeneration is possible even if only one of the fusion partners is capable of regenerating whole plants from protoplasts.

In addition, it has been shown recently that fusion between protoplasts of cultivated tomato, which is incapable of whole plant regeneration, and protoplasts of a wild tomato species, which is capable of plant regeneration, can result in the production of self-fertile somatic hybrid plants. This means that hybrid production by protoplast fusion may be less dependent on plant regeneration capability than was previously thought.

Metabolite production in culture

A special feature of seed plants is the synthesis of metabolic products of great variety, including terpenoids, steroids, alkaloids, essential oils and pigments (Table 10.1). Plant cell

Table 10.1. *Natural products from plants and their associated industries*

Industry	Plant product	Plant species	Industrial uses
Pharmaceuticals	Codeine (alkaloid)	*Papaver somniferum*	Analgesic
	Diosgenin (steroid)	*Dioscorea deltoidea*	Anti-fertility agents
	Quinine (alkaloid)	*Cinchona ledgeriana*	Anti-malarial
	Digoxin (cardiac glycoside)	*Digitalis lanata*	Cardiatonic
	Scopolamine (alkaloid)	*Datura stramonium*	Anti-hypertensive
	Vincristine (alkaloid)	*Catharanthus roseus*	Anti-leukaemic
Agrochemicals	Pyrethrin	*Chrysanthemum cinerariaefolium*	Insecticide
Food and drink	Quinine (alkaloid)	*Chinchona ledgeriana*	Bittering agent
	Thaumatin (chalcone)	*Thaumatococcus danielli*	Non-nutritive sweetener
Cosmetics	Shikonin	*Lithosperum erythrorhizon*	Dye

Modified from Fowler (1983).

culture can be used to elucidate the biochemical pathways needed for synthesis, accumulation and degradation of such metabolic products, and in some cases, for their commercial chemical production.

Until recently, the major research emphasis was on the isolation of cell lines with a capacity for product accumulation that is superior to that of the parent plant. The improvements have depended on the ability to culture cells by using either mass cell culture or immobilization of whole cells.

The biotechnological applications of plant cell culture methods for synthetic purposes can be illustrated by describing the production of the red pigment shikonin, which is the active principle of the purple root of *Lithospermum erythrorhizon*, a perennial herb native to Japan, Korea and China. Shikonin can be used in dyes, ointments and cosmetics. The red colouration in the Japanese flag has been produced for many years with a shikonin dye.

An elegant series of investigations by Mamoru Tabata and Yasuhiro Fujita of Kyoto University in Japan has led to the successful industrial production of shikonin by cell culture methods (Fig. 10.5). The work highlights the importance of coupling basic studies on callus induction, cell selection, and product biosynthesis and localization with applied studies on the improvement of the production medium.

The basic studies of Tabata and Fujita on selection showed that cultured *Lithospermum* cells vary in their shikonin-synthesizing capabilities and that the pigment content of the cells could be increased greatly by selecting cells for high production. Because knowledge of the factors regulating metabolism is as important as the selection of high-producing cells, the researchers systematically investigated the effects of temperature, light wavelength, growth hormones and carbon and nitrogen sources on shikonin production. They showed that it is possible to induce synthesis of the dye in cell suspension cultures by adding agar powder to the culture medium. Agar is relatively expensive, however, and it was economically desirable to devise a production medium in which the cells would produce shikonin in the absence of agar.

Tabata and Fujita found that *Lithospermum* cells will produce shikonin pigments specifically in a medium containing nitrate as the sole inorganic nitrogen source, although cell growth was somewhat impaired. By systematically assessing all the components of this medium, the investigators found that a thirty-fold increase in the concentration of calcium causes a three-fold increase in the yield of shikonin derivatives, but without improving cell growth. They ultimately resolved the cell growth problems by devising a two-stage culture system. In the first stage the cells are grown with agar and then they are transferred to the agar-free medium with high calcium to increase the final yield of product.

Tabata and Fujita estimate that the cell culture procedure for making shikonin is about 800 times more productive than the traditional plant cultivation method. The content of the dye in the cultured cells is about 15 times higher than in the plant material. Moreover, the cell culture method takes less than a month to complete compared with the 48 months necessary for growing the plants.

Although shikonin has recently been chemically synthesized from dihydroxynaphthalene, the final yield was only 0.7 per cent. At present the production of the dye by plant cell culture is far more economical than production by chemical synthesis. Shikonin produced by the large-scale culture of *Lithospermum* cells has been commercially used for cosmetics since 1984.

Assessing the extent to which the type of approach so successfully developed for shikonin synthesis will be applicable to other systems is difficult. For example, stimulating product formation by varying culture conditions, while highly successful for alkaloid production by *Catharanthus roseus* (Madagascar periwinkle) cultures, was largely fruitless for *Papaver* (poppy) cultures, which have shown a limited capacity for accumulation of the alkaloid morphine. However, this could just mean that the correct culture conditions for high production of morphine have not yet been found. It is noteworthy in this regard that temperature stress induces a change in the relative quantities of different alkaloids produced by cultured cells of various *Papaver* species.

Plant transformation with recombinant DNA

The plant cell and tissue culture technologies discussed so far are already providing opportunities for bringing new varieties of improved crop plants to the market place. Nevertheless, there is great interest in applications of culture methods that involve introducing foreign DNA, which is obtained by recombinant DNA technology, into cultured plant cells to bring about their genetic transformation, even though many years may be required before commercially significant plant products are achieved in this way.

Pental and Cocking have pointed out that somatic hybridization by protoplast fusion and genetic transformation are two radically different approaches to the manipulation of plant genomes. In somatic hybridization, as in sexual hybridization, the investigator is aware that the phenotype (the observed characteristic) is related to the genotype (a particular genetic composition), but the reliance is on selecting for a stable and desirable phenotype. By contrast, successful genetic manipulation by recombinant DNA technology is dependent on relating phenotype and genotype, so as to ascertain what biochemical or developmental activity is controlled or modulated by a specific DNA sequence. The success of this approach will depend on how readily the genes coding for particular traits can be identified, isolated from the genome of a plant or other species, and cloned. Such sequences could then be mobilized into the plants that are to be genetically manipulated. Much interest currently centres on the ease with which this can be done, and the use of plant cell and tissue cultures in this respect.

Recombinant DNA technology has made it possible to isolate large amounts of pure genes that might be used to produce better crops. However, John Bingham of the Plant Breeding Institute in Cambridge, England, points out that many plant improvements can be achieved by the conventional breeding methods currently in use, which depend neither on cell and tissue culture methodology nor on the techniques of recombinant DNA. Although goals clearly need to be appraised carefully, significant contributions are nonetheless likely to result from a combination of recombinant DNA and plant and tissue cell procedures with conventional plant breeding.

Two categories of systems are being used to transfer foreign DNA into the genomes of higher plants. One involves the use of bacteria of the genus *Agrobacterium* as vectors for transmitting the genes. Development of this system followed from the demonstration that part of the Ti (for tumour-inducing) plasmid of the organism becomes integrated into the genome of the infected plant cell. A foreign gene that is inserted at the appropriate location in the Ti plasmid can thus be introduced into the plant genome (see also Chapter 11). In addition, DNA of isolated *Agrobacterium* plasmids can be taken up directly by plant protoplasts in culture.

One of the attractions of transformation by *Agrobacterium* is that the procedure is very straightforward for several plant species. Leaf discs are incubated with the bacteria for a few hours. Shoots are formed from the transformed cells of the discs after they are grown for a few weeks in a suitable culture medium. Transformed plants of tobacco and petunia can be obtained within a few months of the infection with the bacteria. This simple method of transferring genes into plants works well for those species that are infected by the *Agrobacterium* bacteria and can be readily regenerated into plants from leaf discs. Such species include the tomato, potato, and several of the forage legumes.

Current success with these species is stimulating attempts to extend this approach to additional crops such as the brassicas and the grain legumes. The cereals are monocotyledons, however, which are much less amenable than dicotyledons to infection by *Agrobacterium*.

Because regeneration of shoots probably occurs mainly from single leaf cells in those species that do not form callus during regeneration, the simple procedure of infecting with *Agrobacterium* has largely replaced the method of transforming protoplasts by culturing them with the bacteria. Nevertheless, such co-cultivation is a highly attractive method of gene transfer, particularly if a reproducible system is available for plant regeneration from the protoplasts. This is the case for tobacco, and when 14 day old protoplasts of this species were co-cultured with *Agrobacterium* cells, between 1 and 3 per cent of them became transformed with Ti plasmid DNA.

The high frequency of transformation in co-cultivation experiments, and also in leaf-disc incubation experiments, contrasts with the low frequency of transformation, usually 0.0001 per cent, that is ordinarily obtained by incubating the isolated Ti plasmids with petunia or tobacco protoplasts. Recently, a combination of procedures that stimulate DNA uptake, including treatment with heat shock, polyethylene glycol, and an electric current, has very significantly increased the frequency of tobacco protoplast transformation with foreign DNA until it is now similar to, or better than, that obtained by co-cultivation with *Agrobacterium*.

As might be expected from the absence of a cell wall on the protoplasts, there so far appears to be no species barrier to the direct uptake methods. Even grasses and cereals such as *Lolium multiflorium*, *Triticum monococcum*, maize, rice and sugarcane, which are all monocots and not readily infected by *Agrobacterium*, are amenable to direct gene transfer into protoplasts.

Direct gene transfer thereby bypasses the limitations inherent in the use of whole *Agrobacterium* bacteria. The major remaining problem limiting gene transfer into cereals is the inability to regenerate whole plants from protoplasts. However, as previously discussed, the reproducible regeneration of plants from rice protoplasts has now been unequivocally shown. It seems likely, provided that emphasis is placed on finding the correct cultural conditions for regeneration by means of similar approaches, that the present problems will be resolved for additional cereals and other species that are currently intransigent to protoplast regeneration.

Fig. 10.5 In this composite, (*a*) shows a shikonin plant (*Lithospermum erythrohizon*) in bloom. In (*b*) are flasks of cultured cells from the plant, before and after they have been induced to produce the red shikonin dye. (*c*) shows cultures of the dye-producing cells in the laboratory and in (*d*) are the larger containers for the commercial-scale production of the dye. (By kind permission of Professor M. Tabata, Kyoto University, Japan.)

Overall it is evident that methods are now well developed for the transformation of plants by foreign DNA. The outstanding challenge is to use the methods to transfer agronomically valuable traits into plants. Only a small number of such traits, including resistance to some herbicides, are under the control of one or a few identified genes. Progress is likely to be slow in most cases because many of the traits agronomists would like to see improved are controlled by many genes. Moreover, the genes have not in most instances been identified, and the factors that control their expression at particular stages of plant development are also unknown.

Similar considerations apply to the transformation of cultured plant cells for the production of alkaloids and other valuable drugs. At least for these applications, however, researchers do not have to be concerned about regenerating whole plants from transformed protoplasts.

The outlook for plant cell and tissue culture in biotechnology.

Plant cell and tissue culture methods will continue to have a profound influence on biotechnology, perhaps more in the longer term than animal cell culture because of the ability to regenerate whole functional plants from cultured cells. Although this chapter has separated the various uses of plant cell and tissue culture for the generation of variation, for hybrid production by protoplast fusion, for metabolite production, and for transformation with recombinant DNA, all these applications are intimately linked.

For example, transformation of plant cells by foreign genes could perhaps enhance secretion of a metabolite product, thereby facilitating product recovery. Or, transformation with genes from *A. rhizogenes*, a relative of *A. tumefaciens* that causes the abnormal root proliferation known as 'hairy root disease', might facilitate the rooting of plants that have been regenerated from cultured cells. Rooting can often be a bottleneck in plant regeneration. Moreover, hairy roots may themselves be a transformed system in which synthesis of a product is enhanced.

Finally, the fusion of plant protoplasts may also be used to yield hybrid cells in which the synthesis of a metabolite product is increased. In biotechnology, how the product is made is of secondary importance. The overriding consideration is that it be produced economically at a price competitive with that of any alternative technology.

Additional reading

Bingham, J. E. (1984). Achievements, prospects, and limitations of conventional plant breeding. In *The Genetic Manipulation of Plants and Its Application to Agriculture. Annual Proceedings of the Phytochemical Society of Europe*, vol. 23, ed. P. J. Lea and G. R. Stewart, p. 1. Oxford University Press, Oxford.

Chaleff, R. S. (1983). Isolation of agronomically useful mutants from plant cell cultures. *Science*, **219**, 676.

Cocking, E. C. (1983). Applications of protoplast technology to agriculture. *Experientia (Supplement)*, **46**, 123.

Cocking, E. C. (1986). The tissue culture revolution. In *Plant Tissue Culture and Its Agricultural Applications*, ed. L. A. Withers and P. G. Alderson, p. 3. Butterworths, London.

Fowler, M. W. (1983). Commercial applications and economic aspects of mass plant cell culture. In *Plant Biotechnology*, ed. S. H. Mantell and H. Smith, p. 102. Cambridge University Press, Cambridge, England.

Gautheret, R. J. (1985). History of plant tissue and cell culture: a personal account. In *Cell Culture and Somatic Cell Genetics of Plants*, vol. 2, ed. I. K. Vasil, p. 1. Academic Press, New York.

Orton, T. J. (1984). Somaclonal variation: theoretical and practical considerations. In *Gene Manipulation in Plant Improvement*, 16th Stadler Genetics Symposium, ed. J. P. Gustafon, p. 427. Plenum Press, New York.

Pental, D. and Cocking, E. C. (1986). Some theoretical and practical possibilities of plant genetic manipulation using protoplasts *Hereditas (Supplement)*, **3**, 83.

Riker, A. J. and Hildebrandt, A. C. (1958). Plant tissue cultures open a botanical frontier. *Annual Review of Microbiology*, **12**, 469.

Tabata, M. and Fujita, Y. (1985). Production of shikonin by plant cell cultures. In *Biotechnology in Plant Science*, ed. M. Zaitlin, P. Day, A. Hollaender and C. M. Wilson, p. 207. Harcourt Brace Jovanovich, Orlando, USA.

11 Improving crop plants by the introduction of isolated genes

Jozef Schell, Bruno Gronenborn and Robert T. Fraley

The goal of the plant breeder is to improve agricultural crop plants by changing their genetic properties. This is usually achieved by combining the genetic traits of different, but related, plants either by sexual mating or by the more recently developed technique of somatic cell fusion (see also Chapter 10). In addition, gene mutations can be induced in plants by treating them with chemicals or radiation. A small percentage of the mutations thus produced will be beneficial.

The success of the classic plant breeding methods is obvious. They were a major contributor to the Green Revolution of the 1950's and 1960's which led to an eventual doubling or tripling of the yields of the major grain crops. Nevertheless, the classic methods have a number of limitations. For one, a particular genetic improvement cannot be attempted unless the trait that the breeder wants to introduce into the target plant is carried by a sexually compatible species, which means that the two plants must be closely related.

For another, the crosses and backcrosses that have to be performed to select the best possible combination of genetic properties usually take a great deal of time – typically 10 to 15 years. This is because sexual mating combines the entire genomes of the two partners, with the result that a very large number of genetic combinations is possible. Only a very few of these will be satisfactory and finding the desired combinations requires considerable time, effort, skill, and a certain amount of flair and luck.

The ability to isolate and clone genes, coupled with the development of reliable techniques for introducing genes into plants, has opened a new route to the genetic improvement of plants that can circumvent the limitations of conventional breeding methods. The new strategy consists of looking for any living organism – whether plant, bacterium, fungus, or even animal – that carries the trait that the investigator wishes to introduce into the target plant, and then using recombinant DNA techniques to isolate the gene that controls the trait. The isolated gene is then 'reprogrammed' so that it can be expressed in plant cells and is introduced into a crop plant by a gene transfer vector.

Once a useful gene has been isolated, it can in principle be transferred to many different crops without a lengthy breeding programme. The main limitation of the transfer methods is that, currently at least, they can only be used for traits that are controlled by one or a very few genes. Such useful traits as resistance to herbicides and disease belong to this category and researchers have already demonstrated that new strains of herbicide- or disease-resistant plants can be produced by gene transfer. Nevertheless, gene transfer techniques cannot replace classical breeding programmes. The transfer methods are instead a powerful, additional tool for the plant breeder.

The development of gene transfer vectors for plants

The use of nature's own gene transfer methods

Plant pathologists have long been interested in the soil bacterium *Agrobacterium tumefaciens* (Fig. 11.1) because of the plant

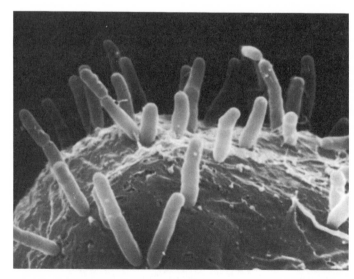

Fig. 11.1 Agrobacterium tumefaciens. In this scanning electron micrograph, the rod-shaped *A. tumefaciens* bacteria are shown attached to cap cells of pea roots. (Photograph by Jerry White, University of Missouri-Columbia. Reprinted by kind permission of Martha C. Hawes, University of Arizona, Tucson.)

disease it causes. Various strains of the organism can readily infect most members of one of the two major subclasses of plants, namely the dicotyledonous plants (dicots) which are so called because the embryos have two seed leaves (cotyledons). Most trees and shrubs are dicots, as are such common crop plants as the potato, soybean, tomato and tobacco. *Agrobacterium* can also infect some members of the second major plant subclass, the monocotyledons (monocots), which have one seed leaf. This subclass includes the grasses and the major cereal crops rice, wheat and corn.

As a result of *Agrobacterium* infection of the dicots the wound tissue proliferates as a cancerous growth, commonly known as a crown gall (Fig. 11.2). Two main properties characterize crown galls: the cells are able to grow in culture without the supplement of hormones that are required for the growth of normal plant cells; and the crown gall cells synthesize a new set of compounds called opines, which are not present in normal cells. Opines are amino acids or sugar derivatives that are used as carbon and nitrogen sources by the bacteria that incite the tumour formation.

Agrobacterium first attracted the attention of molecular biologists when they discovered that a large plasmid, the Ti (for tumour-inducing) plasmid, is the cause of the bacterium's oncogenic capabilities. Tumour formation is the direct result of the transfer of Ti plasmid DNA from the bacterial cells to the infected plant cells, where the plasmid DNA integrates in the nuclear genome. The *Agrobacterium* cell is in effect a miniature genetic engineer that can bring about the stable insertion of foreign genes into the plant cell genome. The bacterium does not cause tumours on cereal plants, but there is indirect evidence that it can introduce DNA into some embryonic tissues of cereals.

Plant molecular biologists have now taken advantage of this ability and have adapted the Ti plasmid as a vector for accomplishing the long-envisaged goal of introducing functional new genes into plants. Before the Ti plasmid could be developed as a successful vector, however, investigators had to learn a great deal about the basic molecular biology of the plasmid and of plant gene expression.

Infection by *Agrobacterium* includes at least two stages. First, the interaction of the bacteria with plant cells initiates a series of events that ultimately leads to the transfer to the plant cells of a specific segment of the Ti plasmid, which is called the T-DNA (for transferred DNA). Then, in the second step, the T-DNA becomes stably integrated into the nuclear genome of the plant cells.

The groups that contributed to this understanding of the molecular biology of the Ti plasmid and the design of the vectors include those of Mary-Dell Chilton, at Washington University in St Louis, and more recently at the Ciba-Geigy Company in North Carolina; Robert Fraley, Robert Horsch and Stephen Rogers, who are at the Monsanto Company in St Louis; Milton Gordon and Eugene Nester at the University of Washington in Seattle; Jozef Schell, Marc Van Montagu and

Patty Zambryski at the State University of Ghent, Belgium, and the Max Planck Institute for Plant Breeding in Cologne, West Germany; Robbert Schilperoort at the University of Leiden in the Netherlands; Jacques Tempé of the Institut National de la Recherche Agronomique in Versailles, France; and Allen Kerr of the University of Adelaide, Australia.

For the T-DNA to be transferred, the *Agrobacterium* must contain 'virulence' genes that enable the bacterial cells to recognize and interact with, and then introduce the T-DNA into, plant cells. Although the virulence genes are located on the Ti plasmid, they are not in the T-DNA segment. Plants that have been wounded in some way are especially susceptible to infection by *Agrobacterium* because the wounded plants release low molecular weight phenolic compounds that activate the bacterial virulence genes.

A second requirement for the introduction of *Agrobacterium* DNA into plant cells is that the DNA to be transferred must be bordered by a set of 'integration sequences' which are 25 base pairs long. *Agrobacterium* cells that harbour the necessary virulence genes can transfer any DNA segment, whether it has a plasmid or chromosomal location, provided that it is bounded by the integration sequences.

The products of some of the virulence genes recognize the integration sequences and cause single-strand breaks in them. This in turn leads to the formation in the *Agrobacterium* cells of single-stranded DNA copies of the DNA between the integration sequences. The single-stranded DNA ultimately gets transferred to the plant cell and integrated into the cellular genome as a double-stranded DNA insert. How the transfer occurs and the mechanism by which the single-stranded DNA is inserted into the plant cell genome are as yet unclear.

Although the T-DNA segment that is normally introduced into plant cells by *Agrobacterium* contains genes that are needed for opine synthesis and tumour induction, these genes are not required for transfer and insertion of the DNA. This design, in which the genetic information for infection and DNA transfer is completely independent of that required for formation of the crown gall tumours, is beneficial with regard to designing vectors for introducing new genes into plants. The genes for tumour induction, which could be harmful to recipient plants, can be removed from the vector without affecting its ability to be transferred by *Agrobacterium*. Any DNA sequence that is flanked by the specific integration sequences can serve as a vector that will be transferred from *Agrobacterium* cells to plant cells. Such vectors can carry at least 40 kilobases of DNA, enough to code for approximately 10 genes.

The first genes that investigators attempted to transfer with vectors derived from the Ti plasmid were bacterial genes coding for resistance to antibiotics. The experiments worked in the sense that the vector DNA entered the plant cells and became integrated in the cellular genome. However, the protein products of the bacterial genes were not made in the plant cells nor were the products of other foreign genes from higher species. For example, the yeast gene coding for the enzyme

Fig. 11.2 Tumour induced on a tobacco plant by the bacterium *A. tumefaciens*. During infection, the bacterium introduces DNA from the Ti plasmid into plant cells. The transferred DNA contains genes that elicit abnormal plant cell growth and tumour formation. In this case, high production of a plant growth hormone results, thereby causing green shoots to sprout from the tumour.

Plant molecular biologists have adapted the *A. tumefaciens* system for use in the genetic engineering of crop plants, by removing the genes that cause tumour formation from the Ti plasmid. Genes coding for some desired trait, such as herbicide or disease resistance, are inserted instead into the plasmid for transfer into the target plant species. (Reproduced by kind permission of Milton P. Gordon and Eugene W. Nester, University of Washington School of Medicine, Seattle.)

alcohol dehydrogenase and the mammalian genes for beta-globin and interferon failed to work. Analysis of the messenger RNA of the plant tissues that contained these foreign genes revealed that none of them were being transcribed into messenger RNA, which is the first step in protein synthesis.

Numerous studies of gene expression have shown that genes have regulatory sequences that are usually located before the start of the protein-coding regions and are necessary for the initiation of transcription. The failure of the bacterial, yeast and mammalian genes to be expressed in plant cells suggested that the regulatory sequences of these genes were not recognized by the plant transcriptional machinery. At that point it became obvious that if foreign genes were to work in plant cells their normal regulatory sequences would have to be replaced by those of plant genes.

Chimaeric genes were then constructed in which the transcription signals from genes known to be expressed in plant cells were attached to the coding sequences of bacterial antibiotic resistance genes. The chimaeric genes were expressed in plant cells, to which they gave a selectable advantage; that is, cells that contained the genes could be identified because of their ability to grow in the presence of the appropriate antibiotic. Current Ti-derived vectors still usually include antibiotic-resistance genes as selectable markers in addition to any other gene the investigator might want to transfer. The *Agrobacterium* system has now been used to introduce new genes into several different types of plants, including tobacco, petunia, tomato, potato, alfalfa and soybean.

Plant viruses as gene vectors and potential gene amplification systems

The application of viruses as vectors for transferring genes into plants is in its infancy compared with the development of the viruses used for introducing genes into bacterial and mammalian cells. Development of the plant viruses as vectors has been slow mainly because the vast majority contain an RNA genome. Only two groups of plant viruses – the caulimoviruses and the geminiviruses – have DNA as their genetic material. Members of both groups have been used to propagate and express foreign genes in plant cells.

Caulimoviruses, of which cauliflower mosaic virus (CaMV) is a typical example, are 'retroid' plant DNA viruses. Viruses normally use the synthetic capabilities of the cells they infect to reproduce themselves. Because CaMV multiplies in terminally differentiated plant cells in which the host DNA synthesis is shut off, the virus has evolved an unusual two-step mode for replicating its DNA. First the CaMV genome, which consists of about 8 kilobases of double-stranded DNA, is transcribed into a genomic RNA inside the nucleus of the host cell. This genomic RNA is in turn transcribed back into double-stranded DNA by a virus-encoded reverse transcriptase.

Not only does CaMV have a unique mode of replication, but it also displays a peculiar translational coupling of at least three genes that imposes some serious constraints on the use of the virus to express foreign genes in plants. Only small genes that do not exceed 300 base pairs in length and are devoid of introns are stably maintained and expressed by CaMV. The only gene of the virus that may be replaced by a foreign gene is the one encoding aphid acquisition factor, a protein necessary for the spread of the virus in nature.

Genes that have been successfully expressed after introduction into plant cells by CaMV are a bacterial dihydrofolate reductase gene and a human interferon gene. In addition the CaMV genome contains two strong promoters of gene expression in plant cells that are often spliced into other vectors, including the Ti plasmid, to drive the expression of foreign genes in plants.

The genome of the second group of plant DNA viruses, the geminiviruses, consists of either one or two circular, single-stranded DNA molecules. The single-stranded viral DNA, which is 2.6 to 3.0 kilobases long, is converted in the nucleus of plant cells into a double-stranded replicative form by an as yet unknown mechanism. Many copies of the replicative form of a geminivirus genome accumulate inside the nuclei of infected cells. There is no evidence to date for a reverse transcription step in geminivirus replication.

Geminiviruses exhibit a much wider host range than do caulimoviruses. This makes the geminiviruses more suitable for general application as vectors for introducing new genes into plants. Although little is known about the geminivirus genes and their functions, recent evidence suggests that the gene encoding the viral capsid protein may be replaced by foreign genes without interfering with the replication of the virus genome.

The cereal geminivirus, wheat dwarf virus, is being developed as a vector for introducing genes into cereals. Although the single chromosome of the wheat dwarf virus is only 2.7 kilobases long, it it nonetheless capable of accommodating and replicating gene inserts up to about 3 kilobases in length. In this way the size of the viral DNA may be more than doubled.

Three bacterial genes that have been inserted into the wheat dwarf virus genome have been successfully replicated and expressed after transfer into cultured cells of the cereals *Triticum monococcum* (one-grained spelt) and *Zea mays* (maize). This illustrates the potential of geminiviruses for serving as the basis of autonomously replicating expression units in plants.

Direct uptake of DNA by plant cells

Gene transfer into protoplasts

The gene vectors that are derived from the Ti plasmid have provided a means of testing the activity of chimaeric selectable marker genes in plants. Once a particular chimaeric gene has been shown to be active in plant cells and to confer a selectable advantage on the recipient cells, it can be used in experiments aimed at developing more direct ways of introducing isolated genes into plants.

For example, chimaeric marker genes have been transferred without the aid of vectors into protoplasts, plant cells from which the rigid cellulose walls have been removed. Various protocols are now available for introducing new genes into the protoplasts of a wide variety of plants, including the dicots tobacco, petunia and carrot, and the monocots wheat and maize.

In all these cases the protoplasts are capable of developing into the undifferentiated type of tissue known as callus. Whole plants that contain the transferred DNA can be regenerated from some of the callus tissue thus formed. This works for the tobacco callus and that of other dicotyledonous plants, but notably not for that of the cereals.

The foreign DNA often inserts at a single chromosomal location in the plant cells, although it less frequently integrates at multiple sites. In most cases inheritance of the transferred DNA follows Mendelian rules. However, the integrated DNA often undergoes various rearrangements, probably before it integrates in the plant genome, and this may make direct transfer of DNA into protoplasts a less desirable means of introducing foreign DNA into plant cells than transfer by *Agrobacterium*.

Transfer of selectable marker genes by injection

Plant molecular biologists would like to apply the new methods of gene transfer to the genetic modificatation of the cereals, which are the major crops of agriculture. The research has been frustrated, however, because *Agrobacterium* did not appear to infect the cereals and because attempts to regenerate whole plants from cereal protoplasts have generally failed – although that situation may now be changing as groups in England, France and Japan have reported the regeneration of rice protoplasts (see Chapter 10).

A number of approaches are currently being explored as alternative ways of transferring genes into cereal plants. These approaches include the use of DNA viruses as gene vectors; the uptake of DNA through the cell walls of germinating pollen; and the injection of DNA into germ cells before they undergo meiosis, the form of nuclear division in which the number of chromosomes is halved as a prelude to the formation of the egg and sperm cells.

Alicia de la Pena of the Max Planck Institute has recently obtained some promising preliminary results on gene transfer into rye by an injection method. The germ cells of the rye plant are located in a flower cluster known as an inflorescence. Materials injected into the inflorescence can penetrate the developing germ cells. For example, injection of the chemicals caffeine and colchicine penetrate the cells and produce developmental abnormalities in them.

To test whether DNA would also penetrate the cells, de la Pena injected rye inflorescences with DNA plasmids consisting primarily of a gene coding for resistance to the antibiotic kanamycin. The seeds that were eventually produced were then screened for kanamycin resistance by germinating them in the presence of the antibiotic. The first batch of about 3000 seeds, which were derived from approximately 70 injected rye inflorescences, produced three seedlings that contained and expressed the kanamycin resistance gene. Although this initial success should be considered preliminary and in need of reproduction, the result indicates that simple methods can probably be developed for introducing foreign genes into cereal plants.

Transposable elements as potential gene vectors

The cloning from maize and other plants of several transposable elements may allow their development as vectors for introducing new genes into plants. Transposable elements naturally move about in the genomes of the host plants, inserting first in one place and then in another. A maize transposable element has recently been shown to be capable of transposition in a new host plant such as tobacco. Experiments are currently in progress to test whether isolated transposable elements that carry selectable marker genes will integrate into plant cell genomes if the elements are introduced into protoplasts or injected into inflorescences.

The methods that are already available for introducing new genes into plants mean that gene transfer into crop species is no longer a rarity. With even more methods under development, it will soon be possible to put new genes into most, if not all, of the major crop plants.

Vectors for studying plant gene expression

Plant genes that are expressed all the time

The protein-coding segments of the genes of higher organisms are flanked by sequences that regulate gene expression. In yeast and animal genes the regulatory sequences needed for the accurate initiation of transcription are usually located upstream from the start or in the non-protein-coding sequences found in most of the genes of higher organisms. In addition, sequences downstream from the end of the coding regions contain signals for transcription termination and for certain modifications of messenger RNAs that are necessary for normal translation into protein.

As previously mentioned, bacterial and animal genes did not work in plants when the genes were transferred with their own regulatory sequences. However, investigators were able to construct functional chimaeric genes in which the coding regions of non-plant genes were flanked with the upstream and downstream regulatory sequences of genes that are active in most, if not all, organs of a wide range of plants. These regulatory sequences were obtained from Ti plasmid genes, such as those coding for opine synthesis, and from genes of the cauliflower mosaic virus. Expression vectors were constructed in which the upstream and downstream regulatory sequences were separated by a linker DNA that could be used to insert the coding region of any desired gene.

Several vectors of this type are now available and have been used to transfer into plants a variety of selectable marker genes, including several antibiotic and herbicide resistance genes. The genes transferred by these vectors are usually expressed in all developmental stages and all tissues of the plants that acquire them. The occasional cases in which the genes are silent are probably the result of integration into a transcriptionally silent portion of the recipient genome or of chemical modifications that inactivate the chimaeric genes.

Plant genes that are expressed only under certain conditions

Many plant genes are not expressed all the time, but are only active in certain tissues or developmental stages, or when turned on by a specific stimulus such as light or heat. As a logical further step in the construction of chimaeric genes, plant molecular biologists are investigating whether the upstream control sequences of these regulated genes will work in their normal fashion when attached to foreign genes that are transferred into plants.

Experiments of this type will not only lead to a better understanding of plant gene control, but can also provide useful information for the eventual genetic engineering of plants. The successful introduction of many new traits into plants will no doubt depend on whether the genes are regulated in a normal fashion by the recipient.

Some of the upstream regulatory sequences have been obtained from genes that are turned on in response to light. Among these are two nuclear genes that code for proteins required for photosynthesis. The *rbcS* gene directs the synthesis of the smaller of the two subunits of ribulose-1,5-bisphosphate carboxylase, the enzyme that catalyses the first step in the pathway that converts carbon dioxide to sugar. The *LHCP* gene codes for the light-harvesting chlorophyll *a/b* binding protein, which is only active after it is transported into the chloroplast where photosynthesis occurs.

The *rbcS* and *LHCP* genes of the pea are only fully active in tissues, such as green leaves, that contain active chloroplasts. The light signal that brings about their activation is conveyed to the two genes by means of the pigment phytochrome. The signal to the *rbcS* gene is also transmitted through a receptor that responds to blue light.

Several studies have shown that the sequences that enable these light-sensitive genes to be regulated in such a precise manner are contained within a few hundred base pairs of DNA, located just upstream from the beginning of the coding regions. When these regulatory sequences from the pea *rbcS* and *LHCP* genes are joined to coding sequences for foreign proteins, the resulting chimaeric genes behave just like the normal *rbcS* and *LHCP* genes. In other words, the foreign proteins are made only in light-activated leaves that have fully developed chloroplasts. The work also shows that the regulatory sequences from the *LHCP* gene work as silencers of gene expression in roots, in contrast to their light-dependent activating effects in leaves.

The chalcone synthase gene, which catalyses one of the key steps in flavonoid synthesis, represents another type of light-regulated gene. The flavonoid compounds, which are highly abundant in plants, have several important functions. In addition to serving as flower pigments, they are part of the defence mechanisms of a plant. Flavonoids can act as anti-microbial agents and as compounds that protect against damage by excess ultraviolet radiation.

The chalcone synthase gene in tobacco is activated by irradiation with ultraviolet light for 20 hours. The regulatory sequences needed for this activation are again located in the upstream flanking region of the gene. Chimaeric genes can be constructed by combining the upstream sequence of a chalcone synthase gene with a foreign coding sequence. Such chimaeric genes are activated by prolonged exposure to ultraviolet radiation just as the normal chalcone synthase gene is.

Some genes are activated not by light, but by elevated temperatures. Among these are a complex of 'heat-shock' genes, the activation of which apparently protects cells against the damage that might otherwise be caused by higher than normal temperatures. When chimaeric genes that contain the regulatory sequence of a heat-shock gene are introduced into plants, they, too, are turned on by increased temperatures.

Expression of many plant genes is limited to specific organs, such as the leaves, seeds, tubers, or the nitrogen-fixing nodules found on legume roots. The upstream flanking regions from such organ-specific genes have also been used to construct chimaeric genes. When the chimaeric genes are transferred into plants they are usually expressed with the same tight organ specificity as are the original genes from which the upstream regulatory regions were obtained.

The organ-specific regulatory regions can be used to change the site of expression of plant genes. For example, the patatin gene codes for one of the major storage proteins of potato tubers, but the tissue specificity of the patatin gene can be altered so that it is expressed in potato or tobacco leaves simply by replacing its upstream flanking sequence with one from a leaf-specific tobacco gene.

These experiments also show that the regulatory sequences are not species specific. The sequence from the leaf-specific tobacco gene works in potato in much the same way that it acts in tobacco itself. Even the phaseolin genes, which code for the storage proteins of bean seeds, are expressed with their normal specificity for seeds when they are transferred into tobacco plants. The protein-synthesizing machinery and its recognition signals must therefore be very similar in the different plant species.

Although most of the genes of higher organisms are located in the nucleus, some organelles – the mitochondria of all cells and the chloroplasts of plant cells – have their own small genomes. The chloroplast and mitochondrial genomes encode some of the proteins of these organelles, whereas others are encoded by nuclear genes. Those that are specified by the nuclear genes are synthesized in the cytoplasm and then have to be transported to their final destinations in the mitochondria or chloroplasts. These proteins carry specific peptide sequences that serve to target them to the right organelle and are then cleaved off.

Foreign proteins can be targeted for transport into chloroplasts by constructing chimaeric genes that carry the coding sequence for the chloroplast transit peptide. The *rbcS* gene, which as already mentioned encodes the chloroplast protein ribulose-1,5-bisphosphate carboxylase, is located in the nucleus and carries a suitable coding sequence for a transit peptide. Attachment of the rbcS transit sequence to a gene coding for resistance to the antibiotc kanamycin causes the product of this chimaeric gene to be transported to the chloroplast as expected.

Having the ability to target the proteins that are produced by foreign genes to specific organelles such as the chloroplast could be very useful. The production of strains of crop plants with increased herbicide resistance is one of the goals of plant genetic engineering. Some genes for herbicide resistance are in the chloroplast. Introducing a foreign gene into the chloroplast genome may be very difficult, but if the gene carries an appropriate transit sequence its product should be transported to the chloroplast even though the gene itself ends up in the nucleus. Attempts to improve photosynthesis might also depend on the ability to have proteins targeted to the chloroplast.

A great deal of work has now shown that expression vectors can be used to construct chimaeric genes that will be expressed when transferred into plants. By incorporating appropriate control sequences into the chimaeric genes investigators can ensure that the genes in the recipient plant will respond as desired to environmental stimuli such as light and temperature and that the products will be made in specific tissues. Use of transit sequences means that the products can even be directed to specific organelles in recipient plant cells.

The engineering of herbicide tolerance in plants

The *Agrobacterium* system is already proving its value in the production of plant strains with new characteristics that could be valuable to farmers. In a recent case in point, researchers have used the system to engineer plants with increased resistance to glyphosate, the active ingredient in the herbicide 'Roundup',which is produced by Monsanto (Fig. 11.3).

Herbicides are widely used in agriculture to control weeds. The application of glyphosate is limited, however. It is currently used mainly for controlling roadside weeds and for other non-

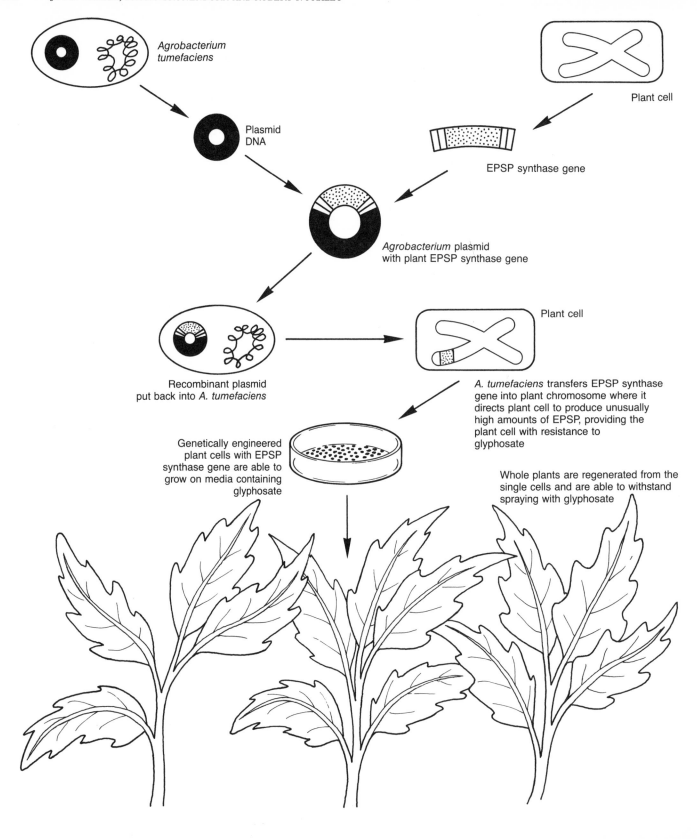

Agrobacterium tumefaciens

Plant cell

Plasmid DNA

EPSP synthase gene

Agrobacterium plasmid with plant EPSP synthase gene

Recombinant plasmid put back into *A. tumefaciens*

Plant cell

A. tumefaciens transfers EPSP synthase gene into plant chromosome where it directs plant cell to produce unusually high amounts of EPSP, providing the plant cell with resistance to glyphosate

Genetically engineered plant cells with EPSP synthase gene are able to grow on media containing glyphosate

Whole plants are regenerated from the single cells and are able to withstand spraying with glyphosate

Fig. 11.3 Gene transfer into plants by *A. tumefaciens*. The diagram shows how the bacterium can be used to produce plants that are resistant to glyphosate, the active ingredient in the herbicide 'Round-up', but other genes can be introduced by similar methods.

Glyphosate kills plants by inhibiting the gene for the enzyme EPSP synthase, which is needed for the synthesis by plants of essential amino acids. To make plants resistant to the herbicide the EPSP synthase gene is first spliced by recombinant DNA methodology into the DNA of the Ti plasmid from *A. tumefaciens*. The recombinant plasmid is put back into the bacterial cells, which infect the cells of susceptible plants such as petunia, tobacco or tomato. As a result of the infection the portion of the Ti plasmid carrying the EPSP synthase gene is introduced into the plant cells. Cells that acquire the new gene make large quantities of the enzyme and can thus grow in the presence of glyphosate concentrations that would kill ordinary plant cells. Whole glyphosate-resistant plants can then be regenerated from the cells. (By kind permission of The Monsanto Company, St Louis, Missouri.)

agricultural applications, because the agent kills crop plants just as effectively as it kills weeds. The usefulness of glyphosate as a weed control agent in agriculture could be enhanced if resistance to the herbicide could be selectively engineered into crop plants. Glyphosate is effective in very low concentrations, is not toxic to mammals, and is rapidly degraded by soil microorganisms.

Mechanisms of glyphosate resistance

Glyphosate exerts its effects in the pathway leading to the synthesis of the aromatic amino acids phenylalanine, tyrosine and tryptophan. If a cell is deprived of these amino acids it will not be able to synthesize proteins and will die. Plants, bacteria and fungi have the pathway and therefore are susceptible to glyphosate's effects; however, mammals do not make their own aromatic amino acids – they must acquire them in the diet – and are thus not susceptible to the herbicide's effects.

In 1980 Nicholas Amrhein and his colleagues at the Ruhr University in Bochum, West Germany, identified 5-enolpyruvyl-shikimate-3-phosphate (EPSP) synthase as the enzyme that is inhibited by glyphosate in the bacterium *Aerobacter aerogenes*. Subsequent studies showed that in higher plants the herbicide inhibits the same enzyme, which catalyses a key reaction in the synthetic pathway for the aromatic amino acids.

These results suggested that plant and other cells could be made tolerant to glyphosate by altering EPSP synthase so that the enzyme would bind the herbicides less efficiently and consequently be less subject to inhibition. Alternatively glyphosate tolerance might be achieved by increasing the production of EPSP synthase. Either way, more herbicide would be required to effect complete inhibition.

Both types of tolerance have been observed. For example, glyphosate-resistant mutants of the bacteria *Salmonella typhimurium* and *A. aerogenes* have been isolated and shown to contain forms of EPSP synthase that are less susceptible to inhibition by the herbicides.

In other cases, overproduction of the enzyme accounts for increased glyphosate tolerance in bacterial and plant cells. Recently, Dilip Shah, Robert Fraley and their colleagues at the Monsanto Company identified a glyphosate-tolerant line of *Petunia hybrida* cells that was established by growing the cells in increasing concentrations of the herbicide. The EPSP synthase made by this cell line is sensitive to the herbicide but is produced in 15 to 20 times the usual amounts.

Using a cloned DNA probe for the EPSP gene, the Monsanto workers showed that the gene had been amplified, with the result that the cells contain 20 times the normal number of copies. The corresponding messenger RNA is also increased by about 20-fold, thereby leading to overproduction of the enzyme protein. Amplification of the EPSP synthase gene may be the underlying molecular mechanism in a number of plant cell lines that have been selected for glyphosate tolerance.

Gene amplification is commonly used by the cells of higher organisms to satisfy the requirement for increased quantities of specific gene products. It may contribute to plant cell tolerance to herbicides other than glyphosate. Howard Goodman and his colleagues at Massachusetts General Hospital in Boston have recently identified an alfalfa cell line that is resistant to the herbicide L-phosphinothricin and have shown that it contains an amplified gene encoding the target enzyme of this herbicide.

Achieving gene transfer of herbicide tolerance

The observations with the glyphosate-tolerant petunia cells suggest that it should be possible to obtain plants that are tolerant to the herbicide by genetically altering them so that they overproduce the EPSP synthase. To accomplish this goal, Monsanto researchers Stephen Rogers and Harry Klee first constructed a chimaeric gene in which the protein-coding sequence of the gene for EPSP synthase was put under the control of a regulatory sequence from the cauliflower mosaic virus. The chimaeric EPSP synthase gene was then introduced into *A. tumafaciens* cells as part of a Ti plasmid from which the tumour-inducing genes had been deleted. The chimaeric gene was transferred by the bacteria into petunia cells in leaf discs, by a method developed by Horsch. Leaf discs that were transformed by acquisition of the chimaeric gene produced callus that could grow on a culture medium containing glyphosate whereas the callus from control leaf discs could not (Fig. 11.4). The tolerance of the transformed callus tissue resulted from overproduction of EPSP synthase.

Plants containing the chimaeric gene were regenerated from the transformed callus tissue and sprayed with glyphosate at a dose equivalent to 0.8 pounds per acre, which is approximately four times the dose required to kill the wild-type plants. The transformed plants survived the herbicide spraying and grew to maturity while the control plants died (Fig. 11.5).

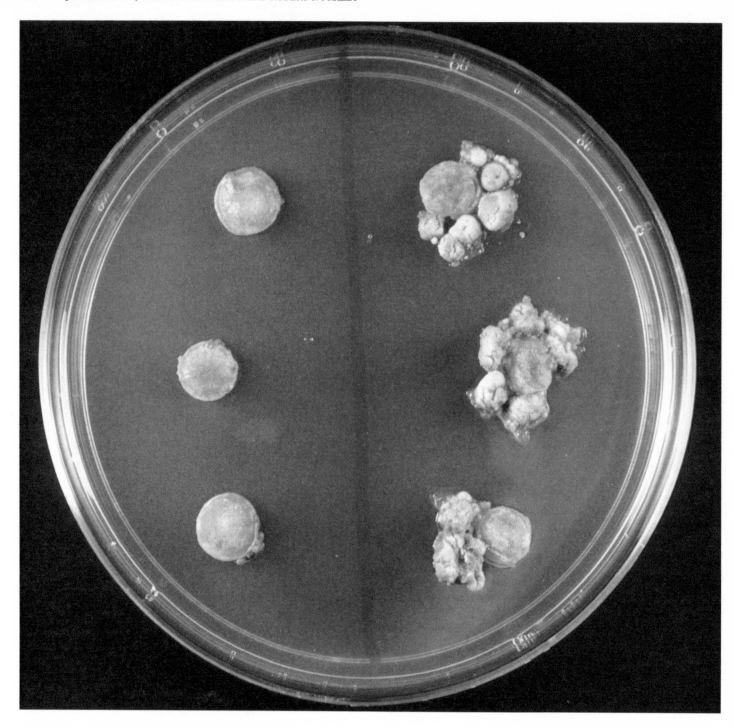

Fig. 11.4 Selection for glyphosate-tolerant callus. The leaf discs on the right were infected with *A. tumefaciens* that carried a vector with the chimaeric EPSP gene. The bacteria transferred the gene to the leaf disc cells, which can form callus in a glyphosate-containing medium. The leaf discs on the left were also infected with *A. tumefa-* *ciens*, but in this case the bacteria did not carry the EPSP gene. The plant cells remain sensitive to the herbicide and do not form callus in its presence. (By kind permission of The Monsanto Company, St Louis, Missouri.)

Fig. 11.5 Glyphosate-tolerant petunia plants. The plants on top, which acquired the chimaeric EPSP gene by transfer from *A. tumefaciens*, survived spraying with glyphosate and grew to maturity, whereas the control plants (below) stopped growing and died. (By kind permission of The Monsanto Company, St. Louis, Missouri.)

Biosynthesis of the aromatic amino acids takes place in the chloroplast. EPSP synthase is made in the cytoplasm of the cell and must therefore be transported to the chloroplast interior. Analysis of the structure of a DNA copy of the messenger RNA for EPSP synthase revealed that when the enzyme is first synthesized it carries a transit peptide that is 72 amino acids long.

Ganesh Kishore and Guy della-Cioppa of Monsanto have shown that newly synthesized EPSP synthase, which contains the transit peptide, is taken up by isolated chloroplasts, whereas the protein without the peptide is not. The investigators propose that the enzyme that is produced by the chimaeric gene in the petunia plants, which should contain the transit peptide, becomes localized in the chloroplasts of the plant cells.

Luca Comai, David Stalker and their colleagues at Calgene, Inc., in Davis, California, have also recently genetically engineered plants with increased glyphosate tolerance. However, these workers used a bacterial gene that specifies a mutant form of EPSP synthase that is less sensitive to the herbicide than the normal enzyme. The bacterial enzyme does not have the transit peptide necessary for transport into chloroplasts, but nonetheless increases the engineered plants' resistance to glyphosate. Presumably the substrate for the enzyme moves from the chloroplast into the cytoplasm where it can be acted upon by the bacterial EPSP synthase, after which the product moves back into the chloroplasts for conversion to the aromatic amino acids.

Regardless of the method used, researchers have now clearly demonstrated that gene transfer can improve herbicide tolerance in plants, at least for those herbicides that inhibit single enzyme reactions.

The production of virus-resistant plants by gene transfer

The production of new plant strains with improved resistances to infectious diseases is another major goal of plant breeders that may be attainable with the help of gene transfer methods. For example, a collaboration was established between researchers at Washington University and Monsanto to use the *Agrobacterium* gene transfer system to produce tobacco and tomato plants with increased resistance to tobacco mosaic virus, a pathogen that causes a disease in which the leaves of infected plants have a mottled (or mosaic) appearance.

The results bear out a prediction that research scientists have been making for the past few years; namely, that it should be possible to produce plants that are resistant to disease by transferring into them one or a few genes. They also hold out the possibility that the gene transfer methods might be successful in improving resistance to the bacterial and fungal diseases of plants.

Mechanisms of disease resistance

At present very little is known about the genetic mechanisms of disease resistance in plants. One possibility is that plants that are resistant to a pathogen carry one or a few specific genes that could be transferred to susceptible plants, thereby improving the recipients' resistance. No one has yet isolated such a gene, however, because the interactions between parasites and their hosts are very complex.

An observation made in 1929 by the late H.H. McKinney pointed to another possible way of increasing the resistance of plants to pathogens. McKinney noted that when a plant has been infected by one strain of a virus, it resists infection by a second strain. This discovery led plant pathologists to embark on a search for viruses that cause mild symptoms, or no symptoms at all, when they infect plants.

Plants that are infected by a mild virus strain are often protected against a more virulent strain, especially when the two viruses are related. This phenomenon, which is known as cross-protection, bears at least a superficial resemblance to the prevention of animal diseases by vaccination. Because cross-protection is most effective for similar viruses, it has been used to determine the degree of relatedness among viruses.

Cross-protection has been applied successfully in agriculture. Notable examples include the protection of glasshouse-grown tomatoes against tomato mosaic virus; the protection of citrus crops against citrus tristeza virus; and the protection of papaya plants against papaya ringspot virus.

Application of cross-protection to large-scale agriculture has its drawbacks, however. First, it requires the widespread distribution of a virus in the growing environment – a practice that many agriculturists find unsatisfactory. Second, a virulent strain may evolve from the mild viral strain and produce greater crop loss than protection. Third, infection by even a mild virus can cause small but significant decreases in crop yields. Finally, cross-protection depends on isolating and characterizing an apppropriate mild viral strain, which often requires considerable effort.

Plant pathologists have suggested a number of different hypotheses to explain the molecular mechanisms of cross-protection. One subset proposes that the first virus somehow interferes with the replication of RNA of the second virus in cross-protected cells. Almost all plant viruses have RNA genomes and such interference would effectively block their life-cycles. The interference might result because the RNAs of the two viruses are so closely related that RNA from the first can hybridize with the small amount of RNA from the second virus that would be present early in infection and prevent either its replication or translation into protein. Alternatively, the first virus might in some way 'use up' the machinery needed for RNA replication, thereby slowing or preventing the reproduction of the second virus.

Another set of hypotheses suggests that cross-protection blocks the establishment of an infection by the second virus. This might be done by preventing the virus from recognizing or binding to its receptor on the cell surface, as a result of which it would fail to enter the cell. Or, the disassembly of the virus and the release of its RNA genome inside the cell might somehow be inhibited.

The production of virus resistant plants

Even though cross-protection is by no means fully understood, Roger Beachy of Washington University, in collaboration with Fraley and Rogers, has shown that gene transfer can be used to induce in plants a state that closely resembles the natural cross-protection conferred by a viral infection. They chose to work with tobacco mosaic virus (TMV) for these experiments because the virus had already been extensively characterized. This enabled the Washington University workers, Patricia Powell-Abel and Barun De, to produce a DNA copy of the RNA genome of TMV and to identify and isolate the gene that encodes the viral coat protein.

They joined the coat protein gene to other DNA sequences that supply the signals needed for the initiation and termination of transcription in plant cells. The resulting chimaeric gene was inserted into a Ti plasmid from which the tumour-inducing genes had been deleted. Sheila McCormick of Monsanto then used *A. tumefaciens* to transfer the Ti plasmid with the chimaeric gene into tomato and tobacco cells.

Plants were regenerated from the cells that had acquired the new DNA. The plants thus produced were allowed to flower and the seeds they yielded were germinated. The seedlings that resulted matured normally and no aberrant characteristics were apparent.

When the very young tomato and tobacco seedlings that had been transformed by the chimaeric gene were inoculated with TMV by Richard Nelson of Washington University, the plants either did not develop an infection or else the disease symptoms developed more slowly than in the control plants (Fig. 11.6). These experiments have now been repeated and confirmed with more than 1000 seedlings that were produced by several different transformed parent plants.

All the experiments so far have been carried out under controlled conditions either in environmental chambers in the laboratory or in glasshouses. Plants grown under the more natural conditions of the glasshouses developed a greater degree of resistance to TMV than did those grown in the environmental chambers, a result suggesting that the resistance of plants in field conditions may turn out to be even better than that already observed.

Still to be addressed is the question of whether the resistance in transformed plants is the same as that found in plants that are cross-protected against virus infection. Although the answer is not known, the two types of resistance show similarities. For one, both cause delays in disease development. In the case of the transformed plants the delay is primarily the result of a 90 to 95 per cent reduction in the number of infection sites on the plant. Some plants escape infection completely. If infection is established, the spread of the virus is slowed in the resistant plants.

For another, resistance can be partially overcome in both the transformed and cross-protected plants by increasing the concentration of the virus in the challenge inoculum. In addition, the resistance can also be partially overcome in both types of plants if the challenge inoculum contains the virus RNA, rather than the complete viral particles.

Finally, resistance acquired by the transformed plants is effective against the common U_1 and a more severe TMV strain, as well as against several strains of tomato mosaic virus. Cross-protection is also most effective against viral strains that are closely related to the first infecting virus and usually diminishes as the degree of relatedness decreases. However, because the studies on the transformed plants have not yet been extended to the more distantly related TMV strains, it is too soon to tell whether the same will be true for the resistance induced by transfer of the TMV gene. Resistance in the plants thus transformed nevertheless appears to have many features in common with the resistance in plants that are cross-protected against virus diseases.

Application of genetically engineered disease resistance to agriculture

Resistance against TMV disease can be produced in tomato and tobacco plants by gene transfer. A similar genetic engineering approach – in which a gene coding for a viral coat protein is introduced into a plant – may be effective in protecting additional plant species against other viruses.

A broader hope is that an analogous approach will also prove effective against fungal and bacterial pathogens. Protection against such agents might be obtainable by transferring a fungal or bacterial gene for virulence into the target plants, although this, too, remains to be demonstrated.

Producing resistance by transferring a viral, fungal or bacterial gene into crop plants will presumably not affect the vigour or fitness of the recipients. This concern can only be evaluated, however, by extensive testing under glasshouse and ultimately field conditions.

Making new strains of insect resistant plants

The bacterium *Bacillus thuringiensis* produces crystalline spores that have a natural insecticidal effect. When the spores are eaten by the larvae of susceptible species, the spore crystals release high molecular weight proteins. When these proteins are partially broken down by enzymes in the insect digestive tract, they release smaller proteins that are toxic to the larvae. The insect larvae that are killed by ingesting *B. thuringiensis* spores thus include those of flies and mosquitos. Many of the caterpillars that damage or destroy crop and foliage plants are also susceptible.

Commercial preparations of *B. thuringiensis* have been available for many years for use in controlling insects such as mosquitos and the gypsy moth. The voracious, leaf-eating larvae of the latter species have had devastating effects on foliage in the northeastern United States and, aided by cars and recreational vehicles, the pest is rapidly spreading through the country.

Using *B. thuringiensis* for insect control has the advantage that the lethal effects of its toxins are limited to susceptible insects. It is therefore safer than chemical insecticides that may poison a broad spectrum of organisms, including birds and other wildlife, and beneficial, as well as harmful, insects. However, insecticides based on *B. thuringiensis* have the disadvantage of being both expensive and readily broken down under field conditions.

Consequently, researchers have begun to explore new ways of using *B. thuringiensis* toxins for insect control. One approach that is now being tried involves introducing the cloned toxin genes into other bacteria that may survive longer in the environment than *B. thuringiensis* itself (see also Chapter 8). Another approach is to construct plants with their own built-in insecticidal toxin. Recently, M. Vaeck, J. Leemans and their colleagues at Plant Genetic Systems N.V. in Ghent, Belgium, have cloned a *B. thuringiensis* crystal gene encoding a protein that has a molecular weight of 130 000. The purified protein is just as toxic to the larvae of the tobacco hornworm (*Manduca sexta*) and the large white butterfly (*Pieris Brassicae*) as are the spore crystals themselves.

When the full-sized protein is exposed to protein-digesting enzymes such as trypsin or chymotrypsin, it releases a fragment with a molecular weight of 60 000 that is relatively resistant to further digestion and is apparently the active toxin. The smaller protein, which consists of almost all of the amino terminal half

Fig. 11.6 Virus-resistant tobacco plants. In (A) the plant on the left carries the chiameric gene for the TMV coat protein; ten days after infection with the virus, its leaves are healthy. In contrast the leaves of the plant on the right, which does not contain the TMV gene, is beginning to show the mottling characteristic of TMV disease. In (B) the plants are shown 14 days after infection. The one on the right has been transformed with the TMV coat protein gene; the one on the left has not. (By kind permission of The Monsanto Company, St Louis, Missouri.)

of the larger molecule, exhibits the complete toxic activity of the original protein.

The Ghent workers have recently used the Ti plasmid to introduce the cloned toxin gene into tobacco plants. The bacterial gene product was made in the tissues of the plants that acquired the foreign DNA. The plants appeared to be normal in all other respects.

The protein produced in the tobacco leaves retains its toxicity for insects. All of the tobacco hornworm larvae that were fed with the toxin-producing leaves died within a week, whereas less than 5 per cent of the larvae fed with leaves from control plants died during the same time period.

Moreover, production of the toxic protein protected whole plants against damage by the insect. In glasshouse tests, 15 freshly hatched tobacco hornworm larvae were placed on each 40-centimetre high tobacco plant. Control plants showed considerable damage within a week and were completely consumed after about 2 weeks. In contrast, all the larvae placed on the genetically engineered plants died within 4 days and damage to the plants was minimal (Fig. 11.7). Further work will be required to determine whether plants that have been genetically engineered to produce the bacterial toxin are also protected from damage under field conditions.

A large number of *B. thuringiensis* strains have been isolated that produce toxins with insecticidal activity against differing spectra of insect species. One day, it may be possible to engineer genetically modified varieties of crop plants that have built-in insecticidal toxins specific for the insect pests that feed on the plants in question.

Conclusions

The *A. tumefaciens* system for introducing new genes into plant cells is proving to be a reliable and efficient means of accomplishing the genetic engineering of a number of crop plants. It has already been used to produce new strains of herbicide and disease-resistant plants. Perhaps the major limitation on the application of the *A. tumefaciens* vector system is the relatively small number of genes that it can carry. Traits that are encoded by many genes will probably be less amenable to transfer by *A. tumefaciens* than those encoded by one or a few genes.

The use of the bacterium for gene transfer is also restricted. Other methods are also being developed for increasing the range of crop plants amenable to gene transfer techniques. The methods include direct DNA injection and the use of plant viruses as vectors. The recent demonstration that whole rice plants can be regenerated from protoplasts is another encouraging development with regard to the genetic engineering of monocots.

Nevertheless, the genetic engineering of new traits into plants cannot be achieved until the genes encoding those traits have been identified and isolated. Studies of plant molecular biology have historically lagged behind those of the molecular biology of bacteria and animals, but that situation is beginning to change. The work on gene tranfer into plants is providing both an impetus to the research on plant molecular biology and new tools with which to approach it.

Fig. 11.7 Genetically engineered insect-resistant plant. The Ti plasmid was used to introduce a toxin gene from the bacterium *B. thuringiensis* into the tobacco plant on the right. The leaves of the plant produce the toxin protein, which kills the larvae of certain insect species when ingested. Fifteen larvae of the tobacco hornworm were placed on each plant. The larvae placed on the genetically engineered plant died as a result of trying to eat the leaves and the plant suffered little damage. In contrast, the unmodified control plant on the left was complely denuded by the leaf-munching larvae. (By kind permission of Plant Genetics Systems N.V., Ghent, Belgium.)

Additional reading

De la Pena, A., Lörz, H. and Schell, J. (1987). Transgenic rye plants obtained by injecting DNA into young floral tillers. *Nature, London*, **325**, 274–6.

Eckes, P., Schell, J. and Willmitzer, L. (1985). Organ-specific expression of three leaf/stem specific cDNAs from potato is regulated by light and correlated with chloroplast development. *Molecular and General Genetics*, **199**, 216–44.

Freeling, M. (ed.) (1985). *Plant Genetics*. UCLA Symposia on Molecular and Cellular Biology, vol. 35. Alan R. Liss, New York.

Horsch, R. B., Fry, J. E., Hoffmann, N. L., Eichholtz, D., Rogers, S. G. and Fraley, R. T. (1985). A simple and general method for transferring genes into plants. *Science*, **227**, 1229–31.

Kaulen, H., Schell, J. and Kreuzaler, F. (1986). Light induced expression of the chimeric chalcone synthase-NPTII gene in tobacco cells. *EMBO Journal*, **5**, 1–8.

Morelli, G., Nagy, F., Fraley, R. T., Rogers, S. G. and Chua, N. H. (1985). A short conserved sequence is involved in the light-inducibility of a gene encoding ribulose 1,5-bisphosphate carboxylase small subunit of pea. *Nature*, **315**, 200–4.

Nester, E. W., Gordon, M. P., Amasino, R. M. and Yanofsky, M. F. (1984). Crown gall: a molecular and physiological analysis. *Annual Review of Plant Physiology*, **35**, 387–413.

Potrykus, I., Paszkowski, J., Saul, M., Krüger-Lebus, S., Müller, T., Schocher, R., Negrutiu, I., Künzler, P. and Shillito, R. D. (1985). Direct gene transfer to protoplasts: an efficient and generally applicable method for stable alterations of plant genomes. In *Plant Genetics*, UCLA Symposia on Molecular and Cellular Biology, vol. 35, ed. M. Freeling, pp. 181–200. Alan R. Liss, New York.

Powell-Abel, P., Nelson, R. S., Barun, D. E., Hoffmann, N., Rogers, S. G., Fraley, R. T. and Beachy, R. N. (1986). Delay of disease development in transgenic plants that express the tobacco mosaic virus coat protein gene. *Science*, **232**, 738.

Rosahl, S., Eckes, P., Schell, J. and Willmitzer, L. (1986). Organ-specific gene expression in potato: isolation and characterization of tuber-specific cDNA sequences. *Molecular and General Genetics*, **202**, 368–73.

Rosahl, S., Schell, J. and Willmitzer, L. (1987). Expression of a tuber specific storage protein in transgenic tobacco plants: demonstration of an esterase activity. *EMBO Journal* (accepted).

Rosahl, S., Schmidt, R., Schell, J. and Willmitzer, L. (1987). Isolation and characterization of a gene from *Solanum tuberosum* encoding patatin, the major storage protein of potato tubers. *Molecular and General Genetics*, **203**, 214–20.

Schreier, P. H., Seftor, E. A., Schell, J. and Bohnert, H. J. (1985). The use of nuclear encoded sequences to direct the light-regulated synthesis and transport of a foreign protein into plant chloroplasts, *EMBO Journal*, **4**, 25–32.

Shah, D. M., Horsch, R. B., Klee, H. J., Kishore, G. M., Winter, J. A., Tumer, N. E., Hironaka, C. M., Sanders, P. R., Gasser, C. S., Aykent, S., Siegel, N. R., Rogers, S. G. and Fraley, R. T. (1986). Engineering herbicide tolerance in transgenic plants. *Science*, **233**, 478–81.

Simpson, J., Schell, J., Van Montagu, M. and Herrera-Estrella, L. (1986). Light-inducible and tissue-specific pea *lhpc* gene involves an upstream element combining enhancer- and silencer-like properties. *Nature*, **323**, 551–4.

Spena, A. and Schell, J. (1987). The expression of a heat-inducible chimeric gene in transgenic tobacco plants. *Molecular and General Genetics*, **206**, 436–440.

Van den Broeck, G., Timko, M. P., Kausch, A. P., Cashmore, A. R., Van Montagu, M. and Herrera-Estrella, L. (1985). Targeting of a foreign protein to chloroplasts by fusion to the transit peptide from the small subunit of ribulose 1,5-bisphosphate carboxylase. *Nature*, (*London*), **313**, 358–63.

Zambryski, P., Herrera-Estrella, L., De Block, M., Van Montagu, M. and Schell, J. (1984). The use of the Ti plasmid of *Agrobacterium* to study the transfer and expression of foreign DNA in plant cells: new vectors and methods. In *Genetic Engineering: Principles and Methods*, ed. A. Hollaender and J. Setlow, pp. 253–78. Plenum Press, New York.

12 Monoclonal antibodies and their applications

Jean L. Marx

Pure research, although it has as its primary purpose the solution of some fundamental scientific problem, often pays off with significant practical benefits. Few discoveries illustrate this premise better than that of monoclonal antibody technology, which has been moving rapidly from the research laboratory into clinical and commercial application since it became available in the mid-1970s.

Monoclonal antibodies are far more specific and reproducible than antibodies made by conventional techniques. These characteristics make them ideal for diagnosing infectious diseases, especially when distinguishing between closely related pathogens is important. The antibodies may also facilitate cancer diagnosis. They are already being used for monitoring the progress of patients undergoing therapy for certain cancers and there are indications that the antibodies detect recurrences earlier than other diagnostic methods.

The US Food and Drug Administration has so far approved more than 150 monoclonal antibodies for diagnostic use. Many of these are for the diagnosis of infectious diseases and some are for monitoring the course of cancer therapy. In addition, the antibodies are used for determining the blood concentrations of therapeutic drugs and of certain hormones. The newer pregnancy tests, including those intended for use in the home, also depend on monoclonal antibodies.

Monoclonal antibodies are beginning to find application in disease therapy as well as in diagnosis. Attempts to treat cancer with them are already under way. The hope is that monoclonal antibodies, with thir great specificity, will behave like the long-sought 'magic bullets' that can home in on and kill cancer cells without significantly damaging normal cells. The results of early clinical trials have generally pointed to the conclusion that this will be a difficult goal to achieve, but the results have nonetheless been sufficiently promising to make the research worth pursuing further.

One clinical application of monoclonal antibodies that is already feasible is immune suppression for kidney transplantation. During the summer of 1986 the US Food and Drug Administration approved the first monoclonal antibody to be given clinically to human patients. This antibody, which is made by Ortho Pharmaceutical Company of Raritan, New Jersey, is directed against the immune cells that mediate graft rejection and is to be administered to kidney transplant patients who are experiencing a rejection episode.

The discovery of monoclonal antibodies

The monoclonal antibody story began in 1974, when Georges Köhler and Cesar Milstein of the Medical Research Council's (MRC) Laboratory of Molecular Biology in Cambridge, England, were studying what was then one of the outstanding unsolved problems of immunology. Antibodies are part of the body's defences against foreign invaders, including disease-causing organisms. When the B lymphocytes of the immune system are stimulated by the presence of an invading pathogen or other foreign material, they produce antibodies, proteins that specifically recognize and bind to the substance that originally elicited their production. This binding of antibody and antigen (antigen being the name given to any substance that triggers antibody production) leads to the destruction of the antigen.

The problem is that the immune system can produce antibodies to an enormous range of antigens. The complete repertoire of antibodies includes perhaps 10 million different varieties, and animal genomes are not sufficiently large to contain individual genes for all possible antibody proteins. This seemed to fly in the face of one of the principal tenets of molecular biology, namely that there is a discrete gene for every protein chain.

At least part of the antibody diversity could be generated, however, if antibody genes were particularly prone to mutation. If the genes were to mutate freely as the B lymphocytes multiply, the result would be a diversification of the genes in the expanded population.

Köhler and Milstein were trying to determine what role, if any, gene mutations play in the generation of antibody diversity. To start with, they wanted a line of cells that could be grown in culture and would produce a single kind of antibody molecule in response to a known antigen. Nothing like that was available at the time. Normal antibody-producing cells could not be maintained in culture for more than a few days. Ordinarily, antibodies were produced by injecting an antigen into an animal and harvesting the resulting antibodies from the animal's blood. This elicits a variable 'polyclonal' mixture of different antibody molecules, even though they all recognize the triggering antigen.

Köhler and Milstein hit on the idea of fusing normal antibody-producing cells with cells from the cancerous tumours known as myelomas. Several years before, Michael Potter of the National Cancer Institute in Bethesda, Maryland, had induced a series of these tumours, which are derived from antibody-producing cells, in mice. Cells from a given myeloma are clonal, that is, they are all descended from the same original parent cell and they all secrete the same type of antibody molecule. For the most part, however, the antigens recognized by the myeloma antibodies were unknown, and trying to identify the correct one from an essentially unlimited number of possibilities would be difficult, to say the least.

Myeloma cells can be induced to grow in culture, however, and Köhler and Milstein thought that if they fused myeloma cells with other antibody-producing cells of known specificity they might be able to obtain a permanent line of cells that made the antibody they wanted. To obtain the antibody-producing cells of known specificity, Köhler injected mice with sheep red blood cells, which are very highly antigenic, harvested the B lymphocytes from the animals' spleens, and then fused the B cells with myeloma cells. When the resulting hybrid cells were cloned, some of the clones proved to produce antibodies that specifically recognize sheep red blood cells (Fig. 12.1).

All hybrids of the same clone produce the same antibody molecules – hence the name monoclonal antibody. The hybrid cells themselves have come to be known as 'hybridomas'. Although the original hybridomas of Köhler and Milstein produced two antibodies, one from the myeloma cell and the other from the spleen cell with which it fused, researchers now use myeloma cells that do not make any antibodies of their own so that the only product of the hybridoma clone is the spleen-cell antibody.

A hybridoma clone can be maintained indefinitely in culture, and the work of Köhler and Milstein has thus made possible the production of virtually unlimited quantities of pure antibodies of known specificity. Although the two investigators may have begun the research to address a purely scientific question, they were certainly aware of the implications of their findings. When Köhler and Milstein reported their results in the *Nature* issue of 7 August 1975, they concluded the paper by saying of the hybrids: 'Such cells can be grown *in vitro* in massive cultures to provide specific antibody. Such cultures could be valuable for medical and industrial use' – a prediction

Fig. 12.1 Production of monoclonal antibodies. A mouse is injected with any desired antigen to elicit antibody production. Antibody-producing cells are subsequently harvested from the animal's spleen and fused with mouse myeloma cells to form hybrid cells, which are grown in culture. Individual hybrids are then grown in culture to form clones, which are screened for the production of antibody that reacts with original antigen. Cells from clones that produce the monoclonal antibody may then be injected into the peritoneal cavity of another mouse where they form a tumour, the hybridoma. The tumor cells secrete the antibody into the fluid that accumulates in the mouse abdominal cavity. The monoclonal antibody is harvested from this fluid. Some cells of the monoclonal-antibody-producing clone can also be frozen for future use.

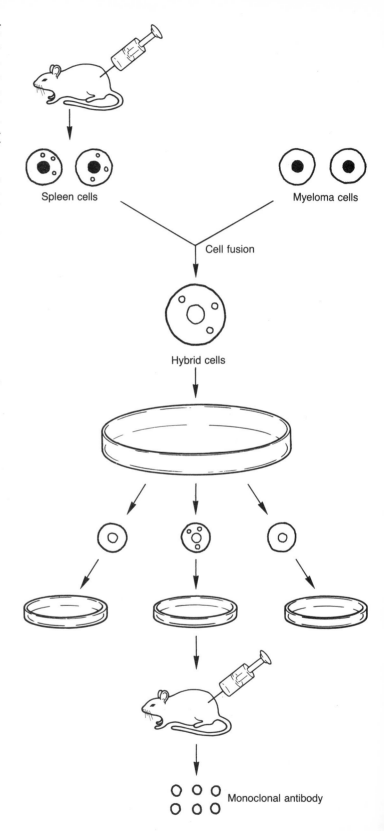

Spleen cells

Myeloma cells

Cell fusion

Hybrid cells

Monoclonal antibody

that is certainly being borne out. For their achievement Köhler and Milstein received the 1984 Nobel Prize for Physiology or Medicine, which they shared with Niels Jerne of the Basel (Switzerland) Institute for Immunology, who has made major theoretical contributions to immunological research.

Applications of monoclonal antibodies

Monoclonal antibodies have proved to be extremely valuable for basic immunological and molecular research. Their many uses include the isolation of proteins and they have also proved helpful in gene cloning. One of the most difficult problems of gene cloning is identifying the cells that contain the desired gene. If a monoclonal antibody that recognizes the gene product is available, it can be used as a probe for detecting those cells that make the product and therefore contain the gene.

In addition, monoclonal antibodies have greatly contributed to the identification of the many different types of cells that participate in immune responses and to the unravelling of their interactions. The immune system contains, in addition to the B cells, another major type of lymphocyte, known as the T cell. Whereas the B cells have as their major function the production of antibodies, T cells have a number of different functions. Some are killer cells that can directly attack cells that they detect as foreign. These include virus-infected cells and possibly cancer cells. T lymphocytes are also important regulators of immune responses, including antibody production by B cells. The regulatory T cells include helpers, which as their name suggests, activate immune responses, and suppressors, which have inhibitory effects.

The use of monoclonal antibodies established that the various types of T cells carry antigens or markers on their surfaces that allow one type to be distinguished from another. The antibodies have also helped to define the changes undergone by the T and B cells during development.

Suppressing immune responses

One of the outgrowths of the research on the T-cell markers is the development of the monoclonal antibody that is now being used to prevent the rejection of transplanted kidneys. This antibody, which is marketed under the name Orthoclone OKT-3, specifically reacts with the 'OKT-3' antigen that is found on all T cells.

To prevent rejection of the foreign tissue, kidney transplant patients are routinely given drugs to suppress the activities of their immune systems. Nevertheless, approximately 60 per cent of the 7000 patients who have kidney transplants every year in the United States experience rejection episodes that threaten the loss of the grafted organ.

Graft rejection is mediated primarily by the T cells. Orthoclone OKT-3 in turn mediates an attack on the cells that are causing the rejection. In a national clinical trial, treatment with the antibody saved the kidneys of 90 per cent of the patients who were having a rejection episode. By comparison, administration of immunosuppressive drugs had only a 75 per cent success rate.

The monoclonal antibody is not only more effective than the drugs in preventing kidney rejection, but it may also be less prone to causing side effects. The antibody suppresses only the T cells, whereas the drugs, which include the antibiotic cyclosporin and certain steriods, suppress all aspects of immune function, thereby leaving the patients more open to infectious diseases. In addition, the drugs may permanently damage the kidneys. The monoclonal antibody is not without side effects of its own, however. It can cause fever, chills, nausea and vomiting, shortness of breath and chest pains, although these effects are reversible.

Several human diseases are caused by an apparent attack of the immune system on the tissues of the body. Such autoimmune conditions include myasthaenia gravis, which is characterized by progressive muscular weakness resulting from the destruction of certain types of nerve cells; systemic lupus erythematosus, in which vital organs including the lungs, kidney or brain can be damaged by the immune attack; and possibly multiple sclerosis, with its progressive degeneration of the central nervous system.

Investigators are attempting to determine whether monoclonal antibodies that are directed against immune cell components that may be involved in triggering the abnormal immune responses can be used to treat autoimmune conditions. For example, researchers at Stanford University School of Medicine in Palo Alto, California, have found that such antibodies can suppress the development in mice of conditions similar to the myasthaenia gravis, multiple sclerosis and systemic lupus erythematosus of humans. This work is still in its early phases, but at least provides an encouraging lead to what may eventually be a new form of treatment for autoimmunity.

Diagnosis of infectious diseases

The diagnostic applications of monoclonal antibodies are by far the most advanced, especially for tests that are performed on body fluids, such as blood and urine samples. Nearly one-third of the 150 diagnostic monoclonal antibodies that have been approved by the US Food and Drug Administration are for pregnancy detection. The urine of pregnant women contains a hormone called human chorionic gonadotrophin, which is secreted by the placenta. The use of monoclonal antibodies to the hormone now permits the diagnosis of pregnancy as early as a week or two after conception.

The diagnosis of venereal diseases is also being improved by the availability of monoclonal antibodies. Gonorrhoea, which is caused by the bacterium *Neisseria gonorrhoeae*, and chlamydia infections, which are caused by the intracellular parasite *Chlamydia trachomatis*, are among the important venereal diseases. There are some 2 million cases of gonorrhoea every year in the United States and perhaps 5 to 10 million cases of chlamydia infections. Moreover, both infections can result in pelvic inflammatory disease and are thus major causes of infertility in women. The two conditions require different treatments, but are hard to distinguish on the basis of their symptoms, which are very similar.

The most accurate method for diagnosing gonorrhoea in women previously required that the causative bacterium be cultured, a procedure that takes 2 to 3 days. Specific diagnosis of chlamydia infections was even more cumbersome, because *C. trachomatis* can only be grown in living cells and completion of the test procedure required from 3 to as many as 7 days. The newer tests, which use monoclonal antibodies, have greatly reduced the time needed to diagnose the two venereal diseases. For example, Robert Nowinski and his colleagues at Genetic Systems Corporation and the University of Washington School of Medicine in Seattle, Washington, have produced monoclonal antibodies that can identify either *N. gonorrhoeae* or *C. trachomatis* in as little as 15 to 20 minutes, thereby allowing the correct therapy to be instituted much more rapidly than in the past.

The Seattle workers have also develped monoclonal antibodies that distinguish between the closely related herpes virus 1 and herpes virus 2 (HSV 1 and 2). Both these viruses can cause painful, recurring inflammatory lesions. HSV 1 usually affects the mouth area, producing the common 'cold sores' and HSV 2 primarily infects the genital regions, but this difference in the preferred location of infection is not absolute. Some 15 to 20 per cent of the genital infections are caused by HSV 1.

Identifying the specific virus is important, because the genital lesions caused by HSV 1 are less likely to recur than those caused by HSV 2 and because individual anti-viral drugs may be more effective against one virus than the other. The viruses can be detected by their ability to kill cultured cells, but this test takes 3 to 6 days. Additional tests are then required to classify the virus detected as either HSV 1 or 2. The monoclonal antibodies can diagnose herpes virus infections and determine the viral type directly from clinical samples that are obtained by scraping cells from lesions. This again reduced the time for diagnosis to 15 to 20 minutes.

Cancer diagnosis

The availability of monoclonal antibodies that recognize immune cell antigens has resulted in improved diagnosis of the leukaemias and lymphomas, cancers of the T and B cells that are sometimes difficult to distinguish. Nevertheless, the prognoses of the different immune cell cancers may be very different. Some are much more amenable to therapy than others and a patient's life can depend on early, accurate diagnosis and the rapid initiation of the therapy appropriate for the particular immune cell cancer. The monoclonal antibodies that recognize the various markers of human immune cells, many of which were developed by Stuart Schlossman and his colleagues at Harvard's Dana-Farber Cancer Institute in Boston, Massachusetts, have provided a means of identifying the specific cell type that gave rise to a cancerous immune cell and therefore of pinpointing the particular type of leukaemia or lymphoma.

The exquisite specificities of monoclonal antibodies are also being applied to the diagnosis of solid tumours, particularly the carcinomas. The carcinomas include the cancers of the lung, breast, colon and rectum, which are the most common cancers in the developed countries. Surgery to remove the primary tumours can sometimes cure these cancers, but if they spread to additional tissues very little can currently be done to eliminate the metastatic disease. The hope is that the great specificity of monoclonal antibodies may allow the earlier detection both of the primary tumours and of cancer recurrences than is now possible, and perhaps lead to a better cure rate.

In the more advanced tests for the solid tumours the antibodies are being used to examine blood, sputum or biopsy samples for cancer cells or for materials that have been released by cancer cells. For several years now, clinicians have monitored the condition of patients who are undergoing therapy for colorectal and other cancers by determining the concentrations of carcinoembryonic antigen (CEA) in their blood. These cancers release CEA into the blood, although its presence there is not an absolutely specific indicator of the presence of a malignancy. It is also produced by normal embryonic tissue, for example.

Nevertheless, CEA has proved valuable as a prognostic indicator for cancer. Its concentration in the blood serum is high in untreated patients and declines after successful surgery. Recurrence of the tumours is usually heralded by increased amounts of CEA in the blood. The concentrations of the antigen were originally determined by ordinary polyclonal antibody preparations but these are now being replaced by the monoclonal variety, which are more sensitive detectors.

Much work over the past several years has been directed at identifying monoclonal antibodies that recognize antigens specific to particular types of tumours – to breast or lung carcinomas, for example. Such antibodies could be used diagnostically or therapeutically. The investigators doing the work have generally come to agree, however that tumour cells are unlikely to produce marker antigens that are absolutely specific for a particular type of cancer. According to Zenon Steplewski of the Wistar Institute in Philadelphia, Pennsylvania, an apparent tumour antigen can always be found on other, normal cells if the researcher looks hard enough.

Although there may be no such thing as 'tumour-specific' antigens, investigators have identified monoclonal antibodies that recognize 'tumour-associated' antigens that occur on particular types of tumour cells but are only rarely found on normal cells. Such antibodies are sufficiently specific for tumour cells to be used for cancer diagnosis.

For example, Steplewski, who works with Hilary Koprowski, also of the Wistar Institute, has produced a monoclonal antibody that reacts with an antigen made by colorectal cancers and by a number of additonal carcinomas of the gastrointestinal system, including cancers of the pancreas, liver and stomach.

In a large study of hundreds of patients who had undergone treatment for colorectal cancer, the Wistar group found that the monoclonal antibody could predict cancer recurrence in certain patients. The serum concentration of the antigen detected by the antibody became elevated some 3 to 18 months before the recurring tumours became clinically apparent.

Robert Bast and his colleagues at Duke University Medical Center in Durham, North Carolina, have identified another monoclonal antibody that shows promise in cancer diagnosis, in this case for ovarian cancers. The Duke workers found that the blood concentration of the antigen that is detected by the antibody is elevated in more than 80 per cent of patients with ovarian cancer, but in only 1 per cent of normal controls. The results also indicate that the monoclonal antibody can be used to monitor the response of the patients to treatment.

More than 120 000 cases of lung cancer occur annually in the United States and 90 per cent of the individuals who get lung cancer die of the disease, which all too frequently spreads before it is detected. John Minna and his colleagues at the National Cancer Institute (NCI) in Bethesda, Maryland, have been working to produce monoclonal antibodies that could aid in the diagnosis of lung cancers.

There are four types of lung cancers. One of the four – small cell carcinoma of the lung (SCLC) – requires a different therapeutic approach from the other three. Because SCLC metastasizes very early in the course of the disease, it is not amenable to surgery but often responds, at least for a while, to chemotherapy. The other three cancers do not respond to chemotherapy, but can be cured by surgery if they are caught before they spread.

Minna and his NCI colleagues, James Mulshine and Frank Cuttitta, have produced two monoclonal antibodies that together identify more than 90 per cent of non-SCLCs but only bind to about 20 per cent of SCLCs. Jeffrey Schlom of NCI, William Johnston of Duke University Medical Center, and their colleagues have also develped a monoclonal antibody that reacts with non-SCLC cells but not with SCLC cells. These antibodies may be useful in distinguishing SCLC from the other forms of lung cancer.

Pleural effusions, which are abnormal fluids located in and around the lungs, and ascites fluids, which are located in the abdominal cavity, can be caused either by infections or by malignancies, especially by those originating in the breast, lung or female genital tract. Recognizing cancer cells in pleural effusions or ascites fluid can be difficult, but monoclonal antibodies may help in this regard, too.

Schlom, Johnston and their colleagues have identified a monoclonal antibody that reacts with cells of several types of cancers, including those of the breast, lung and ovaries. Early studies with the antibody indicate that it reliably detects cancer cells in both pleural effusions and ascites fluid. It can also aid in the identification of cancer cells in biopsy samples that are obtained by withdrawing material from suspected malignant tumours, breast lumps for example, by means of a needle and syringe (Fig. 12.2).

Methods for using monoclonal antibodies to detect cancerous tumours directly in patients are less highly develped than the procedures for analysing blood and other samples that have been withdrawn from the body. Monoclonal antibodies for imaging *in vivo* are tagged with a radioactive material and injected into the bloodstream or elsewhere in the body. The idea is that the specificity of the antibody will enable it to seek out and bind to any cancer cells that bear the appropriate antigen, thereby concentrating the radioactivity at the tumour sites. Clinical methods for determining the localization of radioactive tracers in the body are widely available, although they were not originally designed for use with radiolabelled monoclonal antibodies.

The early results of the *in vivo* imaging experiments have been promising. Several groups have shown that radiolabelled monoclonal anatibodies concentrate in tumours, although more work will be required before such imaging methods can be routinely used in the clinic (Figs. 12.3 and 12.4). In most of this work the monoclonal antibodies have been tagged with iodine-131, mainly because this radioactive isotope is inexpensive and researchers have had a great deal of experience in attaching it to proteins. Moreover, conjugates of the isotope with monoclonal antibodies may be useful for therapy because it emits high-energy radiation.

Nevertheless, iodine-131 is less than ideal for diagnostic imaging. Some antibodies tend to lose it, and the iodine-131 radiation is not efficiently detected by current imaging instruments. Consequently, investigators are beginning to develop methods for labelling monoclonal antibodies with radioactive isotopes that are more suitable for imaging.

Cancer therapy

Ultimately monoclonal antibodies may be used not just to detect cancer cells but to destroy them. The hope is that the antibodies, because of their specificity, will prove to have fewer harmful side effects than more conventional treatments with drugs and radiation, which often damage normal cells as well as the cancerous ones. The specificity of monoclonal antibodies is not likely to be absolute, however, because the antigens they recognize are not completely restricted to tumour cells.

Researchers at a number of medical centres have already begun clinical trials of monoclonal antibody therapy. As is generally the case in the early trials of an experimental therapy, all the patients in these studies have had advanced cancers that have not responded to conventional treatments. Despite this, two promising results have emerged: monoclonal antibodies have induced at least partial remissions in some of the patients; and the individuals have tolerated the antibodies fairly well.

According to Schlom, several hundred individuals have been treated so far at the various medical centres and the side effects of monoclonal antibody administration, which include nausea, vomiting, diarrhoea, fever and chills, have been relatively mild. None of the patients, all of whom were very ill, have died as

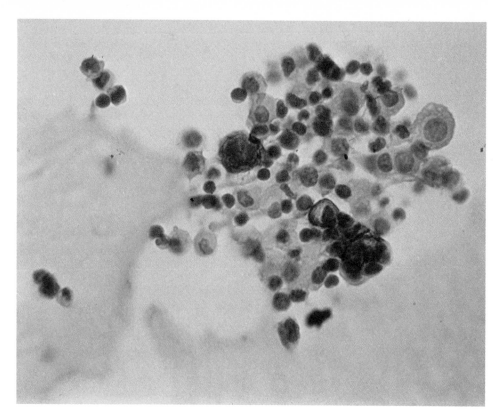

Fig. 12.2 Pleural effusion cells strained with a monoclonal antibody that recognizes cancer cells. The cells were obtained from a patient suspected of having breast cancer. The reddish-brown staining of some cells shows that they have reacted with the monoclonal antibody and are thus carcinoma cells. (By kind permission of Jeffrey Schlom, National Cancer Institute.)

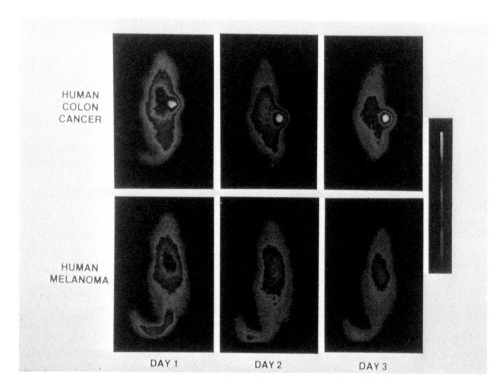

Fig. 12.3 Use of a monoclonal antibody to locate human cancer cells that have been transplanted into mice. The antibody, which is labelled with radioactive iodine, recognizes an antigen that is present on colon cells, but not on melanoma cells. Concentration of the antibody in the colon tumour (white area) can be detected by the first day after injection. It is still there 3 days after injection when less radioactivity is seen in the blood. The antibody does not concentrate in the melanoma. (By kind permission of Jeffrey Schlom, National Cancer Institute.)

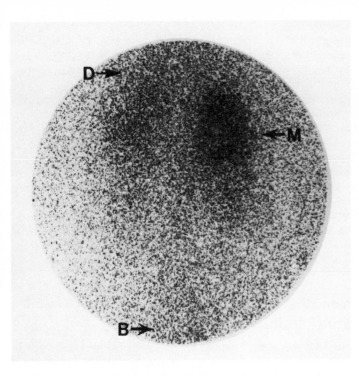

Fig. 12.4 Detection of metastatic cancer in the liver of a human patient. The view on this gamma scan encompasses the entire abdomen from the diaphgram (D) to the bladder (B). The radiolabelled monoclonal antibody has concentrated in the liver (M), thereby indicating the presence of metastatic cancer in that organ.

a result of the experimental therapy. A long experience with clinical trials has shown that potent new therapeutic agents may cause occasional deaths when they are tested in patients weakened by advanced disease, but this has not happened so far with the monoclonal antibodies.

A number of strategies for the therapeutic use of monoclonal antibodies are being explored in the studies. In some the naked antibodies are used. Destruction of the tumour cells then depends on the ability of the antibody to bind to the cells and trigger the normal killing mechanisms of the immune system. In one such trial Henry Sears of the Fox Chase Cancer Centre in Philadelphia has administered the anti-colorectal monoclonal antibody that was made by the Koprowski–Steplewski group to patients with pancreatic or colorectal cancer. Some of the patients experienced regressions of their tumours.

Another approach to using monoclonal antibodies for cancer therapy is to couple the antibodies to cell-killing drugs, toxins or radioactive isotopes. The antibody then serves to concentrate the killing agent in the tumour, while normal cells receive much less exposure. For example, Ralph Reisfeld and his colleagues at the Scripps Clinic and Research Foundation in La Jolla, California, have begun a clinical trial in lung cancer patients

of a monoclonal antibody that is attached to the chemotherapeutic drug methotrexate. The tumours of two of the 11 patients treated thus far shrank following the treatment. Although it is too early to draw any firm conclusions about the therapeutic efficacy of the monoclonal antibody–drug conjugate, its administration is allowing larger than usual doses of methotrexate, which is highly toxic chemotherapeutic drug, to be given to the patients.

Ricin, a protein that is produced by the castor bean plant, is an extremely potent toxin. A single molecule can kill a cell. Several groups, including that of Ellen Vitetta and Jonathan Uhr at the University of Texas Southwestern Medical School in Dallas, and that of Richard Youle at the National Institute of Neurological and Communicative Disorders and Stroke in Bethesda, Maryland, have found that conjugates of recin with monoclonal antibodies can kill leukaemia cells in culture and in experimental animals.

Drugs and toxins must enter cells if they are to exert their lethal effects. Radiation, however, can kill from outside the cell, and may therefore have an advantage over the other agents when delivered by monoclonal antibodies. Efforts to assess the therapeutic potential of conjugates of monoclonal antibodies and radioactive isotopes are under way in a variety of laboratories. Steven Rosen of Northwestern University in Chicago, Illinois, has obtained brief remissions of the skin lesions caused by a T cell lymphoma in two of the five patients he has treated with such a conjugate.

In addition, Stanley Order of the Johns Hopkins University School of Medicine, Baltimore, Maryland, has obtained complete or partial remissions in nearly 50 per cent of the 100 patients he has treated for liver cancer with a complex of radioactive iodine and antibody, although Order uses a conventional polyclonal antibody preparation rather than a monoclonal antibody.

Ronald Levy, Richard Miller and their colleagues at Stanford University School of Medicine have achieved one of the very few cures of a cancer patient by monoclonal antibody therapy. These investigators have devised an unusual strategy for treating B cell lymphoma. The cells of this cancer are derived from antibody-producing cells and therefore carry on their surfaces molecules of the antibody normally made by the parent cell. Moreover, because B cell lymphomas are clonal, all the tumour cells should carry the same antibody. These tumours present an exception to the more general situation in which no absolutely tumour-specific antigens have been found. Every B cell lymphoma carries its own unique tumour-antigen, namely the antibody.

Antibodies are proteins, and like other proteins are capable of eliciting antibody production. The Levy group's approach to treating lymphomas is to make a monoclonal antibody that specifically recognizes the antibody on the lymphoma cells. This strategy, unlike the others being investigated, requires a different monoclonal antibody for every patient. The first patient that was treated with such an antibody went into a complete remission and has remained disease-free for more than 5 years – which is the standard definition for a cancer cure. Levy has since gone on to treat a dozen more patients, but has not been able to repeat the initial success. Some of the patients have had temporary partial remission, but there have been no more cures.

All the monoclonal antibodies that have been used so far in clinical trials are of mouse origin and, because they are foreign proteins, often induce antibody formation in the recipient patients. The patients' antibodies may diminish the effectiveness of the therapeutic mouse monoclonal antibodies and this may have accounted in part for the failure of most of Levy's patients to respond.

The main problem probably originates in the nature of the antibody genes, however. Each antibody molecule consists of four proteins: two identical heavy chains with a molecular weight of about 50 000 each, and two identical light chains with a molecular weight of about 23 000 each (Figs. 12.5 and 12.6). Each light chain and each heavy chain can be subdivided into a constant region, which as its name suggests is the same in all chains of the same type, and a variable region, which differs from one chain to the next. The combined variable regions of the light and heavy chains form the two antigen-binding sites of an antibody molecule. The ability of antibody molecules to recognize so many antigens thus depends on the generation of a very large number of variable regions.

In the years since Köhler and Milstein initiated their experiments on the role of mutations in generating antibody diversity, a great deal of evidence has shown that certain regions of the gene sequences coding for the antibody variable regions are prone to mutation and that these mutations account for a portion of antibody diversity. As it happens, the monoclonal antibodies used for therapy by Levy and his colleagues were specifically directed against the mutation-prone regions of the target antibodies on the lymphoma cells. Because the mutations change the shape of the target antibodies, they can no longer be recognized by the therapeutic monoclonal antibody. Consequently, not all of the tumour cells are killed and the progress of the disease is only temporarily checked, if at all. Apparently the first patient was lucky.

Patients who have leukaemias or lymphomas that do not respond to chemotherapy or radiation may be given bone marrow transplants in a last effort to save their lives. Before such transplants are performed the patient's own bone marrow is destroyed with drugs or radiation to eliminate the source of the malignant cells. This treatment also eliminates the patient's normal blood-forming ability, an effect which would be lethal without a transplant of functional bone marrow.

Bone marrow transplants can be extremely hazardous. The principal danger is 'graft-versus-host' disease, in which the immune cells produced by the transplant attack the cells of the recipient. The result can be widespread tissue damage and even death. To minimize the risks of graft-versus host disease the antigenic composition of the donor cells must be carefully matched to that of the cells of the recipient. Another way of trying to prevent graft-versus-host disease is to treat the bone marrow before it is transplanted with a monoclonal antibody that reacts with the cells that cause the immune attack on the recipient tissue.

Suitable donors cannot be found for every patient who might benefit from a bone marrow transplant, however. Monoclonal antibodies might help in this regard, too. The idea here is to remove some of the patient's bone marrow before he or she is exposed to the lethal, marrow-destroying dose of chemicals or radiation. The bone marrow is then purged of leukaemic cells by treatment with an appropriate monoclonal antibody preparation and reintroduced into the patient. In one early trial of this therapy, Lee Nadler, who is at Dana-Farber Cancer Institute, was able to induce remissions in all of the 40 lymphoma patients treated so far. About 60 per cent have remained disease-free for up to 4 years.

Potential limitations on therapy with monoclonal antibodies

Although the clinical studies have provided some encouraging findings, they have also pointed up a number of problems that will have to be solved before therapies using monoclonal antibodies can be considered anything but highly experimental. There are five major classes of antibodies, for example, and researchers still need to determine whether some of these are more suited to a given application than others. Other basic questions that need to be answered concern the size of the doses that must be given to be effective and how best to administer the antibodies to ensure that they are efficiently delivered to the tumour sites.

The heterogeneity of tumour cells constitutes what may be the biggest obstacle to the use of monoclonal antibodies for cancer therapy. A large body of evidence has shown that the cells of individual tumours differ with regard to a wide range of properties. They can vary in their ability to metastasize, to secrete or respond to hormones, in their susceptibility to chemotherapeutic drugs and radiation – and in the composition of the antigens on their surfaces (Fig. 12.7). Some cells may carry the antigen recognized by a monoclonal antibody whereas others may not.

Moreover, antibody binding may cause an antigen to disappear from the cell surface. In that event, the surviving cells will not be susceptible to further antibody treatments. Effective cancer therapy requires that all cancer cells be killed and that none be left to seed the formation of new tumours. But cells that lack the target antigen for whatever reason will escape an antibody's effects.

Investigators hope that it will be possible to circumvent the heterogeneity problem, perhaps by using a mix of monoclonal antibodies that react with several tumour-associated antigens. Conjugating the antibodies with radioactive isotopes may also help, because an isotope's lethal effects can extend beyond the cell that binds the antibody.

As already mentioned, mouse monoclonal antibodies have been used for all the clinical trials. Administration of foreign proteins to human patients, especially when done repeatedly, has the potential of triggering the severe – and even fatal – allergic reaction called anaphylaxis. This possible hazard proved less serious than might have been expected in the monoclonal antibody trials. Although some patients have experienced allergic reactions to the mouse monoclonal antibodies, the reactions were treatable and none proved fatal.

Even if severe allergic reactions have not been a problem with the mouse monoclonal antibodies, immune reactions to the foreign proteins may nonetheless diminish their therapeutic effectiveness. Investigators have been trying for years to produce stable hybridomas that yield human monoclonal antibodies but have so far had little success.

Hybridomas are made today much as they originally were by Köhler and Milstein. Mice are immunized with the desired antigen and then the antibody-producing cells are isolated from the animals' spleens for fusion with myeloma cells. Antibody producing cells clearly cannot be obtained from human beings by this method.

There might be a way around this problem because many tumours shed antigens into the circulation, which might well elicit an antibody response. Fusing myeloma cells with antibody-producing B cells from the blood of cancer patients could in that event yield hybridomas that produce human monoclonal antibodies that react with the tumour cells. Schlom's group, for one, has found that this approach works. However, the hybrids, which were made by fusing B cells from human breast cancer patients with mouse myeloma cells, were unstable, as is frequently the case with hybrids between cells of different species. Human myeloma cells that are suitable for hybridoma production have never been developed.

Chimaeric antibodies

Recent work that marries monoclonal antibody and recombinant DNA technologies may provide a means of making antibodies that are less likely to provoke a counteracting immune response in humans than the mouse antibodies. In nature a complete gene for an antibody light chain is assembled by combining three separate DNA segments (Fig. 12.8). Two of these make up the coding sequence for the variable region and the third encodes the constant region of the protein chain. Assembly of the heavy chain genes is similar except that three DNA sequences must be joined to form the gene segment coding for the variable region. This assembly from separate DNA segments is another of the mechanisms that allows the immune system to generate such a great diversity of antibody molecules.

With the aid of recombinant DNA technology researchers

are now capitalizing on nature's plan for antibody synthesis to construct antibodies to their own specifications. Among the 'designer antibodies' being made are chimaeric antibodies in which the constant regions are of human origin and the variable regions are of mouse origin. The general assumption is that immune responses to foreign antibodies are directed mainly against the constant regions of the molecule. An individual's immune system is already exposed to an enormous number of variable regions, which are presumably not recognized as foreign as readily as the constant regions.

The investigators who have contributed to the development of chimaeric antibody technology are Sherie Morrison of Columbia University College of Physicians and Surgeons in New York City, who works with Vernon Oi of Becton-Dickinson Monoclonal Center in Mountain View, California; Marc Shulman and Nobumichi Hozumi of the University of Toronto, Canada; and Michael Neuberger of the MRC Laboratory of Molecular Biology in Cambridge, England.

To construct a chimaeric antibody the genes for the light and heavy chains are made individually by recombinant DNA methods. Cloned gene sequences are already available for the constant regions of the two classes of human light chains and the five classes of human heavy chains. The gene sequences

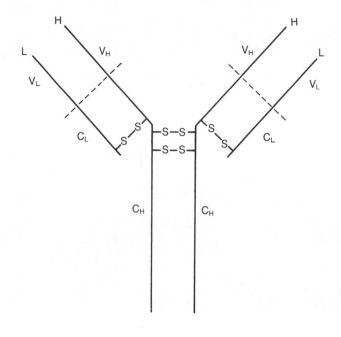

Fig. 12.5 Diagram of an antibody molecule. Antibody molecules are composed of four protein chains: two identical heavy chains (H) and two identical light chains (L). The chains are held together as shown by disulphide links between residues of the amino acid cysteine. Every light and heavy chain contains a variable region (V_L and V_H) and a constant region (C_L and C_H). An antibody molecule contains two antigen-binding sites, which are composed of the variable regions of the light and heavy chains.

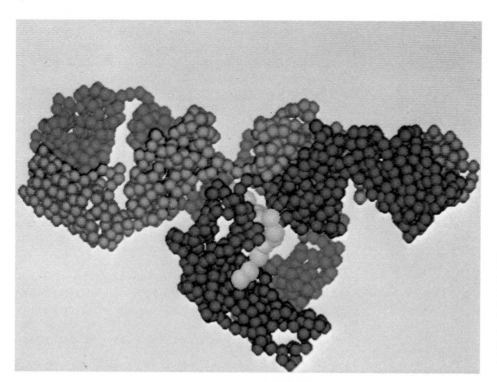

Fig. 12.6 Model of an antibody molecule. In this figure the heavy chains are shown in blue and orange and the light chains are in green. The two arms of the Y-shaped molecule contain the antigen-binding sites. The yellow indicates carbohydrate residues that are attached to the antibody protein. (From E. W. Silverton, M. A. Navia and D. R. Davies (1977) *Proc. Nat. Acad. Sci., USA*, **74**, 5140, with permission.)

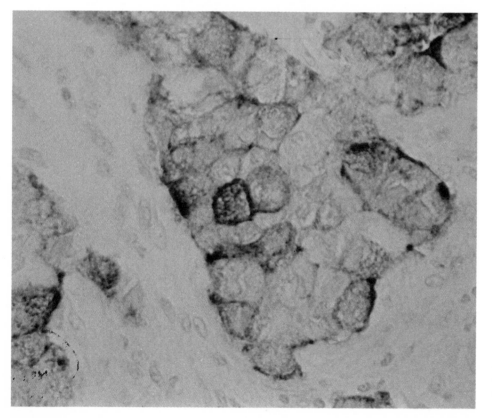

Fig. 12.7 Tumour cell heterogeneity. The cells of human tumours can differ from one another in several ways, including in the kinds of antigens that they carry. When this human breast carcinoma was stained with monoclonal antibody, some cells became highly stained (dark brown), others only lightly stained, and some did not bind the antibody at all. Such heterogeneity could be a problem in attempts to treat cancers with monoclonal antibodies. Cells that do not bind antibody might well escape its lethal effects and remain to cause a recurrence of the malignancy. (By kind permission of Jeffrey Schlom, National Cancer Institute.)

Fig. 12.8 Assembly of complete antibody genes. (*a*) Light chain genes are assembled from three separate coding segments, designated V for variable, J for joining and C for constant. The complete variable region of the light chain protein is encoded in the V and J segments, which are joined during the maturation of antibody-producing cells. During this joining, any of the several V segments can be connected to any of the J segments, thereby contributing to the diversity of antibody molecules. The joined V–J segment remains separate from the constant region segment until the whole stretch of DNA is copied into messenger RNA, at which time the RNA between the V–J and C segments is spliced out.

(*b*) The assembly of complete heavy chain genes resembles that of the light chain genes except that three DNA segments – V, J and D (for diversity) – are needed to encode the variable region of a heavy chain. In addition, there are several different constant regions. The final type of antibody produced is determined by which of the constant regions becomes joined to the completed V–D–J segment.

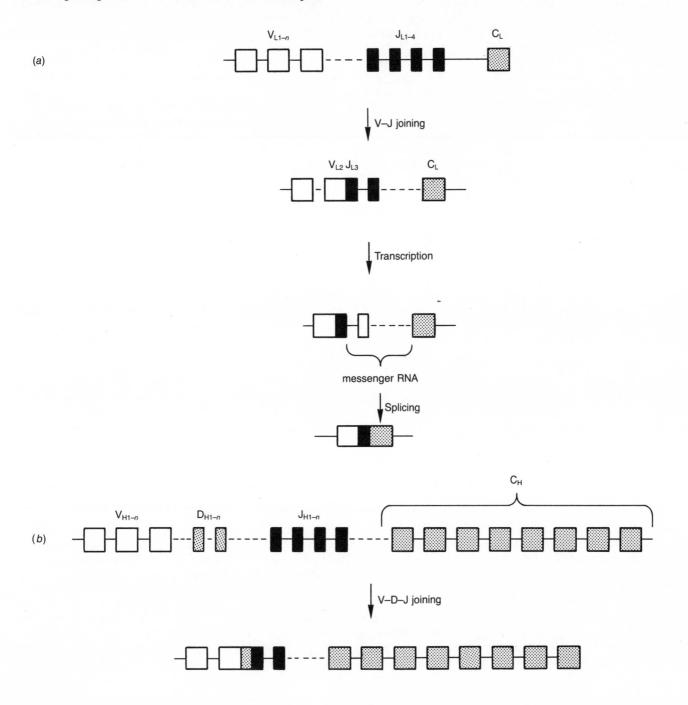

encoding mouse variable regions are cloned from a hybridoma cell line that makes a monoclonal antibody of the desired specificity. The hybridoma itself is generated in the standard manner.

The cloned sequences for the light chain constant and variable regions are then joined, as are those for the heavy chain, to form the complete genes. Once the genes for the chimaeric light and heavy chains are in hand they are introduced together into mouse myeloma cells where the proteins are made and the complete antibody molecules are assembled. Although the procedure for making chimaeric antibodies sounds cumbersome, the individual steps are now routine.

According to Morrison the types of antibodies that can be made are limited primarily by the imagination of the investigator. In addition to producing mouse–human chimaeric antibodies, for example, researchers can make antibodies of a desired specificity and class, which is determined by the structure of the heavy-chain constant region. In monoclonal antibody production researchers have no control over what class of antibody a hybridoma will produce, and this is important because the different classes have different functions.

Anti-idiotype antibodies

Antibody variable regions may not be as effective as the constant regions in eliciting immune responses, but the variable regions are not completely devoid of that capability. Antibody molecules carry regions in or near their antigen-recognition sites that can act to stimulate antibody production. These regions – called 'idiotypes' in the jargon of immunology – are characteristic for each antibody. The antibodies used by the Levy group for treating B cell lymphoma are in fact directed against the idiotypes of the antibodies on the cancer cells.

One of the contributions for which Jerne was awarded the Nobel Prize was his 'network' theory of antibody regulation. He proposed that the immune system contains an interlocking network of antibodies that are directed at one another's idiotypes. Under normal circumstances the concentration of each antibody is kept in check by the antibodies to its idiotype, that is, by the 'anti-idiotype' antibodies. Introduction of an antigen perturbs this system, leading to increased production of the antibodies that recognize the antigen. This in turn leads to increased production of the corresponding anti-idiotype antibodies and ultimately to a diminution in production of the first antibodies. Moreover, the anti-idiotype antibodies further trigger the production of the antibodies that recognize their own idiotypes (the anti-anti-idiotypes) and so on. The perturbation thus spreads through the network much as ripples spread over the surface of a pond.

Whether Jerne's network operates as proposed in regulating antibody production is still a matter of discussion, but there is no question that anti-idiotype antibodies exist. Recently investigators have turned to anti-idiotype antibodies as a possible new type of vaccine. The approach offers a way of vaccinating against a disease-causing organism without using the organism itself. This could be an advantage if the pathogen is difficult to grow or dangerous to handle, or if injecting the pathogenic material could cause harmful side effects.

An antigen and its antibody have complementary shapes. They fit together much as do a lock and key, although the antigen and antibody molecules are perhaps not so rigid. An anti-idiotype antibody binds to the same region of the antibody as does the corresponding antigen. This implies that the antigen and the anti-idiotype antibody have similar three-dimensional shapes. If that is the case, then vaccinating with the anti-idiotype antibody should lead to the production of antibodies that can recognize both the anti-idiotype antibody and the antigen (Fig. 12.9) and that may confer protective immunity against the antigen.

Attempts to develop such anti-idiotype antibodies as vaccines are just beginning, but a number of studies indicate that the approach can work. For example, Ronald Kennedy and Gordon Dreesman of the Southwest Foundation for Biomedical Research, San Antonio, Texas, injected rabbits with human antibodies to the surface antigen of hepatitis B virus. The human antibodies elicited the production of anti-idiotype antibodies in the rabbits, which were then purified and used to immunize chimpanzees, the only non-human animal to get hepatitis B. The animals thus immunized made antibodies to the hepatitis B surface antigen and were protected against the disease when challenged with the virus. In addition, the Koprowski group is developing an anti-idiotype vaccine to protect against the rabies virus, and the approach is being investigated as a possible way of making a vaccine against the dangerous AIDS virus.

Anti-idiotype antibodies may also provide protection against cancer. Ingegerd Hellstrom, Karl Erik Hellstrom and their colleagues at the University of Washington Medical School in Seattle found that they could induce immunity in mice to a human cancer antigen by injecting the animals with anti-idiotype antibody. And Koprowski suggests that the induction of anti-idiotype antibodies in some of the cancer patients that were treated with his group's mouse monoclonal antibody may have contributed to the regression of their tumours.

The future of monoclonal antibodies

Monoclonal antibodies have more than proved their value in basic research and have begun to move from the laboratory into the clinic. The diagnostic applications will continue to grow rapidly and will progress from tests that can be done on blood and other materials that have been withdrawn from the body to the direct imaging of cancerous tumours within patients. The therapeutic applications will develop more slowly, but researchers stress that they are learning more about how to use monoclonal antibodies therapeutically with every clinical experiment.

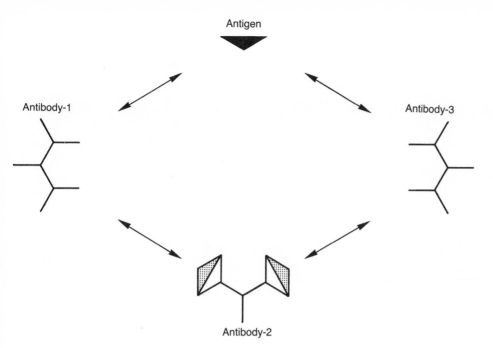

Antigen

Antibody-1

Antibody-3

Antibody-2

Fig. 12.9 Simplified anti-idiotype network. The antibody (antibody-1) produced in response to an antigen can itself elicit the production of anti-idiotype antibodies (antibody-2). The three-dimensional structure of the portion of the anti-idiotype antibody that binds to the first antibody should resemble the three-dimensional structure of the original antigen. Immunizing with the anti-idiotype antibody should therefore elicit the production of antibodies that recognize both the anti-idiotype-antibody and the antigen.

One approach to cancer therapy that has so far received little attention concerns whether monoclonal antibodies can be used to prevent cancer cells from receiving some essential biochemical stimulus. For example, if the growth of the cells depends on the presence of a growth factor or of the product of an activated oncogene, then a monoclonal antibody that blocks the activity of the growth factor or the oncogene product might produce a regression of the tumour.

Minna and his colleagues have evidence that such a strategy might work. The SCLC cells (see p.149) produce bombesin, a protein that acts as a growth factor for this virulent type of tumour. The Minna group has generated a monoclonal antibody to bombesin and shown that it prevents the growth of SCLC tumours that have been transplanted into mice. The investigators are planning a preliminary trial of the antibody in human patients. The receptors to which growth factors such as bombesin must bind to exert their effects constitute another potential target for monoclonal antibody therapy.

The marriage between monoclonal antibody and recombinant DNA technologies, which enables researchers to make chimaeric and other types of designer antibodies, has already been noted. Investigators are still learning how best to join toxins, drugs or radioactive isotopes to antibodies without altering the antibodies' specificity or activity. When this information becomes available, the chimaeric antibody technology may allow appropriate binding sites for the various agents to be built into antibody molecules.

Monoclonal antibody technology is also merging with that for making site-directed antibodies (see chapter 14). Short peptides elicit the production of antibodies that recognize and bind to the same peptide sequence when it is carried in intact proteins. This permits researchers to select in advance the protein site with which the antibody is to react, an ability which is useful when the goal is to interfere with some specific function of a protein, such as the catalytic activity of an enzyme or the binding of a virus to its target. In the past, site-directed antibodies have been made by the classic methods, but investigators are beginning to move to the production of site-directed monoclonal antibodies in yet another demonstration of the versatility of monoclonal antibody technology.

Additional reading

Jerne, N. K. (1985). The generative manner of the immune system. *Science*, **229**, 1057.

Köhler, G. and Milstein, C. (1975). Continuous cultures of fused cells secreting antibody of predefined specificity. *Nature (London)*, **256**, 495.

Köhler, G. (1986). Derivation and diversification of monoclonal antibodies. *Science*, **233**, 1281.

Milstein, C. (1986). From antibody structure to immunological diversification of immune response. *Science*, **231**, 1261.

Mulshine, J. L., Cuttitta, F. and Minna, J. D. (1985). Lung cancer markers as detected by monoclonal antibodies. In *Monoclonal Antibodies in Cancer*, ed. S. Sell and R. A. Reisfeld, pp. 229–46. Humana Press, Clifton, New Jersey.

Nowinski, R. C. and others (1983). Monoclonal antibodies for diagnosis of infectious diseases in humans. *Science*, **219**, 637.

Schlom, J. (1986). Basic principles and applications of monoclonal antibodies in the management of carcinomas. (The Richard and Hinda Rosenthal Foundation Award Lecture.) *Cancer Research*, **46**, 3225.

Sears, H. F., Herlyn, D., Steplewski, Z. and Koprowski, H. (1984). Effects of monoclonal antibody immunotherapy on patients with gastrointestinal adenocarcinoma. *Journal of Biological Response Modifiers*, **3**, 138.

Vitetta, E. S., Krolick, K. A., Miyama-Inaba, M., Cushley, W. and Uhr, J. W. (1983). Immunotoxins: a new approach to cancer therapy. *Science*, **219**, 644.

13 Site-directed antibodies in biology and medicine

Thomas M. Shinnick and Richard A. Lerner

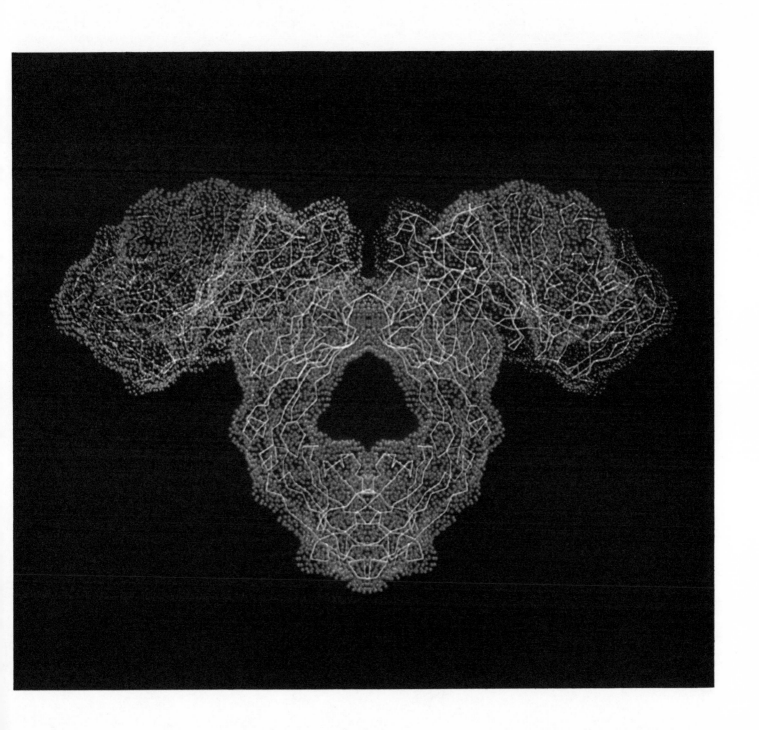

A powerful new approach to making antibodies is not only providing important research tools for the immunologist, molecular biologist and protein chemist, but also holds great promise for such medical applications as the diagnosis and prevention of disease. The new approach is the outgrowth of an observation a few years ago that antibodies that will react with almost any region of a protein can be elicited by a short peptide corresponding to that region. Since then research with the antibodies has generated novel insights into the immunochemistry and structure of proteins in solution. In addition, efforts are under way to use the antibody-producing technology to make safe, chemically defined vaccines and specific new reagents for diagnostic tests.

The discovery that peptides in general can elicit antibodies that react with proteins took the immunological community somewhat by surprise in the early 1980s. F. Anderer of the Max Planck Institute had shown 20 years previously that a short peptide that corresponds to the first six amino acids of the coat protein of the tobacco mosaic virus elicits antibodies that bind to the intact protein. Nevertheless, the prevailing view was that this was not a widespread phenomenon and that most peptides would be incapable of eliciting protein-reactive antibodies.

This view was largely the result of two considerations. First, the antibodies elicited by an intact protein react only with a few, small sites on the protein surface. For example, less than 25 per cent of the amino acids of the well-characterized proteins lysozyme and myoglobin appeared to be directly involved in interactions with antibodies. This finding implied that without previous knowledge of the antibody-binding sites of a protein, random selection of a peptide that contains such a site is unlikely.

Second, the ability of an antibody to bind to a protein is critically dependent on the three-dimensional structure (conformation) of the protein. Antibodies elicited by the native structure react poorly, if at all, with a protein that has had its shape altered by denaturing treatments. These results suggested that to obtain protein-reactive antibodies the immunizing agent not only had to have the same amino acid sequence as the protein but also had to display the same conformation. Peptides in solution are much more flexible, and can assume many more conformations, than proteins. To many, these considerations clearly indicated that the use of peptides to generate protein-reactive antibodies would be feasible only in special cases and could not form the basis of a general technology.

However, the considerations arose mainly from the study of antibodies that had been produced in response to intact proteins. The specificities of antibodies that are elicited by peptides could well be quite different from those made against intact proteins. The antibodies generated by proteins are made against one of the most conformationally restricted presentations of the molecule, whereas the peptide-elicited antibodies are made against a much less restricted – possibly even random – presentation of the molecule. In other words, it is not possible to conclude that antibodies elicited by a peptide will be unable to bind to an intact protein just because antibodies generated by the protein fail to bind to the peptide. Immunogenicity does not equal antigenicity when considering the two types of molecules.

Immunogenicity refers to the ability of a substance to elicit antibodies that bind to it, whereas antigenicity is the ability of a substance to be recognized and bound by a particular antibody preparation. There has often been some confusion over these terms, possible because antibodies are typically generated by the same substance with which they are to react. In that case, an immunogenic site directly corresponds to an antigenic site.

However, antibodies elicited by one substance often bind not just to that substance but also to another, even though the converse is not true. Antibodies elicited by the second substance bind only to it and not to the first. For example, antibodies elicited by intact pneumococcal bacteria can bind to the bacteria and to the polysaccharide coat purified from the bacteria, but antibodies generated by the purified coat do not react with the intact bacteria. With respect to antibodies that bind to the bacteria, the purified polysaccharide coat is therefore antigenic but not immunogenic.

Immunology of proteins

A great deal of effort over the past several decades has gone into defining the immunogenicity and antigenicity of proteins. The typical approach has been to elicit antibodies with the intact protein and then to fragment the protein by various chemical or enzymatic means and determine which fragments retain the ability to react with the antibodies. These studies have produced a well-defined picture of the antigenic sites that are recognized by antibodies that have been generated by intact proteins. Such sites are small – typically only four to seven amino acid residues interact directly with the antibody – and few in number (Figs. 13.1, 13.2 and 13.3). Moreover, antibody binding is critically dependent on the conformation of the protein antigen.

Caption to previous page

Fig. 13.1 Computer graphic representation of the structure of the antibody molecule. The antibody molecule is a Y-shaped protein that consists of four protein chains. The heavier two proteins (blue surfaces) extend from the stem of the Y into the arms. The two lighter chains (green surfaces) are confined to the arms. Each of the four chains has a constant region (white and yellow skeletons) and a variable region (red skeleton). The combined variable regions from the light and heavy chains form the antigen-recognition sites of the antibody. (By kind permission of Arthur Olson, T. J. O'Donnell and Michael Connelly, Research Institute of Scripps Clinic.)

Studies of the antigenic structure of the enzyme lysozyme, which is obtained from the whites of chicken eggs, played a major experimental role in the development of the concepts of protein antigenicity and immunogenicity that prevailed in the late 1970s. Lysozyme was used as a model for globular proteins because it is highly immunogenic and small, containing only 129 amino acid residues in a single polypeptide chain. Moreover, the complete three-dimensional structure of the protein had been determined by X-ray crystallographic methods.

In most of the lysozyme studies the antibodies were obtained by immunizing a goat or rabbit with intact, conformationally native protein. The studies were thus defining the antigenic sites recognized by antibodies that were elicited by native lysozyme.

Four disulphide bonds play a major role in holding lysozyme in its native three-dimensional structure. The first indication that binding of the antibodies to the corresponding antigenic sites was critically dependent on the conformation of those sites came from experiments in which the lysozyme molecule was made to unfold by treating it with denaturing agents and then splitting the disulphide bonds with reducing agents. This treatment almost completely abolished the ability of the antibodies that were originally generated by the native protein to bind to the molecule. Conversely, antibodies that were elicited by the denatured protein bound readily to the denatured molecule but poorly, or not at all, to conformationally native lysozyme. These results indicated that antibodies do not simply 'read' amino acid sequences but must recognize and bind to something special about the conformation of the target protein.

To define and map the antigenic sites of lysozyme that bind the antibodies generated by the native enzyme, M. Atassi of the Baylor College of Medicine in Houston, Texas, and his coworkers fragmented the enzyme and measured the binding of the antibodies to the various pieces. In this way they defined three antigenic sites that bound more than 90 per cent of the antibodies. Site 1 includes amino acid residues 5 to 14 and 125 to 128, which are brought together on the lysozyme surface by a disulphide bond between cysteine residues 6 and 127. Site 2 contains residues 60 to 80 and 87 to 97, which are held in close contact by the disulphide bonds between cysteines 64 and 80 and between cysteines 76 and 94. Site 3 includes residues 113 to 116 and 30 to 34, which are linked by the disulphide bond between cystine residues 30 and 115.

To define which particular residues within the three sites directly interact with the antibodies, Atassi and his colleagues chemically synthesized short peptides in such a way as to reproduce the conformations of the amino acids in those regions. They assumed that in solution antibodies would react only with amino acid residues on the surface of the target protein. From the known X-ray crystallographic structure of lysozyme, the investigators determined which residues of the three antigenic sites of lysozyme were exposed on the surface, and therefore likely candidates for residues. By using the amino acid glycine as a 'spacer', they then synthesized peptides in which the exposed amino acids were held in the correct positions relative to one another.

Analysis of the ability of a series of such 'surface simulated' peptides to bind antibodies that were elicited by native lysozyme revealed that the residues required for antibody binding to site 1 are arginines 5, 14 and 125, glutamic acid 7, and lysine 13; in site 2 they are tryptophan 62, lysines 96 and 97, asparagine 93, threonine 89, and aspartic acid 87; and in site 3 they are lysines 33 and 116, asparagine 113, arginine 114, and phenylalanine 34. Atassi and his colleagues designated these 16 amino acids as the 'critical' residues for antibody binding. However, the residues that dictate the folding of the protein molecule, although not themselves in contact with a bound antibody molecule, are also critical in the sense that they are needed to maintain the correct conformation of the protein.

Each of the three sets of critical amino acids forms a spatially continuous portion of the protein, although the individual amino acids may be distant from one another in the linear lysozyme sequence. Given that the critical residues are in the proper positions to interact with antibodies only in the correctly folded protein, it is not surprising that denaturation essentially abolishes antibody binding to the lysozyme molecule.

Experiments that were carried out in the late 1960s and early 1970s in the laboratories of Michael Sela and Ruth Arnon, both of the Weizmann Institute in Rehovot, Israel, and Christian Anfinsen at the National Insitututes of Health in Bethesda, Maryland, also suggested that the antigenic and immunogenic determinants of protein are conformation-dependent. Residues 64 to 80 of lysozyme form a loop structure on the surface of the protein that is held together by a disulphide bond between cysteines 64 and 80. This loop can be obtained either by cleavage from the protein or by chemical synthesis and was used by Sela, Arnon, Anfinsen and their colleagues to test whether opening the loop alters its immunogenicity and antigenicity.

Antibodies to native lysozyme or to the closed loop reacted strongly with the intact protein and the closed loop, but poorly or not at all with the open loop. In contrast, antibodies made against the open loop could react only with the open loop. The experiments showed once again that antibodies react with something unique to the conformation of the antigens. But more than that, the studies suggested that it would be necessary to reproduce complex conformations to produce synthetic antigens that could bind to antibodies elicited by intact proteins.

Such a task would be difficult at best and only feasible where the complete three-dimensional structure of the protein had been determined. The two general conclusions about antigenic determinants, namely that they are relatively few in number and discontinuous and that they are highly dependent on conformation, did not bode well for efforts aimed at developing a general technology for generating antibodies to any region of a protein.

Recently, however, the notion that a few sites determine the entire antigenicity of a protein has come under attack. Studies of lysozymes that carry amino acid substitutions at positions

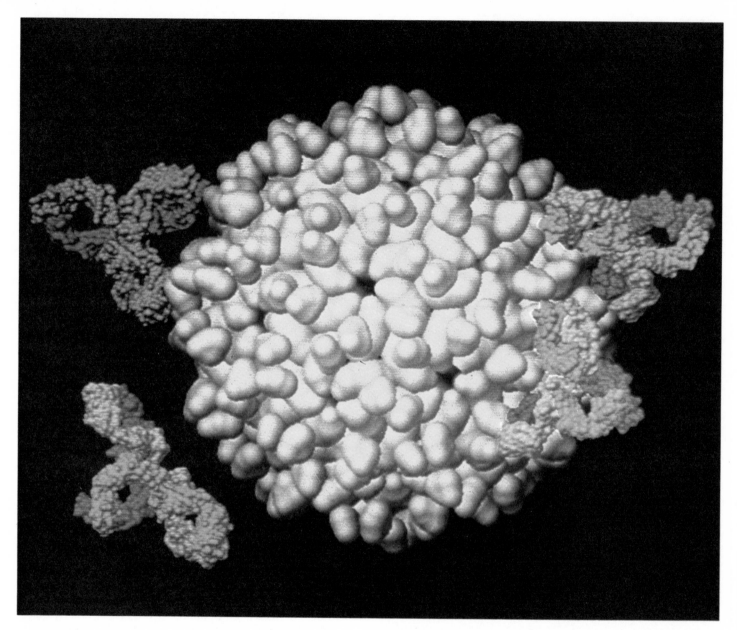

Fig. 13.2 Computer graphic model of four antibody molecules interacting with a hypothetical viral particle (shown in gold). The heavy chains of the antibody are blue and the light chains are magenta. The model shows the relative sizes of the antibodies and a typical viral target. Each antibody molecule contacts only a small portion of the viral surface. The interactions of antibodies with viral proteins are also restricted to small segments of those proteins. (By kind permission of Arthur Olson and Michael Connelly, Research Institute of Scripps Clinic.)

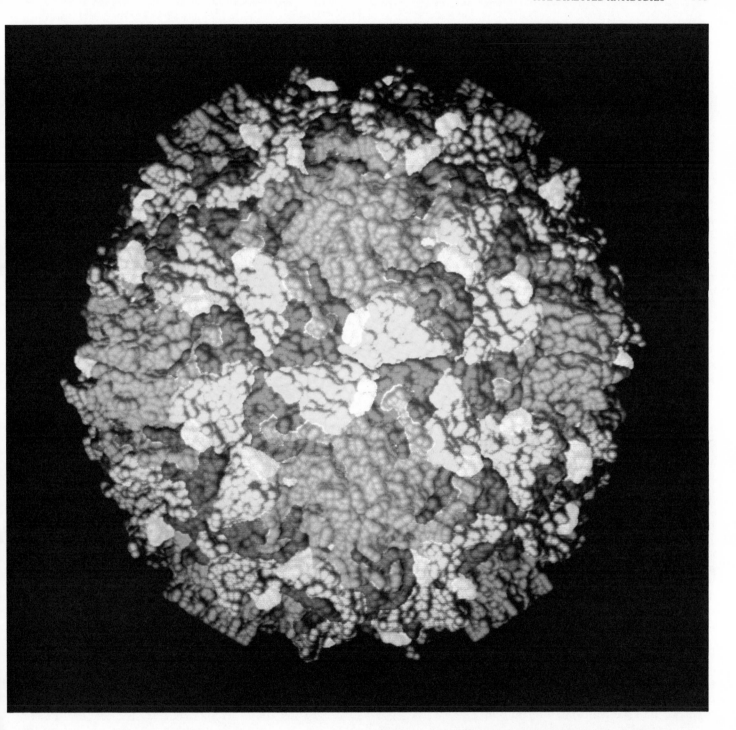

Fig. 13.3 Computer graphic model of the poliovirus. The blue, yellow and orange colours designate the different proteins on the viral surface. The white regions indicate the antibody-binding segments of the viral proteins, again constituting only a small portion of those proteins. (By kind permission of Arthur Olson and James Hogle, Research Institute of Scripps Clinic.)

outside the antigenic sites that were defined by the Atassi group showed, for example, that the different lysozymes could be distinguished by their antibody reactivity. This result indicates that there are more antigenic sites than previously thought. It is, of course, still possible that one of the original determinant sites could be altered by an amino acid change at another position.

Studies by Sandra Smith-Gill and her colleagues at the National Institutes of Health give further support to the suggestion that there are additional antigenic sites on lysozyme. When these investigators made a monoclonal antibody (see Chapter 12) to a chicken lysozyme they found that it reacted completely with lysozymes from seven different species of galliform birds, partially with lysozymes from two other galliform species, and not at all with a duck lysozyme.

By comparing the amino acid sequences of the proteins, the Smith-Gill group determined that the likely antibody-binding site includes arginine 45 and arginine 68 and extends into the cleft between arginine 45 and arginine 114. This site is outside those defined by Atassi.

It may be unrealistic to assume that the complete immunogenicity and antigenicity of a protein can be defined with a single antibody preparation. What is immunogenic may depend as much on the mode of presentation of the antigen and the species immunized as on the structure of the protein antigen.

Synthetic peptide immunogens

In 1980 additional signals began to appear that challenged the line of reasoning that held that relatively small peptides are unlikely to elicit antibodies that can react with intact proteins containing those peptide sequences. When Richard Lerner and his colleagues at the Research Institute of Scripps Clinic in La Jolla, California, completed the determination of the nucleotide sequence of murine Moloney leukaemia virus, they encountered a 'mystery' region that was located at the right-hand end of the viral *env* gene, which encodes the proteins of the viral envelope. The region in question had the potential to encode a protein, but the predicted protein did not appear to correspond to any of the known viral proteins.

To analyse this segment of the *env* coding region the Scripps workers chemically synthesized a peptide corresponding to the carboxyl end of the predicted protein. They made antibodies to the peptide and used them to look for viral proteins that contain the sequence. The antibodies reacted with and precipitated two viral proteins from infected cells. One was the large protein that is the precursor from which the envelope proteins are cut. The second turned out to be a smaller protein that contained both the envelope protein (designated p15E) and the sequence encoded in the mystery region of the rival genome.

As it turned out, the mystery protein did not correspond to any of the known viral proteins because it is cut off from p15E

protein when the leukaemia virus buds from cells. However, the precursor that contains the p15E and mystery proteins could be detected by the antibody, thereby solving two problems at once. The product of a potential protein-coding region in the viral genome was identified and the fate of a small portion of a protein was specifically traced during virus maturation.

At about the same time, workers from Russell Doolittle's laboratory at the University of California at San Diego made antibodies to synthetic peptides corresponding to the carboxy and amino-terminal regions of the large tumour antigen of simian virus 40 and showed that the antibodies reacted specifically with the intact protein. Taken together, these two studies revived the possibility that peptide immunogens might be used to generate protein-reactive antibodies – but a large conceptual hurdle still had to be overcome.

In both studies described above, the antibodies had been made to peptides corresponding to the ends of proteins, which may often be disordered and perhaps more easily mimicked by short peptides. As far back as Anderer's studies in 1963 there had been some success in using the ends of proteins as immunogens, but few attempts had been made to move beyond this. Consequently, there was still little certainty that peptides other than end fragments could be used for generating antibodies to proteins. If the technology was restricted to the use of the end fragments, peptide immunogens would certainly not be a general way of inducing the formation of site-specific immunological reagents for studying proteins.

A full investigation of whether peptides corresponding to sequences in proteins can generally elicit antibodies that react with the intact proteins required a protein that met certain stringent criteria. Its amino acid sequence and three-dimensional structure had to be known and its antigenic or immunogenic sites had to have been identified and mapped on the structure of the molecule The haemagglutinin protein of influenza virus, which is located on the surface of the viral particle, met all these criteria.

The haemagglutinin molecule contains two different polypeptides, which are designated HA1 and HA2. During the course of an immune response to a natural infection by influenza virus or to purified haemagglutinin, antibodies are made that bind essentially to just four antigenic sites – according to Donald Wiley and his colleagues at Harvard University, who have mapped the locations of the sites, all of which are on HA1

Nicola Green and her colleagues in Lerner's laboratory synthesized 20 different peptides that together represented about 75 per cent of the HA1 primary sequence. The investigators did not make any attempt to mimic the three-dimensional structure displayed by the corresponding regions of the protein. Moreover, the individual peptides correspond to regions that display diverse secondary structures, including alpha-helices, beta-pleated sheats, loops and extended conformations. The peptides were coupled to carrier proteins and injected into rabbits to elicit antibodies.

Of the 20 peptides, 18 elicited antibodies that reacted with the full-length haemagglutinin molecule or with the intact virus. A key result was that the 18 peptides were not just from the four antigenic sites that the Wiley group had identified, but were scattered throughout the entire HA1 amino acid sequence. The only requirement for a peptide to generate protein-reactive antibodies was that the peptide should represent a region on the surface of the protein. The results of the experiment indicate that the information carried within a relatively short linear peptide is sufficient to elicit reactivity against a much larger protein molecule with a complex three-dimensional structure – and suggest that chemically synthesized peptide immunogens can be used to generate antibodies that react with most regions of proteins.

The study also demonstrates that the immunogenicity of an intact protein is less than the sum of the immunogenicity of its pieces. In other words, there are regions of a protein that are not immunogenic when presented to the immune system as part of the intact protein, but are immunogenic when presented as peptides. One possible explanation for this difference is that the apparently non-immunogenic regions of proteins are capable of eliciting antibody production but that these antibodies are not detected by the neutralization assay normally used for mapping antigenic sites in intact proteins. In that event, the total collection of antibodies to a protein such as the influenza haemagglutinin would have a reactivity pattern much broader than that observed in the neutralization studies.

This is probably not the case for haemagglutinin, however, because potent antibody preparations that were generated by the intact haemagglutinin protein or by the virus do not cross-react with any of the synthetic peptides. Rather, most of the immune response to native haemagglutinin appears to be directed against determinants that are not mimicked by the short peptides. This finding is consistent with the body of data that shows that antigenic determinants on intact proteins are largely conformational in nature and that such determinants are rarely mimicked by small peptides.

The high frequency with which peptides can elicit antibodies that react with intact folded proteins raises the question of how a relatively disordered peptide can generate antibodies that interact with highly ordered proteins. Two models have been put forward to explain this. The first evokes a stochastic argument. The idea here is that a peptide in solution can adopt a variety of conformations, some of which mimic those in the native protein, and that the antibodies made against that small fraction of peptide conformations are the only ones that react with the folded protein. Then the success of the technology might be more a testimony to the sensitivity of the immunological assays that can detect a small percentage of the antibodies in a mixture than to something more fundamentally interesting.

However, peptides in solution can adopt thousands or even hundreds of thousands of conformations, and as judged by spectroscopic means, no single conformation is preferred over the others. Assuming then that the conformations that elicit antibodies in a particular animal are chosen at random and that

an individual animal expresses only a small portion of the repertoire of antibodies that could potentially react with an antigen, a great deal of animal-to-animal variation might be expected in the antibodies induced by a peptide. But this is not the case. When different animals are immunized with the same peptide they tend to produce the same amounts of protein-reactive antibodies.

To test the stochastic model in a formal way, Henry Niman of the Lerner group used monoclonal antibodies as a way of estimating the frequency with which small peptides induce antibodies that react with the native protein. He made monoclonal antibodies to each of six peptides from three different proteins and then assayed the antibodies for the ability to bind to the appropriate intact protein. Seventeen of the 21 antibodies that were elicited by one peptide from the haemagglutinin protein of influenza were capable of reacting with the intact haemagglutinin molecule – a frequency at least 10 000 times greater than that predicted by the stochastic model. Similar results were obtained with the other five peptides.

Antigenicity and the mobility of protein segments

If the stochastic model is incorrect, then antibodies may be reacting with protein conformations that are different from those that have essentially been seen frozen in the crystal structure. One protein might induce a conformational change in a small region of another protein simply by binding to it – a type of induced-fit model that had been previously seen in the interactions between enzymes and their substrates. A difficulty with this view is that it is somewhat circular in nature in that antibodies can only bind to, and distort, a region that they can recognize. It seems unlikely that the antibodies could recognize a region with a three-dimensional structure that is only distantly related to the correct binding conformation.

A second possibility is that a given site in a protein might be somewhat mobile and be able to adopt a number of related conformations. Binding would then take place when the site passes through a conformation that the antibody recognizes. The antibody would thus act as a sink, trapping the protein in that conformation. This model in essence greatly increases the number of conformations shared by the peptide and the intact protein. It predicts that the peptides that elicit antibodies with the capability of recognizing intact proteins are located in areas of great conformational mobility in the intact protein.

As a test of these ideas, Lerner and his Scripps colleagues John Tainer and Elizabeth Getzoff, studied the antigenicity of segments of proteins as a function of their atomic mobility. The test protein that they selected for the experiments was myohaemerythrin, an oxygen-carrying pigment that occurs in lower animals, including certain marine worms. The three-dimensional structure of myohaemerythrin has been solved to sufficient resolution to provide accurate indicators of the atomic

Fig. 13.4 'Glowing coal' model of the myohaemerythrin protein. The more mobile and antigenic regions of the protein are depicted by the lighter colours. (By kind permission of John Tainer, Elizabeth Getzoff and Michael Connelly, Research Institute of Scripps Clinic.)

mobility throughout the molecule. Moreover, the protein is completely foreign to the animals used to produce the antibodies. Animals do not normally make antibodies to their own proteins and might also fail to respond with antibody production to peptides from other sources if the peptides have structures similar to those in indigenous proteins. Myohaemerythrin was chosen to avoid the problem of unresponsiveness as a result of this type of tolerance.

The Scripps workers synthesized 12 peptides representing segments of the myohaemerythrin molecule with varying degrees of atomic mobility. Eleven of the peptides elicited strong immune responses and the antibodies thus produced were assayed for their reactivity with the intact, conformationally native protein. Antibodies generated by peptides corresponding to regions of high atomic mobility reacted strongly, whereas antibodies elicited by peptides corresponding to regions of low atomic mobility displayed little or no reactivity with the intact protein (Fig. 13.4). For some percentage of the time then, the more mobile regions of a protein apparently can adopt or be induced to adopt conformations different from those seen in the static crystal structures. Such behaviour might explain the ability of relatively disordered peptides to elicit protein-reactive antibodies at a high frequency.

The myohaemerythrin retained its native conformation during the assays for reactivity with the antibodies; its characteristic visible spectrum, which is very sensitive to changes in the three-dimensional structure of the protein, did not alter. Therefore peptide immunogens can elicit antibodies that react with native proteins. These same immunogens can often also generate antibodies that react with completely or partially denatured proteins. A mixture of antibodies that can react with both native and denatured proteins may have greater utility in the laboratory than a conventional antibody preparation that can typically react only with an antigen in the same conformational state as the original immunogen.

The correlation between antigenicity and mobility has been found in other proteins in addition to myohaemerythrin. For example, Aaron Klug of the MRC Laboratory of Molecular Biology in Cambridge, England, M.H.V. Van Regenmortel of the CNRS Institute de Biologie Moleculaire et Cellulaire in Strasbourg, France, and their colleagues came to a similar conclusion concerning the coat protein of tobacco mosaic virus. In addition, the correlation has been found by the Klug–Van Regenmortel group and the Lerner group for lysozyme and myoglobin, and by the Lerner group for haemoglobin, leghaemoglobin, cytochrome c, ribonuclease and insulin.

Recent studies with peptides have also revealed that the immunogenic sites of a protein may actually be quite numerous and not located just on the surface of the molecule. For example, the Lerner group has mapped the immunogenic sites of myohaemerythrin by using short linear peptides. They immunized seven rabbits with the intact protein and then assayed the resulting antisera for reactivity with over 2500 peptides that comprised all of the possible six-residue fragments of myohaemerythrin.

The immunogenicity of the protein varied considerably from one region to another and the patterns of the antibody responses in the seven rabbits also varied greatly from one animal to another. Overall, the more immunogenic sites tended to be in regions of high atomic mobility, decreased packing density, convex shape, and significant electrostatic potential. However, the combined responses of the seven rabbits produced antibodies that react with a set of peptides encompassing all 113 amino acid residues of the myohaemerythrin sequence. This means that under some circumstances each residue of the protein appears in an immunogenic region of the molecule. Once again the data suggest that proteins in solution are not the static structures seen by X-ray crystallography, but are dynamic molecules.

Potential applications

The usefulness of peptide immunogens is greatly enhanced by their ability to generate a population of antibodies, some of which react with the native protein and some of which react with the unfolded molecules. Consequently, the antibodies should be useful in almost any assay for detecting specific proteins.

Analysis for specific proteins

Two facets of the peptide immunogen approach make it particularly valuable. The first is that essentially the only requirement for making antibodies that will react with a given protein is the amino acid sequence of the protein. This is particularly important nowadays because most amino acid sequences are obtained by translation of the nucleotide sequences of the corresponding genes. In fact, occasionally the only thing known about a putative gene product is its amino acid sequence. But peptides corresponding to portions of that sequence can be chemically synthesized and used to generate antibodies that can serve as probes for identifying the protein in cells.

In addition to identifying the protein products of genes, such antibodies can be used to investigate the biological function of proteins by determining their cellular locations and analysing their possible enzymatic activities in the test tube. The antibodies are also valuable for purifying proteins by immunoaffinity chromatography. In this type of chromatography the antibodies, which are attached to some kind of solid support, specifically bind to the protein they recognize, thereby removing it from a complex mixture such as a cell extract. The specifically bound proteins can then be gently removed from the antibodies by use of an excess of the original immunizing peptide. Enzymes can often be recovered in active form by this method.

The second facet of the peptide immunogen approach is that the antibodies react with a small region of the protein that can be chosen in advance by the investigator who selects the peptide to be used for antibody generation. The antibodies can thus be said to have predetermined or preselected specificities. In some studies, such as those in which protein detection is the only goal, the site where the antibodies bind is not too important. But in others, targeting the antibodies to a particular, small site can be essential.

For example, in the previously described studies on the synthesis of the envelope proteins of murine Moloney leukaemia virus, identification of the step in which the carboxyl terminal segment of the p15E precursor is cleaved off would have been very difficult in the absence of antibodies that specifically react with that portion of the precursor protein. Targetted antibodies are excellent reagents for following the fate of a particular portion of a protein through processing pathways; for tracking which coding sequences of genes are used when alternative choices are possible, as in the rearranged genes that code for antibody proteins themselves; and for producing antibodies that can distinguish closely related proteins.

Using peptides as immunogens may be the easiest, most practical way to make completely specific antibodies that can distinguish closely related proteins. The Thy-1 antigen, which is carried on all T lymphocytes, is a case in point. This antigen exists in two forms that differ by only one amino acid out of a total of 112. The form designated Thy-1.1 is thought to have an arginine residue at position 89, whereas the Thy-1.2 form is thought to have a glutamine residue at that position. Because the two proteins differ by only a single amino acid they offered a good system for testing whether anti-peptide antibodies can discriminate between closely related proteins.

Hannah Alexander and her colleagues in the Lerner group synthesized six peptides, four of which corresponded to constant regions of Thy-1.1 and Thy-1.2. The other two spanned the region from amino acids 79 to 98. One had an arginine at position 89, as in Thy-1.1, and the other had a glutamine at position 89, as in Thy-1.2. Antibodies to the peptides from the conserved regions of the proteins reacted both with Thy-1.1 and Thy-1.2, whereas antibodies to the peptides that spanned the variable region reacted only with the predicted variant. By targeting the antibodies to the region that differs, the investigators were able to produce antibodies specific for the particular Thy-1 variants, which would have been difficult to achieve by conventional means.

Antibodies that behave like enzymes

It may be possible to take the predetermined specificity of anti-peptide antibodies one step further and make antibodies that bind to the same structures in proteins as enzymes do. The catalytic effectiveness of enzymes may depend on their ability to stabilize certain high-energy states – the transition states as they are called – of the substrate molecules. Antibodies that recognize those same states might carry out catalytic functions. In that event, the immunological antibody repertoire may serve as source of any kind of enzyme provided that the enzyme's substrate is sufficiently large to be immunogenic.

There is a real impetus to test these notions because, given the current ability to make antibodies to almost any position on a protein, success would be tantamount to being able to produce enzymes that split proteins at specific sites. Such enzymes would be analogous to the restriction enzymes that cut DNA at specific sites, except that the specificity of the enzymes that act on the proteins could be chosen by the experimenter.

But how could this be accomplished? One approach might be based on the assumption that anti-peptide antibodies work by an induced-fit mechanism and thus distort the site at which they bind. If additional energy were supplied, the protein might be split at the binding site. An alternative strategy is to make antibodies to intermediates in the catalyzed reaction to tip the equilibrium in favour of hydrolytic splitting of the peptide bond. The principle, then, is one of immunological catalysis.

To test whether immunological catalysis might be possible, Alfonson Tramantano, Kim Janda and Lerner attempted to obtain monoclonal antibodies to a substance that mimics the structure of the transition state in the hydrolysis of an ester. The investigators immunized mice with an ester that was designed to be a stable analogue of the transition state for the hydrolysis of carboxylic esters and isolated the monoclonal antibodies that could bind to the analogue. The antibodies were then tested for the ability to hydrolyse an ester. Three of them accelerated the rate of hydrolysis of the ester by factors of 10^3 to 10^4.

In a related development, Peter Schulz and his colleagues at the University of California at Berkeley began with a pre-existing antibody and showed that it could catalyse the hydrolysis of a compound that is structurally similar to the antigen of the antibody.

These results demonstrate the feasibility of deriving enzymatic activities from immunological specificities. The applications of such a method in protein chemistry, biochemistry and medicine could be profound. For example, instead of engineering the immune system to produce antibodies that simply bind to viruses or tumour antigens, it might be possible to evoke antibodies that inactivate the viruses or kill tumour cells by catalysing specific protein cleavages. The antibodies would essentially act directly without depending on help from accessory factors such as the proteins of the complement system. Regardless of whether these concepts are correct in detail, antibodies of predetermined specificity may soon take on roles that transcend their simple binding functions.

Medical applications of peptide-elicited antibodies

The specificity of peptide-elicited antibodies might also be exploited for the detection, treatment and prevention of human disease. The antibodies might form the basis of immunodiagnostic procedures to detect the presence of pathogenic organisms or of antibodies to the pathogens in clinical samples. Quite often the immunodiagnosis of diseases such as tuberculosis has relied upon rather crude preparations of antigens from the pathogenic organisms. Not only is it hazardous to grow and handle the large amounts of pathogen required to produce clinically useful amounts of the antigens, but it is difficult to standardize reactivities from one batch to the next. Also, the crude preparations of antigens often do not allow infections by closely related pathogens to be distinguished.

For example, the tuberculin skin test is commonly used to detect tuberculosis, which is caused by infection with *Mycobacterium tuberculosis*. Tuberculin is an acid precipitate of *M. tuberculosis* bacteria that have been maintained in culture for several weeks. When this antigenic material is injected into the skin it will trigger an immune reaction that is characterized by redness and swelling of the injected area in persons who have previously been exposed to the bacterium.

Tuberculin's usefulness is limited, however, because it cannot distinguish between responses caused by previous exposure to *M. tuberculosis* and those caused by exposure to closely related bacteria. In some regions of the world, perhaps as many as 30 per cent of the tuberculin-positive reactions result from contact with other mycobacterial species, such as *M. intracellulare* or *M. fortuitum*, and not from contact with *M. tuberculosis*.

A more specific skin-test antigen would help to detect tuberculosis with greater accuracy. Such a skin test might be based on the use of synthetic peptide reagents that are chemically defined and easy to standardize. The Lerner group has synthesized a 13 residue peptide that corresponds in sequence to a portion of a protein from *M. tuberculosis* that has a molecular weight of 10 000. The peptide can be used as a skin-test antigen to detect the bacterium in experimental animals and has proved to be much more specific in this regard than tuberculin.

Peptides and peptide-elicited antibodies might also form the basis of immunoassays for detecting infections. Here it might be possible to exploit the exquisite sensitivity of these antibodies to distinguish between closely related pathogens, such as the different serotypes of a virus. Serotypes are virus variants that are detected because they give rise to somewhat different antibody responses in infected individuals. For example, neutralizing antibodies to the hepatitis B virus are primarily directed against the surface antigen designated HBsAg. This protein also carries the three markers that define the four possible serotypes of the virus, two of the d type (adw and adr) and two of the y type (ayw and ayr). Several studies have shown that the surface antigens from the y and d serotypes can differ in as few as two of the 226 amino acid residues of the HBsAg sequence.

To produce antibodies that could distinguish between two proteins that differ by so little would be quite difficult if conventional techniques were used. However, with peptide immunogens the antibodies can be targeted to the region that contains the key amino acid differences. Two 13-residue peptides were synthesized to correspond to either the y- or d-specific sequence. Antibodies generated by the 'y-peptide' reacted only with the surface antigen from virus of serotype y and those elicited by the 'd-peptide' reacted only with the surface antigen from d-serotype virus. Such peptides might form the basis of a clinically useful immunoassay for serotyping hepatitis B virus. The possible advantages of an assay of this type include the ease with which serotype-specific antibodies could be produced without having to use the pathogen itself and the ready availability of the synthetic peptide as a stable, positive control.

One of the more exciting potential applications of peptide immunogens is in the production of safe, chemically defined vaccines for preventing disease. Such a vaccine could be made without having to grow or handle large quantities of the pathogenic agent, as is now required for the production of most conventional vaccines, and the vaccine would be free of any biological contamination. Moreover, peptides have excellent

stability at room temperature, a characteristic that would be particularly valuable in the developing countries where lack of refrigeration for storing vaccines can be a problem.

Although no peptide vaccine has yet made its way into clinical application, several have shown great promise either in tissue culture or in animal models. These include peptides that can confer some level of immunological protection against the hepatitis B, herpes simplex, feline leukaemia, foot-and-mouth disease, polio and rabies viruses, and also against a diarrhoea-causing strain of the bacterium *Escherichia coli*, and against the malaria parasite.

Work with hepatitis B virus, the cause of serum hepatitis, illustrates attempts to use synthetic peptides in a vaccine. Infection with this virus is characterized by a severe inflammation of the liver and afflicts more than 200 million people worldwide. Moreover, infection early in life has been correlated with a high incidence of liver cancer in adulthood.

Neutralizing antibodies, which bind to a virus and thereby trigger its destruction, are the important ones for immunological protection. As mentioned previously, most of the neutralizing antibodies to hepatitis B are directed against the surface antigen. Variation in these antibodies may be the cause of the serotype variation in the immune response to hepatitis B virus.

On the basis of this assumption, Lerner's group with that of John Gerin at Georgetown University School of Medicine, Washington, DC, asked whether the synthetic peptides that correspond to the d/y serotype region could confer immunological protection on chimpanzees that were challenged by infection with hepatitis B virus. Immunization with a peptide containing 28 amino acids that span the sequence specific to the y serotype resulted in partial or complete protection of the chimpanzees that were infected with the serotype y virus. This result, together with those on the other pathogens mentioned, indicates that a short peptide is capable of conferring immunological protection against a large, complex pathogenic agent.

However, because only partial protection was achieved in several animals, the results also indicate that much more must be learned about how to elicit the best protective response with peptide immunogens. The questions that must be addressed concern the most effective peptide and adjuvant to use and the optimum mode of administration of the vaccine. The answers to these questions are being actively pursued in a number of laboratories around the world in attempts to achieve the degree of vaccine efficacy required for widespread clinical application.

Although the specificity of the antibodies that are elicited by synthetic peptides is an advantage for some applications, it raises some questions about the utility of peptide vaccines on a worldwide basis. A change of two amino acids in the target region of the hepatitis B surface antigen is apparently capable of causing a serotype switch and thereby presumably enabling the virus to escape destruction by the serotype-specific neutralizing antibodies. Genetic variation is always present in a pathogen population and vaccination of very large numbers of people with a serotype-specific peptide might provide a positive selective force for the emergence of other serotypes and as a result render the vaccine less effective.

This problem can be circumvented, however, either by including peptides corresponding to all possible serotypes in the virus or by targeting the peptides and antibodies to regions of the protein that are functionally important and hence less likely to mutate. For hepatitis B virus, a vaccine that contains peptides corresponding to the y and d serotypes might be sufficient to protect against this virus, because all hepatitis B virus surface antigens are of either the y or d serotype.

For viruses that show a great deal of serotypic variation, such as the one that causes foot-and-mouth disease, including all possible variants might not be feasible. For these pathogens it might be better to concentrate instead on the sequences that are conserved among the various serotypes. These sequences are likely to include the functionally important regions of the molecule that cannot tolerate change and might thus make good targets for neutralizing antibodies.

Robert Neurath and his colleagues at the New York Blood Center in New York City have, for example, been attempting to target anti-peptide antibodies to a functionally important, conserved region of the hepatitis B virus. The surface antigen of the virus is encoded by the *env* gene, which can specify up to 400 amino acids. The 226 amino acids on the carboxyl end of this sequence constitute the mature surface antigen and the corresponding coding sequence is called the S region or S gene. The *env* DNA that precedes the S gene is called the pre-S sequence or gene. The pre-S protein is present in the infectious virus particle and appears to be involved in the binding of polymerized albumin to the virus. The supposition is that this binding site plays a role in the attachment of the virus to liver cells, which is necessary for infection to occur.

A peptide consisting of 26 amino acids from the amino terminal of the pre-S protein contains a dominant antigenic determinant for the protein and the virus. Antibodies to this peptide are detected early in the course of a hepatitis B virus infection in humans and may provide the basis of an immunoassay for diagnosing serum hepatitis.

When the 26-amino-acid peptide is used as an immunogen, it elicits antibodies that bind to the pre-S protein and inhibit binding of the virus particles to polymerized albumin. This result suggests that the antibodies might be able to interfere with virus binding to liver cells, thereby preventing the initial step in infection. Immunization of chimpanzees with the peptide can confer immunological protection against a subsequent challenge with virulent hepatitis B virus.

The pre-S peptide represents an antigenic determinant that is shared by all hepatitis B viruses so that immunization with the peptide should confer protection against essentially any serotype of the virus. Moreover, if genetic drift changed the target sequence in the virus so that the peptide-generated antibodies could no longer bind to it, the sequence might well be altered to the point where it could no longer bind to liver cells either. In other words, mutants that escape immune surveillance by altering the amino acid sequence of the pre-S peptide region may turn out to be non-infectious.

Summary

Peptides and peptide-elicited antibodies have become important research tools for the molecular biologist and protein chemist despite some initial reluctance on the part of the immunological community to accept peptide antigens as a general way of producing protein-reactive antibodies. The impact of these reagents on basic research endeavours continues to grow, as evidenced by the increase each year in the amount of published work that uses them in some manner. Peptides and the antibodies they elicit are providing important insights into the structure of proteins in solution and the relation between protein structure and function.

The reagents also open up new approaches for studying the workings of the immune system. They can help define the precise chemical nature of the determinants recognized by T cells and of antigenic and immunogenic variation, the mechanism of immune tolerance, and the regulation of immune responses. In addition, peptides are now being used as immunogens for producing monoclonal antibodies. In this way, the two technologies are being combined to achieve the easy generation of large amounts of pure antibodies that react with preselected, precisely defined targets.

In the laboratory, the peptides are easy to synthesize and the protein-reactive antibodies relatively easy to produce. However, the procedures used for generating the antibodies in laboratory animals are in general too harsh to use for immunizing humans or domestic animals. Consequently, a number of questions must be answered before peptide immunization technology can be applied to preventing clinically important diseases. These questions concern the design of the most efficacious peptide, the choice of the optimum carrier and adjuvant, and the determination of the vaccine doses, schedules and routes of administration that are necessary for disease prevention.

The ability of peptides to elicit the appropriate immune responses to provide protection and the effects of the various genes that regulate those responses must also be considered. However, because peptide immunogens display such features of an ideal vaccine as specificity, stability, ease of manufacture and lack of potential biological contamination, the questions are receiving high priority in research laboratories around the world. Much progress towards the goal of a clinically useful, peptide-based vaccine can be expected over the next few years. Of even greater promise are the possible uses of peptide reagents for the immunodiagnosis of disease. This goal is much more readily attainable than a vaccine, and a peptide-based immunoassay for use in the clinic can be expected within the next few years.

Additional reading

Attassi, M. Z. (1978). Precise determination of the entire antigen structure of lysozyme: molecular features of protein antigen structures and potential of 'surface simulation synthesis' – a powerful new concept for protein binding sites. *Immunochemistry*, **15**, 909–33.

Benjamin, D. C., Berzofsky, J. A., East, I. J., Gourd, F. R. N., Annum, C., Leach, S. J., Margoliash, E., Michael, J. G., Miller, A., Pager, E., Reichlin, M., Sercarz, E. E., Smith-Gill, S. J., Todd, P. E. and Wilson, A. C. (1984). The antigen structure of proteins: a reappraisal. *Annual Review of Immunology*, **2**, 67–101.

Brown, F., Chock, R. M., and Lerner, R. A. (eds.) (1986). *Vaccines 86*. Cold Spring Harbor Laboratory, Cold Spring Harbor, New York.

Green, N., Alexander, H., Wilson, A., Alexander, S., Shinnick, T. M., Sutcliffe, J. G., and Lerner, R. A. (1982). *Cell*, **28**, 477–88.

Niman, H. L., Houghten, R. A., Walker, L. E., Reisfeld, R. A., Wilson, I. A., Hogle, J. M. and Lerner, R. A. (1983). Generation of protein-reactive antibodies by short peptides is an event of high frequency: implications for the structural basis of immune recognition. *Proceedings of the National Academy of Sciences, USA*, **80**, 4949–53.

Pollack, S. J., Jacobs, J. W. and Schultz, P. G. (1986). Selective catalysis by antibody. *Science*, **234**, 1570–3.

Shinnick, T. M., Sutcfliffe, J. G., Green, N. and Lerner, R. A. (1983). Synthetic peptide immunogens as vaccines. *Annual Review of Microbiology*, **37**, 425–6.

Tainer, J. A., Getzoff, E. D., Paterson, Y., Olson, A. J. and Lerner, R. A. (1985). The atomic mobility component of protein antigenicity. *Annual Review of Immunology*, **3**, 501–35.

Tramontano, A., Janda, K. D. and Lerner, R. A. (1986). Catalytic antibodies. *Science*, **234**, 1566–70.

14 New methods for the diagnosis of genetic diseases

Yuet Wai Kan

Recombinant DNA technology has revolutionized the study of the human genetic diseases. Many of these diseases are caused by gene mutations that result either in the complete absence of a functional protein product or in a defective protein that does not work as it should. Recombinant DNA technology is providing new methods of detecting such gene lesions, thus permitting precise identification and localization of the defects that cause hereditary disorders. As a consequence, many of the conditions can be diagnosed in the fetus at an early stage of pregnancy. Moreover, the previously undetectable carrier states of some genetic abnormalities can now be identified, with the result that prospective parents can make informed genetic choices about whether to begin or continue a pregnancy.

For diseases in which the nature and precise locations of the genetic defects are not yet known, recombinant DNA methods are being used to determine the chromosomal regions where the affected genes are located. These diseases include cystic fibrosis and Huntington's disease, and intense efforts are currently under way to identify the particular genes in question. This will not only help in diagnosis, but may also lead to a better understanding of the biochemical bases of the conditions and possibly to more effective therapies. Finally, the technology has been very useful for understanding the natural history of genetic diseases. It allows the tracing of the origins and migrations of mutant genes and the quantitation of the number of times that the same genetic lesions have arisen in different populations.

Sickle cell anaemia and thalassaemia

The power of DNA analysis was first apparent with sickle cell anaemia and the thalassaemias, inherited disorders that are characterized by abnormal haemoglobin production. Although there are many different human genetic diseases, those involving haemoglobin are among the most important health problems in the world. Haemoglobin, which is the predominant protein in the red blood cells, functions primarily in carrying oxygen from the lungs to the tissues. The haemoglobin molecule is made up of four protein subunits, each of which carries an iron-containing haem molecule that serves to bind the oxygen. The protein chains are called globins.

Six major types of globins, which are designated alpha, beta, gamma, delta, epsilon and zeta, have been found in normal human haemoglobins (Table 14.1). These globin molecules are

Table 14.1. *The major globin types in human haemoglobin*

Haemoglobin	Globin types	Produced by	Percentage in adult
A	$\alpha_2\beta_2$	Adult	97
A$_2$	$\alpha_2\delta_2$	Adult	2.5
F	$\alpha_2\gamma_2$	Fetus and adult	<1
Portland I	$\xi_2\epsilon_2$	Embryo	—
Portland II	$\xi_2\gamma_2$	Embryo	—

produced in a regulated manner during embryonic, fetal and adult life. The embryo first produces the epsilon- and zeta-globins and the haemoglobin molecule then consists of two of each of these chains. The epsilon- and zeta-globins are replaced at around the tenth week of gestation by the alpha- and gamma-globin chains, which combine to form haemoglobin F, the major haemoglobin of fetal life. Just before birth, production of the gamma-globin chain decreases and it is replaced by synthesis of beta-globin. The principal type of adult haemoglobin is called HbA and consists of two alpha and two beta chains. There is also a minor adult haemoglobin, HbA2, which contains two alpha and two delta chains and constitutes about 2.5 per cent of the total (Fig. 14.1).

Genetic abnormalities can affect the synthesis of any of the globin chains, but because HbA is the major haemoglobin of adult life the aberrations that affect the production, structure or function of the alpha and beta chains cause the clinically important disorders. These disorders can be broadly separated into two categories: the haemoglobinopathies, in which abnormal proteins are produced, as in the case of sickle cell anaemia; and the thalassaemias, in which one of the globin chains is either not produced at all or is produced in inadequate amounts. Its structure remains normal, however.

These two groups of diseases are caused by recessive mutations and are inherited according to the Mendelian laws. The individual must be homozygous and have the defective gene

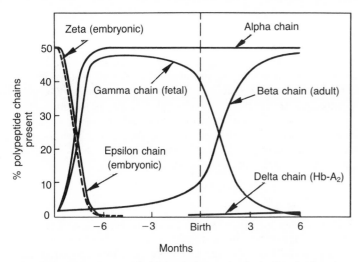

Fig. 14.1 The pattern of globin chain synthesis during fetal and postnatal life.

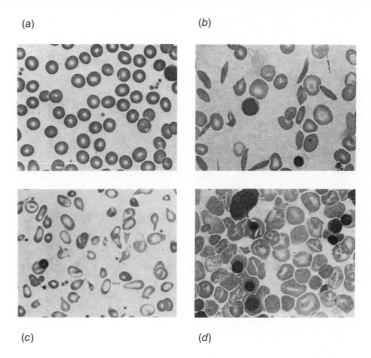

Fig. 14.2 Red blood cells from a normal individual (*a*) and from patients with sickle cell anaemia (*b*), homozygous *beta*-thalassaemia (*c*) and homozygous alpha-thalassaemia (*d*).

on both members of the appropriate chromosome pair for the genetic disease to be manifest. If both parents are heterozygous carriers that have the mutant gene on just one chromosome, each child has a 1 in 4 chance of inheriting two copies of the mutant gene and consequently developing sickle cell anaemia or a thalassaemia. Carriers do not have overt disease symptoms, but can be detected by blood tests. Sickle cell anaemia and the thalassaemias primarily affect people in Africa, the Mediterranean area, the Middle East and Asia.

Sickle cell anaemia

Of the hundreds of mutations that can affect either the alpha or beta chains of haemoglobin, the most important is the one that causes sickle cell anaemia. The sickle cell mutation occurs at the sixth amino acid of the 146 that make up the human beta-globin chain. Glutamic acid, the normal amino acid at this position, is changed to valine as a result of a single base change in codon 6 of the corresponding gene. This single amino acid substitution changes the properties of the haemoglobin molecule and produces serious clinical consequences.

When the abnormal haemoglobin loses its oxygen to the tissues, its molecules stack together to form fibrils that impart a rigidity to the normally pliable, biconcave red blood cells and distort them into sickled shapes (Fig. 14.2*a* and *b*). The cells can no longer flow freely through the small blood vessels, which become blocked by the cells. Affected individuals are plagued by painful crises, vital organs are damaged, and haemolytic anaemia results because the rigid sickle cells are prematurely destroyed in the circulation.

Thalassaemia

The thalassaemias are divided into two main groups, the alpha and beta thalassaemias, according to which globin chain is

synthesized in abnormal amounts. In the homozygous form of altha-thalassaemia the alpha-globin chains are absent and the primary haemoglobin in the fetal blood is composed of four gamma chains, a form of haemoglobin that does not function physiologically in oxygen transport. Immature red blood cells are found in the circulation (Fig. 14.2*d*) and the fetuses are extremely anaemic. They suffer from a condition known as hydrops fetalis, and almost always die during the last three months of pregnancy or within a few days of birth.

Because beta-globin does not replace the gamma-globin chain in haemoglobin until about the time of birth, the clinical manifestations of beta-thalassaemia do not become obvious until after birth. The red cells of affected individuals are smaller than normal and contain decreased amounts of haemoglobin (Fig. 14.2*c*). Regular blood transfusions are required to maintain their haemoglobin concentrations at adequate levels. To compensate for the anaemia the bone marrow mass of the patients expands and they absorb iron from the diet with abnormally high efficiency. The increased absorption, coupled with the iron overload that results from the blood transfusions, causes heavy deposits of iron in the tissues and damages many organs, including the heart and liver.

Until recently, patients with beta-thalassaemia generally died during childhood from infections, or in their early teens from heart failure as a result of the heavy iron load. Their prognosis has improved greatly during the last 15 years, largely as a result of a treatment regimen that consists of frequent blood transfusions plus daily injections of the drug desferroxamine to remove iron from the tissues. Some patients are now in their thirties.

The clinical diagnosis of sickle cell anaemia or thalassaemia is accomplished by examination of the blood. Under conditions of low oxygen concentration sickled red blood cells are easily discernible under the microscope. The small size of the red blood cells and the decreased haemoglobin content that are characteristic of thalassaemia are also clearly visible in a microscopic examination. The diagnoses can be confirmed by examining the haemoglobin itself. Sickle cell and normal haemoglobins can be readily distinguished as can the characteristic alterations in the alpha-or beta-globins of thalassaemia.

Prenatal diagnosis

With a few exceptions, the underlying defects of genetic disease cannot currently be corrected. Most treatments are therefore directed towards prevention or early treatment of the complications. In sickle cell anaemia, for example, infections are a common cause of death in infancy. Early diagnosis is desirable so that affected babies can be given antibiotics to help prevent infections and medical care at the first symptoms if an infection does develop. Although the growth and development of individuals who have beta-thalassaemia have been greatly improved by the regimen of transfusions and daily injections of desferroxamine, the patients nonetheless require constant medical treatment and hospitalization for complications.

One way of controlling genetic diseases is to prevent the birth of affected children. Members of a population with an increased risk of having a genetic disease can be educated about the nature and prevalence of the particular defect and tested systematically to see which individuals are carriers. With this information, couples who are at risk of having a child with the disease can be identified and counselled so that they can make informed reproductive choices. They have the option of preventing pregnancy entirely or, if they still want to have a baby, of having prenatal diagnosis and aborting an affected fetus. Over the past 10 years safe and accurate tests for the prenatal diagnosis of sickle cell anaemia and thalassaemia have been developed.

Fetal blood analysis

The first attempts at prenatal diagnosis of thalassaemia and sickle cell anaemia were straightforward. Because the diseases are manifested in the blood, a small sample of fetal blood could be obtained and tested for the presence or absence of the abnormal haemoglobins. Methods for obtaining fetal blood were developed in the mid-1970s and this early approach to prenatal diagnosis was quite successful. Fetal blood sampling is still the primary method of prenatal diagnosis for beta-thalassaemia in some European countries, but it has several drawbacks.

The procedure is usually performed fairly late in pregnancy – around the eighteenth gestational week – and a diagnosis is not reached until the twentieth week. If the parents decide to terminate the pregnancy then, the abortion must be performed in late pregnancy when the physical risks to the woman are greater and the psychological effects more pronounced. Furthermore, withdrawing fetal blood is not a routine procedure and can only be done by a few experienced obstetricians. The procedure is associated with some risks to the fetus, including bleeding and infection, although in experienced hands these risks are small. In a few cases of beta-thalassaemia the technique used to analyse the fetal blood cannot distinguish with certainty between the carrier and the disease states.

DNA analysis

Amniocentesis has been used to acquire fetal tissues for the prenatal diagnosis of birth defects since the 1960s. In this procedure a hollow needle is inserted through the abdominal wall of the pregnant woman and into the fluid filled amniotic sac that contains the fetus. A sample of the amniotic fluid, which contains cells of fetal origin, can then be withdrawn through the needle. The amniotic cells are derived from the amnion, the fetal skin, or the fetal respiratory and gastrointestinal tracts.

To detect chromosomal defects, such as the one that causes Down's syndrome, the cells are grown in culture and the chromosomes of the dividing cells are analysed. Analysis of the enzymes of the cultured fetal cells yields information about hereditary enzyme deficiencies, such as the one that causes Tay–Sachs disease. Analysis of the chemicals in the amniotic fluid can give information about the presence of some developmental abnormalities. For example, in spina bifida, a serious condition in which the spinal column does not form normally, the concentration of alpha-fetoprotein is elevated in the amniotic fluid. However, the amniotic fluid does not provide red blood cells and, consequently, amniocentesis could not at first be used for diagnosing sickle cell anaemia and the thalassaemias, nor could it be used for detecting genetic diseases for which the affected proteins are unknown. The advent of DNA analysis for the direct detection of gene mutations, which occurred towards the end of the 1970s, greatly expanded the possible scope of prenatal diagnosis.

Another development that helped in this regard is the use of chorionic villus biopsy, a new method for obtaining fetal cells and DNA. Amniocentesis cannot be attempted until the fifteenth week of gestation and the culturing and analysis of the cells requires another 2 weeks. This again means that if abortion were to be elected it would have to be done fairly late in pregnancy. Chorionic villus biopsy has made prenatal diagnosis possible at a much earlier stage of gestation. The chorion, which is another of the membranes enveloping the fetus, is of fetal origin. The early embryo is surrounded by microscopic projections of the chorion, which are called villi (Figs. 14.3 and 14.4). A few villi can be sampled, either by biopsy with forceps or by suction with a catheter, to provide DNA for analysis. Because chorionic villus biopsy is performed between the

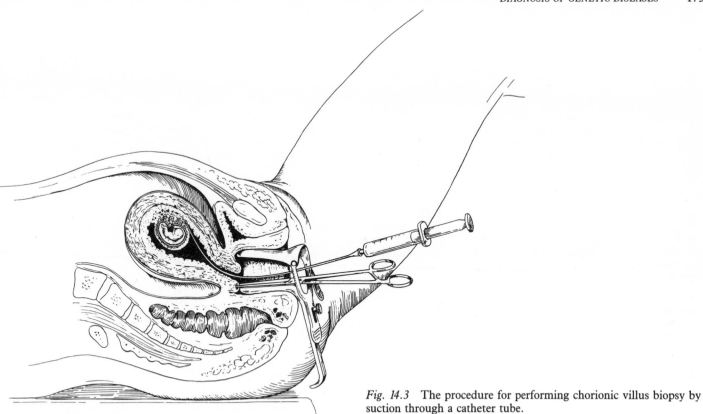

Fig. 14.3   The procedure for performing chorionic villus biopsy by suction through a catheter tube.

Fig. 14.4   Microscopic view of chorionic villi removed during biopsy (a), and a higher-magnification view of a single villus (b).

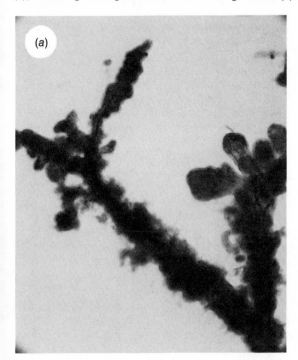

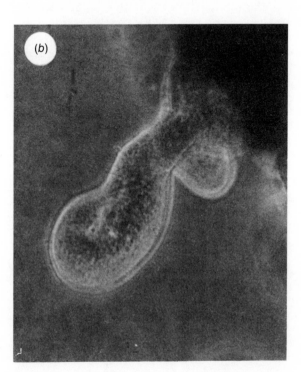

seventh and tenth gestational weeks and does not require lengthy periods for cell culture, it permits a fetal diagnosis within 3 months of pregnancy.

The technique was used in the Scandinavian countries in the early 1970s, in China and Russia in the mid-1970s, and was recently revived in Europe and the United States. Preliminary results of a cooperative study currently in progress in several medical centres to assess the safety of chorionic villus biopsy indicate that the risks of the procedure are small.

Development of DNA analysis

In 1976 alpha-thalassaemia became the first genetic disease to be successfully diagnosed by prenatal analysis of DNA. The functional alpha-globin genes are duplicated in humans and an individual has four copies of the gene per cell, two on each member of chromosome pair 16. In the lethal homozygous form of alpha-thalassaemia all four alpha-globin genes are deleted from the chromosomes. Diagnosis can be achieved by determining whether the DNA from the cells that were obtained by amniocentesis contains any alpha-globin genes.

In the early studies on the prenatal diagnosis of alpha-thalassaemia, Yuet Wai Kan and his colleagues at the University of California in San Francisco used a hybridization method in which a radioactive copy of the alpha-globin gene serves as a probe for detecting the gene in the subject's DNA. When the radioactive probe and the subject's DNA are mixed together in solution the probe will adhere to the DNA of the alpha-globin gene. Under carefully controlled conditions the degree of this hybridization to the subject's DNA is related to the number of copies of the alpha-globin gene that are present and can thus indicate whether the individual has the homozygous form of alpha-thalassaemia (Fig. 14.5). Although the hybridization method is feasible, it is technically difficult and requires large amounts of DNA that can only be obtained by culturing the cells of the amniotic fluid for 3 to 4 weeks.

The discovery of restriction enzymes (see Chapter 1) and the development of the Southern blot analysis have greatly simplified the prenatal diagnosis of the haemoglobin disorders. These methods can detect qualitative changes in genes, including mutations of single nucleotides, in addition to the quantitative changes in gene number that are caused by gene deletions. The development of these methods can be illustrated by their application to the detection of the three major types of mutations in the globin system, that is, those resulting in alpha-thalassaemia, sickle cell anaemia and beta-thalassaemia.

Prenatal diagnosis of alpha-thalassaemia has been accomplished by Southern blot analysis, a widely used method for analysing DNA that has been named after its originator, Edwin Southern of the University of Edinburgh. A sample of fetal DNA is digested with a restriction enzyme that cuts it at specific nucleotide sequences. The fragments are then separated by electrophoresis on a gel support and the fragment containing the alpha-globin genes is detected with a probe consisting of

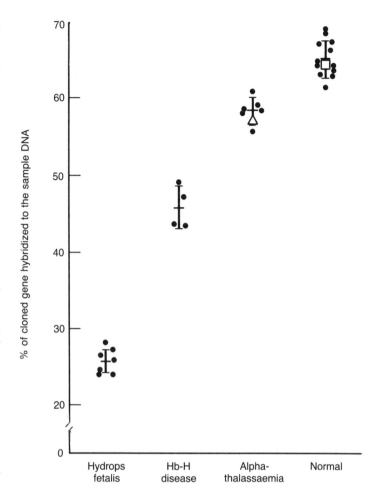

Fig. 14.5 Measurement of the number of alpha-globin genes by hybridization with a cloned radioactive alpha-globin gene. The first lane on the left indicates hybridizing to DNA from subjects with hydrops fetalis, who have zero copies of the alpha-globin gene; the second lane represents hybrizing to DNA from subjects with haemoglobin H disease, who have one copy of the gene; the third lane shows hybridizing to DNA from carriers of alpha-thalassaemia, who have two copies of the gene; and the right-hand lane shows hybridization to normal DNA samples, with four copies of the gene.

the radioactive alpha-globin gene. The fragment that normally contains the genes will be absent if the fetus has homozygous alpha-thalassaemia (Fig. 14.6).

Kan and his colleagues have recently developed a greatly simplified method for the prenatal diagnosis of alpha-thalassaemia. The fetal DNA is spotted on a piece of filter paper without first undergoing digestion or electrophoresis and a specific probe for the alpha-globin gene is applied. The absence of alpha-globin genes in fetuses with homozygous alpha-thalas-

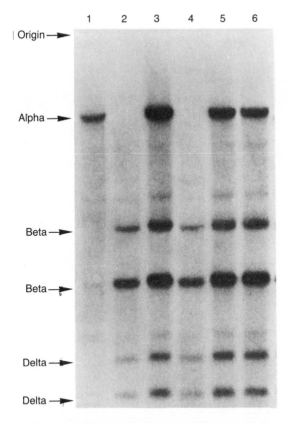

Fig. 14.6 Southern blot analysis of human DNA samples. Lanes 1, 3, 5 and 6 are control DNA samples that contain intact alpha-globin genes. Lanes 2 and 4 contain samples that were obtained from patients with homozygous alpha-thalassaemia.

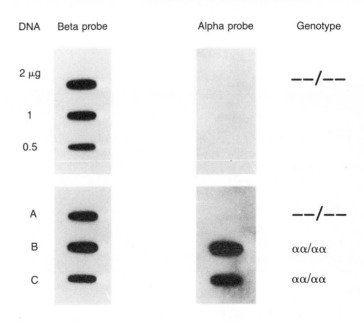

Fig. 14.7 Dot-blot analysis of DNA for alpha-thalassaemia. Decreasing quantities of a DNA sample from an infant with hydrops fetalis who had no alpha-globin genes were spotted on the upper pair of filters. The beta-globin probe indicates the presence of that gene but the alpha-globin genes are missing. Duplicate samples of amniotic fluid from three different subjects (A, B and C) were applied to the lower pair of filters. Samples B and C showed the presence of alpha- and beta-globin genes, but the alpha-globin genes were missing from sample A.

saemia is clearly discernible (Fig. 14.7). This method is now widely used for the prenatal diagnosis of alpha-thalassameia in China, where the disease is common.

DNA polymorphism and linkage analysis

The evolution of the prenatal test for sickle cell anaemia illustrates how rapidly progress has occurred in DNA diagnostic technology. Initially, the sickle cell gene could only be identified indirectly by linkage analysis with DNA polymorphisms, which are variations in DNA sequences that occur naturally in populations. In the original work the mutant gene was found to be linked to a particular variation that was located close to, but outside, the beta-globin gene. Within a few years this approach was replaced by a method of analysis that could directly pinpoint the single nucleotide mutation in the DNA of the gene itself. Linkage analysis with DNA polymorphisms is still used to study and diagnose many genetic disorders, however.

When Kan and his colleagues used the restriction enzyme designated *Hpa*I to cut samples of human DNA they found that the patterns of the fragments thus produced may vary from sample to sample. A probe for the beta-globin gene, for example, showed that the gene is normally located on a fragment that is 7.6 kilobases long. However, *Hpa*I digestion of the DNA from certain individuals yields a longer, 13-kilobase fragment that contains the beta-globin gene. The longer fragment results from a single nucleotide mutation that abolishes the first *Hpa*I site to the right of the gene. Consequently, the length of the DNA fragment is extended by about 5 kilobases, the distance to the next normally occurring site for the restriction enzyme (Fig. 14.8). The 13-kilobase fragment occurs mainly in people of African origin and is frequently associated with the sickle cell mutation of the beta-globin gene in the American black population.

The change in the *Hpa*I site does not affect the function of the beta-globin gene, but can serve as a useful genetic marker for the sickle cell mutation. In some families determination of the *Hpa*I fragment length can indicate whether or not a chromosome contains the sickle cell mutation. This type of DNA polymorphism, which results in fragments of different lengths when the DNA is digested with restriction enzymes, is known as 'restriction fragment length polymorphism' (RFLP).

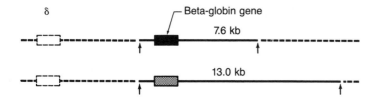

Fig. 14.8 The *Hpa*I sites (arrows) that are associated with the normal beta-globin gene (upper line) and the sickle cell gene (lower line). The site to the right of the normal gene is missing from the DNA with the sickle cell gene. As a result digestion of that DNA with the restriction enzyme generates a 13-kilobase (kb) fragment instead of a 7.6-kilobase fragment.

Analysis of RFLPs has application to many genetic disorders in addition to sickle cell anaemia.

Evolution of the sickle cell gene

Family studies, in which inheritance of a genetic disease can be linked to a particular RFLP, are a prerequisite for the prenatal diagnosis of the disease. In sickle cell anaemia, for example, not all sickle genes are associated with the 13-kilobase *Hpa*I fragment, and not all 13-kilobase fragments contain the sickle gene. Analysis of the *Hpa*I and other polymorphic restriction sites in the beta-globin gene cluster have provided insights into the origins and migration patterns of the sickle cell mutation and strongly suggest that it has occurred several times in world populations. This finding explains why in the United States only some of the sickle cell mutations of the beta-globin gene are associated with the 13-kilobase fragment.

The mutation of the *Hpa*I site that gave rise to the 13-kilobase fragment containing the beta-globin gene apparently first occurred in Upper Volta in West Africa. Subsequently, a second mutation – the one giving rise to the sickle cell gene – arose on the same chromosome that already contained the *Hpa*I mutation. The sickle cell mutation protects against malaria, which is endemic in that part of Africa, and the frequency of the 13-kilobase fragment with the sickle gene increased as a result. Because there is no selective advantage for the 13-kilobase fragment with the normal beta-globin gene, its frequency remained low. Consequently, in West African countries, such as Ghana and Nigeria, the majority of the 13-kilobase *Hpa*I fragments contain the sickle cell genes whereas only a few carry the normal beta chain gene (Fig. 14.9).

Other regions that have a high prevalence of sickle cell anaemia include East Africa, Saudi Arabia and India. In these areas the chromosomes containing the sickle cell mutation have retained the normal *Hpa*I site and yield the 7.6-kilobase fragment when digested with the enzyme. This result implies that the sickle cell mutation in these areas arose independently of that in West Africa. In fact, more recent studies in which Ronald Nagel of Albert Einstein College of Medicine in New York City,

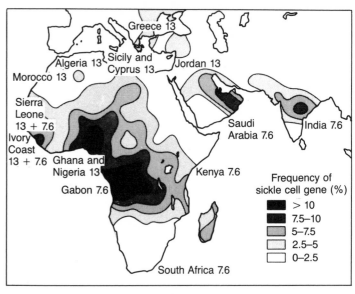

Fig. 14.9 The distribution in the Old World of the 7.6- and 13-kilobase restriction fragments that are generated by the *Hpa*I enzyme. The frequency of the sickle cell gene is also depicted.

Dominique Labie of INSERM's Institut de Pathologie et Biologie Cellulaires et Moleculaires in Paris, and their colleagues analysed several different restriction sites in the vicinity of the beta-globin gene suggest that the East African, Saudi Arabian and Indian genes also arose independently of one another. In each region the mutant genes reside on chromosomes with different restriction patterns.

Approximately 70 per cent of the sickle cell genes in the American black population are associated with the 13-kilobase *Hpa*I fragment; the remaining 30 per cent are associated with the 7.6-kilobase fragment. This distribution probably reflects the fact that most American blacks had their origin in those regions of western Africa where the 13-kilobase fragment is prominent. Although linkage analysis with the *Hpa*I polymorphism only works for prenatal diagnosis of sickle cell anaemia in about 70 per cent of the American black population, the use of polymorphisms for additional restriction enzyme sites increases the percentage of cases amenable to prenatal diagnosis.

Direct analysis of a point mutation

A restriction enzyme cuts DNA wherever a sequence corresponding to its recognition site occurs. The amino acid sequence at positions 5 to 7 of the normal beta-globin gene is proline – glutamic acid – glutamic acid, which is encoded by the nucleotide sequence CCT–GAG–GAG. The restriction enzyme *Mst*II cleaves at the recognition sequence CCTNAGG, where N equals any of the four nucleotides in DNA, and will therefore cut the normal beta-globin gene at precisely the site affected by the sickle cell mutation.

Digestion of normal DNA with the enzyme yields two fragments from this region: one that extends 1.15 kilobases to the left of the recognition site and another that extends 0.2 kilobases to the right of the site (Fig. 14.10). The cause of the substitution of valine for glutamic acid at position 6 in beta-globin is a change in the corresponding DNA sequence of the gene to CCT–GTG–GAG, which is not a cleavage site for *Mst*II. In sickle cell anaemia then, the two DNA fragments fuse into one of 1.35 kilobases. By using a specific radioactive DNA probe for the region of the beta-globin gene that includes the *Mst*II recognition site, it is possible to analyse directly for its presence or absence and therefore to detect the sickle cell gene.

DNA analysis has completely replaced fetal blood sampling for the prenatal diagnosis of sickle cell anaemia. Digestion with *Mst*II is now the most commonly used technique for detecting the sickle cell mutation, although the mutation can also be detected by another method of direct DNA analysis, which used oligonucleotide probes for determining the presence or absence of specific DNA sequences.

Beta-thalassaemia

Prenatal diagnosis of beta-thalassaemia is more complicated than that of alpha-thalassameia or sickle cell anaemia. More than 50 different mutations that affect the beta-globin gene and give rise to deficiencies of the beta-globin protein have thus far been described. Deletions account for about 15 of these, but the deletions are rare and do not cause a significant number of cases. The clinically important beta-thalassaemias are caused by more than 35 point mutations that alter just one or a few nucleotides.

The different point mutations that have been described disrupt the normal function of the beta-globin gene in various ways (Fig. 14.11). Some disrupt one or another of the control regions just before the start of the genes that are necessary for initiation of the transcription of the gene into messenger RNA, thereby preventing messenger RNA synthesis. In addition, mutations occasionally affect the termination signal at the end

Fig. 14.10 Direct detection of the sickle cell mutation in the beta-globin gene. When the *Mst*II site, which spans codons 5, 6 and 7 of the gene, is lost as a result of the sickle cell mutation, digestion with the enzyme produces a 1.35-kilobase fragment instead of the normal 1.15-kilobase fragment. The lanes marked AS contain DNA samples from two carriers of the sickle cell gene; these have both types of fragments. The DNA in the lane marked AA is from a normal individual, who has only the normal adult beta-globin gene. And the lane marked SS contains a DNA sample from a sickle cell patient with only the mutated gene.

Fig. 14.11 The different types of mutations that may disrupt globin gene expression and give rise to beta-thalassaemia. The upper line shows the organization of the beta-globin gene with the protein-coding exons in black and the non-coding introns represented by the dotted lines. Deletions (1) can remove part or all of the gene, but are uncommon. The clinically important mutations affect just one or a few nucleotides. If these occur in the promoter (2), a control region just before the beginning of the gene, they can prevent transcription into messenger RNA. Mutations in the introns (3) or in the polyA tail (4) can interrupt beta-globin synthesis by preventing the normal processing of the messenger RNA. Mutations that block translation of the processed messenger into protein include changes in the initiation codon (5), and nonsense and frameshift mutations (6) that generate termination codons in the coding regions. Mutations in the normal termination codon (7) can also block translation. Finally, some mutations lead to production of an unstable globin (8) that is destroyed prematurely.

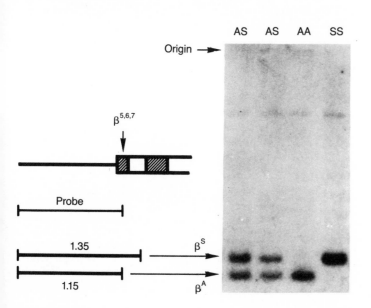

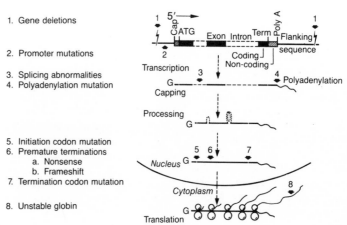

1. Gene deletions

2. Promoter mutations

3. Splicing abnormalities
4. Polyadenylation mutation

5. Initiation codon mutation
6. Premature terminations
   a. Nonsense
   b. Frameshift
7. Termination codon mutation

8. Unstable globin

of the gene and result in abnormal termination of the messenger RNA, which consequently becomes deficient in amount.

Other mutations prevent the normal processing of the messenger RNA. The globin genes, like most of the other genes of higher organisms, consist of protein-coding regions (exons) that are separated by non-coding intervening sequences (introns). The globin genes have three exons and two introns. Both the exons and introns are transcribed into messenger RNA, but the introns are spliced out before the messenger RNA is translated into protein. Point mutations at, or in the vicinity of, an intervening sequence may cause it to be processed abnormally. The result would be a deficiency in beta-globin synthesis because abnormally processed messenger RNAs are often unstable and usually non-functional.

Several types of point mutations also interfere with the translation of the beta-globin messenger RNA into the protein. Mutations in the initiation codon can affect the beginning of translation. Coding sequence mutations may change a normal amino acid codon into a stop codon, thereby causing premature termination of the chain synthesis. Other point mutations may add or delete nucleotides and shift the reading frame, thereby changing all the codons after the mutation. None of the beta-globin genes with translational mutations produce functional proteins. Finally, some mutations produce haemoglobins that are extremely unstable because of amino acid substitutions in critical regions of the beta-globin chain.

Only a few of the point mutations in beta-thalassaemia happen to create a new restriction site or abolish an existing one and are thus detectable by restriction enzyme analysis. Two approaches are available for detecting the rest of the mutations. The first is an indirect method that depends on linkage of the mutation with polymorphic restriction sites. The second is hybridization with oligonucleotides that specifically bind to the gene in the region of the mutation.

Polymorphic restriction sites along the beta-globin gene cluster

The beta-globin gene is one of a family of related genes that are located in a cluster on chromosome 11. The family includes the epsilon- and gamma-globin genes, the non-functional pseudo-beta-globin gene, and the delta-globin genes, in addition to the gene for beta-globin.

Since the discovery of the *Hpa*I polymorphism, about 20 additional restriction sites in the beta-globin gene cluster have been found to be polymorphic, that is, present in some individuals but absent in others. The different polymorphic sites can be detected by using probes for the various genes located in the cluster. By analysing for the presence of these sites on a specific chromosome, a pattern can be established and the chromosomes can be classified into haplotypes according to their patterns.

Detailed studies of Italian beta-thalassaemia patients and their families showed that nine haplotypes commonly occur in this population (Fig. 14.12). Each of the different beta-thalassaemia mutations became linked to the haplotype of the chromosome on which it originated. Of course, chromosomes with the same haplotype can have more than one mutation. Conversely, the same mutation may be linked to more than one haplotype, either because the mutation occurred more than once or because two chromosomes exchanged genetic material. Consequently, knowing the haplotype of a particular chromosome does not definitively point to the presence of a specific beta-thalassaemia mutation.

Nevertheless, in a given family, the polymorphic restriction sites are effective as markers to indicate which chromosomes contain an abnormal beta-globin gene. By choosing strategic restriction sites one or two restriction enzymes will usually be sufficient to mark the chromosomes and identify those whose chromosomes contain the normal and mutant genes within a family known to be at risk for the hereditary disease. This technique can be employed successfully for prenatal diagnosis of beta-thalassaemia in families once the linkage between a haplotype and a mutation has been established. However, these family studies are tedious and time-consuming, and may be impossible if no family members who are homozygous for the mutant gene are available.

Fig. 14.12   The beta-globin gene cluster. The arrows indicate the positions of seven polymorphic restriction sites; the restriction enzyme for each site is identified below the arrow. The patterns and frequencies of the nine haplotypes found in the Italian population are shown. Plus signs indicate which sites are present, and minus signs indicate which are absent, for each haplotype.

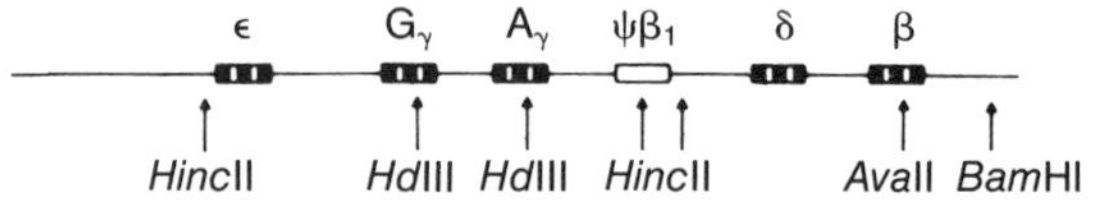

Haplotypes	ε HincII	Gγ HdIII	Aγ HdIII	ψβ1 HincII	ψβ1 HincII	δ AvaII	β BamHI	Overall frequency (%)
I	+	−	−	−	−	+	+	47
II	−	+	+	−	+	+	+	17
II	−	+	−	+	+	+	−	8
IV	−	+	−	+	+	−	+	1
V	+	−	−	−	−	+	−	12
VI	−	+	+	−	−	−	+	6
VII	+	−	−	−	−	−	+	6
VIII	−	+	−	+	−	+	−	1
IX	−	+	−	+	+	+	+	3

Direct analysis with oligonucleotide probes

A more precise means of analysing point mutations in beta-thalassaemia is with oligonucleotide probes – DNA sequences that usually contain 18 to 20 nucleotides and are tailored to match any desired target sequence in the genome. To analyse for a point mutation a pair of radioactive probes is synthesized, one corresponding to the normal sequence and the other to the mutated sequence. The two probes are hybridized to the test DNA on separate Southern blots. Under appropriately controlled conditions the normal probe will only hybridize with normal DNA, and the mutant probe only with the thalassaemic DNA (Fig. 14.13). This technique can distinguish completely normal individuals who have two unaffected genes from carriers who have one normal and one mutant gene and from patients with two mutant genes. It has been used very successfully with beta-thalassaemia and other genetic disorders.

The task of choosing the appropriate probe for a given family at risk seems formidable because so many different point mutations give rise to beta-thalassaemia and each mutation needs its own unique pair of oligonucleotide probes. However, recent studies of beta-thalassaemia in Italy indicate that although there may be many different mutations in the population, only a few occur frequently. Three of the nine mutations described in Italy account for more than 80 per cent of the beta-thalassaemia lesions.

Furthermore, the mutations have distinctive geographic distributions. Two mutations predominate in the Ferrara area, for example, while one mutation accounts for 95 per cent of the lesions on the island of Sardinia. Consequently, if the part of Italy where a patient's family originated is known, appropriate probes can be chosen to test for the beta-thalassaemia mutation.

Other genetic disorders

Approximately 3000 genetic disorders are caused by mutations affecting single genes. Defective genes that are carried on the X chromosome primarily affect males, who have only one X chromosome. Males have a 50 per cent chance of inheriting a mutant gene from a female carrier and developing the disease in question. Females have two X chromosomes and would have to have the defective gene on both to develop the symptoms. Haemophilia, a disease in which the blood fails to clot normally, is an example of such an X-linked disease.

Genetic disorders that are caused by genes that are carried on the autosomes, that is, on chromosomes other than the sex chromosomes, can be either dominant or recessive. Autosomal dominant diseases are evident when only one of the two gene copies is affected; Huntington's disease (which is also called Huntington's chorea) is a case in point. Autosomal recessive disorders, which include most hereditary diseases, are usually symptomatic only when both gene copies are defective. During the past 5 years DNA analysis has uncovered the affected genes in increasing numbers of these disorders.

In those for which the defective gene products have been identified, the approach to DNA analysis is similar to that for the globin genes. The gene DNA is isolated and, if the mutation causing the defect has not yet been delineated, is used as a hybridization probe for detecting the abnormal gene by linkage to polymorphic restriction sites. If the genetic lesions are known, direct detection with restriction enzymes or oligonucleotide probes can be performed. If the defective gene products are unknown, as is the case for Huntington's disease and cystic fibrosis, the strategy is to use random fragments of human DNA as probes to search for linkages to the mutant gene.

Fig. 14.13 Oligonucleotide probe analysis for a beta-thalassaemia mutation. The mutation converts the thirty-ninth codon of the beta-globin gene from CAG to TAG, thereby changing a glutamine codon to a stop codon. This mutation is common in Italy. The ß^A oligonucleotide probe detects the normal beta-globin gene sequence and the ßth probe detects the mutated sequence. In the family tested the parents (I-1 and I-2) are heterozygous for the mutation; their DNA hybridizes with both probes. The DNA from their child (II-1), who is homozygous for the mutation, hybridizes only with the mutant probe.

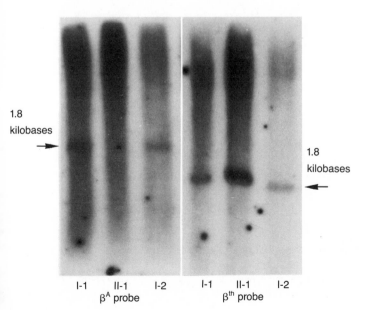

1.8 kilobases

1.8 kilobases

I-1   II-1   I-2
ß^A probe

I-1   II-1   I-2
ßth probe

Diseases in which abnormal gene products are known

Phenylketonuria is an autosomal recessive disease that is most commonly caused by a defect in the enzyme phenylalanine hydroxylase, which converts the amino acid phenylalanine to tyrosine. As a result of the deficiency in the enzyme activity, high concentrations of phenylaline accumulate in the blood and other tissues. The elevated phenylalanine concentrations have a harmful effect on brain development. No effects of the disease are seen at birth, but as soon as the newborn begins to consume foods rich in phenylalanine – such as milk – the symptoms become increasingly evident. Left untreated, 98 per cent of the patients will have IQs below 70 (100 is considered average for the population). Other manifestations of phenylketonuria include convulsions, psychotic behaviour, eczema and loss of skin pigmentation.

The condition can be readily detected at birth because of the high concentrations of phenylalanine that appear in the urine. Current treatment consists of maintaining the affected child on a diet low in the amino acid. Careful monitoring of the blood phenylalanine is necessary because low concentrations can be just as dangerous as high concentrations. Abnormally low levels of the amino acid can lead to poor growth, retarded bone development, an enlarged liver, infection, low blood sugar and neurological disorders.

Patients who adhere rigidly to their diet will develop fairly normally, although they tend to have IQs on the low end of the average range. Maintaining a child on a low phenylalanine diet can be difficult, however, because many popular foods must be avoided. Even when treated, therefore, the disease can pose management and psychological problems.

Savio Woo and his colleagues at Baylor College of Medicine in Houston, Texas, have recently cloned the gene for phenylalanine hydroxylase. The cloned DNA, which is a copy of the corresponding messenger RNA, is approximately 3 kilobases long. The entire genomic sequence spans nearly 100 kilobases of DNA. Because the gene is so large, it contains many polymorphic restriction sites. These sites serve as markers for linkage analysis for carrier identification and for prenatal detection of the phenylalanine hydroxylase deficiency. The mutations that give rise to phenylketonuria are just beginning to be understood. So far the results indicate that only a few mutations cause the disorder, at least in northern Europe. Once these mutations have been defined, direct detection with specific oligonucleotides or restriction enzymes will be possible.

Another hereditary disease in which the protein defect is known is the alpha-1-antitrypsin deficiency. Alpha-1-antitrypsin is the body's principal inhibitor or the protease enzymes that break down proteins. The disease is quite variable in severity, but affected individuals suffer liver damage because the protease inhibitor fails to be released from the liver, and emphysema because the alpha-1-antitrypsin is the primary inhibitor of the protein-splitting enzyme elastase in the respiratory tract. As a result of the overactivity of elastase, the lungs undergo the characteristic damage of emphysema.

The alpha-1-antitrypsin protein is remarkably polymorphic – more than 30 variants have been identified to date. The most important is the Z type, which arises from a mutation of glutamic acid to lysine at amino acid position 153. The Z gene can now be differentiated from the normal M gene by hybridization with synthetic oligonucleotide probes, and Woo's group

has devised a successful method for the prenatal diagnosis of the alpha-1-antitrypsin deficiency.

The most common hereditary bleeding disorders are the haemophilias. Haemophilia A, which is caused by lack of clotting factor VIII, accounts for 90 per cent of the cases, and haemophilia B, which results from a deficiency of clotting factor IX, accounts for the remaining 10 per cent. The degree of bleeding suffered by the patients varies according to the amount of clotting factor remaining in the blood. Some patients require regular infusions of factor VIII or IX to maintain concentrations sufficient to prevent bleeding; others have less severe deficiencies and require fewer infusions.

The two haemophilias are sex-linked disorders, the genes for which are located on the long arm of the X chromosome. A female carrier has a 50 per cent chance of having a son with the disease and also a 50 per cent chance of having a daughter who is a carrier. Spontaneous mutations – those that appear in a family with no previous history of haemophilia – have also been delineated. Queen Victoria, for example, apparently carried such a spontaneous mutation. Prince Leopold, the youngest of her four sons, had haemophilia and two of her daughters, Alice and Beatrice, were carriers. Alice's daughter eventually married Tsar Nicholas II of Russia and became the Empress Alexandra. Their son Alexis proved to have haemophilia, a circumstance that contributed to the eventual overthrow of the Russian royal house in the revolution of 1917.

Two groups of investigators, one from Genentech, Inc., in South San Francisco and one from the Genetics Institute in Boston, Massachusetts, have cloned the gene for factor VIII. George Brownlee of Oxford University in England, Earl Davies of the University of Washington in Seattle and their colleagues have cloned the factor IX gene. Many mutations, including both point mutations and partial or complete deletions, affect the functions of the two genes. In addition, defects that prevent normal splicing of the messenger RNA for factor IX have been identified.

Several of the mutations of the factor VIII gene affect restriction recognition sites and are detectable by direct enzyme analysis. There are also a number of polymorphic sites in and around the clotting factor genes that can be used to track haemophilia in cases where direct detection of the mutation is not possible. A probe made from a random DNA fragment detects a highly polymorphic site that is closely linked to the factor VIII gene, and can be used for linkage analysis of haemophilia A.

DNA probes can also be used for carrier detection. Although a haemophilia patient can easily be diagnosed by measuring the level of factor VIII or IX activity in his blood plasma, identifying female carriers of the haemophilia gene is difficult. Their clotting factor concentrations average 50 per cent of normal, but show sufficient variability to cloud positive identification of some carriers. DNA markers can determine which chromosome carries the abnormal gene and can therefore identify female carriers.

Diseases with undefined genetic loci

The protein defects underlying many hereditary disorders are unknown, although the conditions cause well-defined clinical symptoms. Even the chromosomes on which defective genes are located are often unidentified, unless the pattern of inheritance of the disease places the gene on the X chromosome. The chromosomal location might also be identifiable if the clinical symptoms occur in patients who have deletions of specific chromosome segments, in which case the affected gene can be inferred to lie on the deleted segment.

DNA analysis offers a way of identifying the genes affected in those hereditary diseases for which there is no information about the biochemical defect. Some DNA sequences appear only once in the human genome, while others are repeated hundreds or thousands of times. To study diseases involving unknown protein products, individual unique sequences are first used to identify restriction sites that display polymorphism. Such a sequence can then be used to examine the DNA of members of a population that is affected by a hereditary disease to see whether any particular variant of the polymorphic site is inherited together with the disease. If such a linkage is found, then the disease gene and the polymorphic site are presumably near one another on the same chromosome. The identity of that chromosome can then be determined by any of several techniques.

Individuals who have Huntington's disease are healthy until they are about 40 years of age. At that time degenerative changes in the central nervous system cause uncoordinated and involuntary movements. As the disease progresses, patients become mentally and physically debilitated, usually dying within a few years of the onset of symptoms.

The children of a patient with Huntington's disease have a 50 per cent chance of inheriting the gene. Because the gene is an autosomal dominant one, all of those who inherit it will develop the disease. Previously, those who inherited the Huntington's gene could not be identified until symptoms developed, but James Gusella of Harvard Medical School and his colleagues have recently identified a marker for Huntington's disease. They did this by using several random DNA probes to study a large Venezuelan kindred that is affected by the condition. One of the probes identified a polymorphic site that is closely linked to the disease and subsequently localized the site to the tip of chromosome 4. Efforts are under way to pinpoint the gene itself.

Duchenne muscular dystrophy is the most common form of muscular dystrophy. Affected males suffer from weakness resulting from gradual muscle degeneration that begins *in utero* and continues after birth. The disease is first evidenced by weakness of the limb muscles, but progresses to all the muscles, including the heart. Affected individuals usually die during childhood.

Some patients with Duchenne muscular dystrophy have a rearrangement of the short arm of the X chromosome and the affected gene is presumed to lie in this region. Several research groups have concentrated their attention on this chromosome region and have found a probe for a site very close to the gene for Duchenne muscular dystrophy. The probe enables researchers to trace the inheritance of the disease with 95 per cent certainty.

Cystic fibrosis is a common autosomal recessive disorder that occurs in 1 in every 2000 live births in the Caucasian population. Carriers of the defective gene have no symptoms, but individuals who inherit two copies of the gene have a severe illness. The secretory glands produce a thick mucus. When this accumulates in the bronchial tubes of the lungs the result is chronic obstructive lung disease and frequent infections. Secretions from the pancreas are also abnormal, with the result that food is not properly digested and absorbed from the intestines. Affected individuals have cirrhosis of the liver and males are often infertile. The sweat glands secrete abnormally high concentrations of sodium chloride, which can be used as a diagnostic test for the disease. The course and prognosis of cystic fibrosis vary, depending on how severely the lungs are affected.

Lap Chi Tsui of the University of Toronto, Canada, and Robert Williams of St Mary's Hospital in London, England, used random DNA probes to search for the genetic locus for cystic fibrosis. They found DNA probes that identify sites close to the gene, which has been localized to the long arm of chromosome 7. Two of the sites are so closely linked to the gene that this disease can be tracked with a confidence level greater than 95 per cent. The probes are being used successfully for carrier detection and prenatal diagnosis.

Ethical issues raised by prenatal diagnosis

The number of genetic diseases amenable to detection by the new DNA technology is increasing rapidly. It has made prenatal diagnosis possible for diseases that could not previously be detected before birth and has permitted the identification of carriers of defective genes. These recent advances will no doubt benefit increasing numbers of families that are affected by genetic diseases.

Nevertheless, scientific progress often raises questions and several important ethical issues have arisen as a consequence of the availability of prenatal tests. Because cures are not available for most genetic disorders, identification of an affected fetus primarily gives parents the option of terminating the pregnancy. Whether this action is justified for every inherited disease is the subject of intense debate. Sometimes the choices are clear-cut. A fetus with alpha-thalassaemia will inevitably die before or at birth. Early termination of the pregnancy saves the mother from 20 weeks of gestation as well as the potential complications of toxaemia, difficult labour and post-partum haemorrhage.

The current treatment of homozygous beta-thalassaemia patients with daily injections of desferroxamine and monthly blood tranfusions enables them to lead normal, or nearly normal, lives. However, the long-term effects of the treatment cannot be assessed yet because it has only been in use since the 1970s. Futhermore, the treatment regimen requires significant effort on the part of the patient and the parents. The high cost of the therapy – up to $50 000 per year – places a severe strain on the individual families and on the economies of those developing countries in which the disease is common.

The availability of prenatal diagnosis and elective abortion has already influenced the reproductive patterns of individuals who are carriers of the beta-thalassaemia gene. Many such individuals, who have already had to watch the suffering of an affected child and would previously not have had more children, are now electing to do so. In Italy, Greece and Cyprus the number of homozygous births has been significantly reduced, or even eliminated (Fig. 14.14).

Prenatal diagnosis of sickle cell anaemia has had a mixed reception, primarily because of the variability of the clinical course of the disease. Most patients with sickle cell anaemia can expect to have some of the symptoms, but some people are affected worse than others. The reasons why patients have such variable prognoses are not entirely known and it is impossible to predict whether an affected individual will develop severe clinical symptoms. Preliminary experience with prenatal diagnosis indicates that roughly half of the parents select abor-

tion of an affected fetus, whereas the other half continue the pregnancy.

The availability of prenatal diagnosis for phenylketonuria raises a different issue. This disorder can be diagnosed at birth with a simple blood test. If the affected child follows a strict diet during infancy and childhood, many of the complications can be avoided. However, some of the children who have the disease have residual brain damage, and some patients experience severe psychological and disciplinary problems because of the strict dietary regime they must follow. Consequently, some parents may still choose prenatal diagnosis and elective abortion.

The ability to identify those who have inherited the gene in conditions such as Huntington's disease that have a late onset presents a different set of problems. The DNA probes allow for their identification many years before the disease actually strikes. Because no treatment exists to alter the course of the disease, identifying these individuals at an early age could impose an enormous psychological burden on them. It could also help them to make an informed decision about whether to have children and risk passing the gene on to another generation, however. Moreover, it could provide a tremendous relief to the 50 per cent of potential Huntington's victims who turn out not to have inherited the gene.

Fig. 14.14 The number of children born with homozygous beta-thalassaemia has declined in some countries since the introduction of prenatal diagnosis.

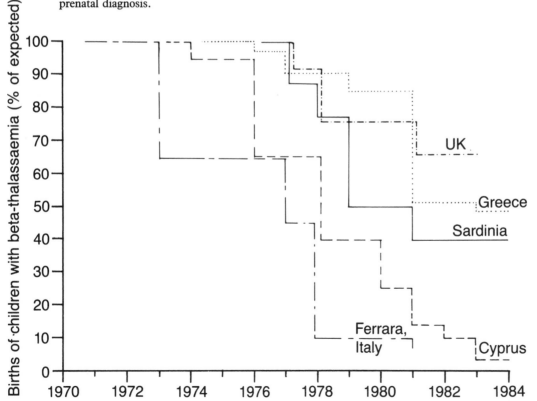

Future directions of DNA research in genetic diseases

The use of DNA probes will ultimately enable investigators to define the exact molecular defects in such diseases as Duchenne's muscular dystrophy, cystic fibrosis and Huntington's disease. Methods for 'walking' or 'jumping' along a chromosome allow researchers to move relatively quickly from one point to another on a chromosome and should eventually allow them to move from the markers they have identified for these hereditary diseases to the affected genes.

The development of some diseases may be partially influenced by genetic susceptibilities, even though the conditions are not completely genetic in origin. Another active area of research concerns the use of DNA probes to identify persons who are susceptible to such diseases, which may include heart diseases and at least some cancers. There is some indication, for example, that certain DNA polymorphisms that are associated with a gene for a cholesterol-carrying protein may indicate a propensity to heart attacks. Similar efforts are under way to identify persons who are susceptible to cancer.

Perhaps the most exciting prospect offered by the DNA technology is the eventual possibility of curing genetic disorders by inserting cloned normal genes into the human genome (see Chapter 15). In one strategy, viral vectors are used to carry the normal gene, although the point of insertion in the genome cannot be controlled with these vectors. Nevertheless, investigators are moving toward trying this approach to correct an enzyme deficiency that causes a severe immune deficiency disease. Lesch–Nyhan syndrome and Gaucher's disease are also candidates for gene therapy by this procedure.

Another strategy, which is still in the very early stages of development, aims at replacing the abnormal gene with a cloned normal version. This approach may be most effective with sickle cell anaemia and the thalassaemias, in which the genes must be accurately controlled during normal development. Results with cells growing in culture show that a gene can be inserted into its normal location, although the insertion must occur with a much greater frequency than is currently possible before this approach could be effective. Nevertheless, a few years ago the prospect of gene therapy would have been merely a dream. Thanks to recombinant DNA technology, it may soon become a reality.

Additional reading

Alter, B. P. (1984). Advances in the prenatal diagnosis of hematologic diseases. *Blood*, **64**, 329.

Bunn, H. F. and Forget, B. G. (1986). *Hemoglobin: Molecular Genetic, and Clinical Aspects*. W. B. Saunders, Philadelphia.

Cao, A., Pirastu, M., and Rosatelli, C. (1986). The prenatal diagnosis of thalassaemia. *British Journal of Haematology*, **63**, 215.

Jackson, L. G. (1985). First-trimester diagnosis of fetal genetic disorders. *Hospital Practice*, **20**, 39.

Kan, Y. W. and Dozy, A. M. (1980). Evolution of the hemoglobin S and C genes in world populations. *Science*, **209**, 388.

Kazazian, H. H. Jr (1985). The nature of mutation. *Hospital Practice*, **20**, 55.

Lawn, R. M. and Vehar, G. A. (1986). The molecular genetics of hemophilia. *Scientific American*, **254** (March), 48–54.

Stanbury, J. B., Wyngaarden, J. B., Fredrickson, D. S., Goldstein, J. L and Brown, M. S. (1983). *The Metabolic Basis of Inherited Disease*, 5th edn. McGraw-Hill, New York.

15 The prospect of gene therapy for human hereditary diseases

Jean L. Marx

Perhaps the most profound impact of the new ability to manipulate DNA will come in the realm of therapy for the human hereditary diseases, which are caused by defective genes. Many of these conditions are relatively rare. For example, cystic fibrosis, which is most common genetic disease of Caucasians, afflicts about 1 in every 1800 white persons in the United States. Sickle cell anaemia primarily affects blacks and occurs in 1 in every 500 black babies born in the United States. Other genetic diseases may affect just a few individuals. Nevertheless, the total impact of gene defects is large. According to current estimates there are at least 3000 genetic diseases.

Moreover, the defects are all too often lethal or at least debilitating. An abnormal gene may either produce no protein at all or may direct the synthesis of one that does not work as it should. A few hereditary diseases can be treated simply by giving the patient the protein that the gene fails to make normally. The type of dwarfism that is caused by inadequate production or release of growth hormone by the pituitary gland is a case in point. Injections of human growth hormone, which is becoming more readily available now that the gene has been cloned, can restore normal growth if they are given early enough.

Most genetic diseases are not so amenable to treatment, however. This may be true even when the gene defect is well understood, as is the case for sickle cell anaemia. Good medical care can help sickle cell patients to live for decades, but they will be subject to painful and incapacitating 'sickle-cell crises'. Individuals with cystic fibrosis rarely live much past their twentieth or thirtieth birthday. Other genetic diseases, such as Tay–Sachs disease and certain severe hereditary immune deficiencies, kill even earlier, after just one to a few years of life.

The ability to isolate and clone specific genes has opened the way for a dramatic new approach to treating human hereditary diseases – not just by supplying the missing protein but by actually giving the patient a good new gene to take over where the defective one has failed. Since the beginning of the 1980s researchers have made rapid strides towards developing methods for introducing new genes into animal cells. The genes have proved to work well, producing their active protein products, in cells that are grown in laboratory dishes. By the end of 1984 the first clinical trials of gene therapy in human patients were considered imminent. An unexpected obstacle arose, however, when investigators found that genes introduced into the cells of living animals by methods similar to those envisioned for human gene therapy more often that not failed to produce significant amounts of active proteins.

Attempting to use gene transfer on human patients would be pointless, and even unethical, without a reasonable expectation that the new gene would work. Nevertheless, not all the gene transfer experiments gave discouraging results. In a few, the newly introduced genes were found to be active in experimental animals, even if the amounts of proteins made were not considered sufficient to correct a genetic defect. Moreover, transferred genes were found to correct a gene defect in human cells that were grown in culture. By mid-1986 researchers were becoming optimistic again that the expression problems could be overcome and human gene therapy ultimately attempted.

Candidates for gene therapy

The hereditary diseases that are currently the best candidates for gene therapy are those in which the defect can be corrected by introducing a good new gene into bone marrow cells. Bone marrow, which is the soft, spongy tissue in the cavities of most bones, forms both the oxygen-carrying red blood cells and the various types of white blood cells that are needed for mounting effective immune responses. Marrow cells can be readily removed from an individual by withdrawing them through a hollow needle inserted into the back of the hip bone. They can be maintained in laboratory dishes while undergoing the manipulations required for introducing a new gene and then simply be injected into the patient's bloodstream, from where they can make their way back into the bone marrow itself.

Among the conditions that might be corrected by gene transfer into bone marrow cells are two enzyme deficiencies, both of which result in severely crippled immune responses and death at an early age. The two enzymes, adenosine deaminase (ADA) and purine nucleoside phosphorylase (PNP), are part of the same biochemical pathway for synthesizing the purine bases, which are needed, among other things, as building blocks of the nucleic acids (Fig. 15.1). Lack of either of the enzymes results in the accumulation of compounds that, although normally present in low concentrations in the cell, are toxic when their concentrations are excessive.

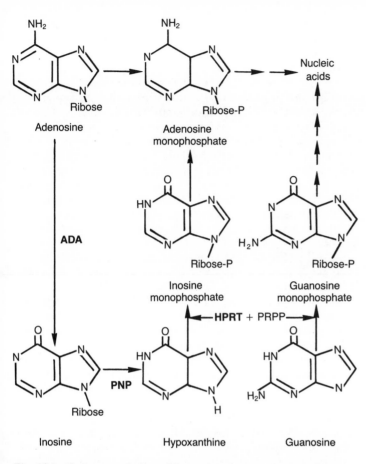

Fig. 15.1 Purine metabolism. The enzymes adenosine deaminase (ADA), purine nucleoside phosphorylase (PNP) and hypoxanthine-guanine phosphoribosyl transferase (HPRT) are all part of the pathway for purine metabolism. ADA converts adenosine to inosine by removing the amino group from the purine base; PNP removes the ribose from inosine, thereby producing the free purine base hypoxanthine; and HPRT transfers ribose phosphate from phosphoribosyl pyrophosphate (PRPP) to hypoxanthine and guanine, thus producing inosine monophosphate and guanosine monophosphate.

The expectation is that gene therapy will be attempted first for the ADA deficiency, which afflicts about 100 children worldwide. The PNP deficiency is very rare – only 10 cases or so have been identified throughout the world – and is receiving less attention in gene therapy research.

Although all the cells of individuals who are deficient in ADA or PNP fail to make the enzyme in question, the only types that appear to suffer deleterious effects are the T and B lymphocytes, which are needed for making normal immune responses. The T and B cells die and, as a result, the patients cannot fight off infections and succumb to the deficiencies within the first few years of life. Because the T and B lymphocytes are among the cells that originate in the bone marrow,

the ADA or PNP deficiencies ought to be curable if a functional copy of the corresponding gene can be introduced into the precursors of the T and B lymphocytes in the bone marrow.

No gene therapy could be attempted without the genes themselves. This requirement has been met for the ADA and PNP deficiencies. The human ADA and PNP genes have been cloned and are therefore available for introduction into the bone marrow cells. Moreover, an indication that such a strategy will work is given by the success of bone marrow transplants in curing patients with severe hereditary immune deficiencies, including those caused by ADA or PNP deficiency. This shows that having bone marrow capable of making healthy T and B cells is sufficient to reverse the patient's symptoms. Manipulations of other types of body cells are not required.

Although bone marrow transplants can cure the hereditary immune deficiencies, alternative therapies are still needed. The application of the transplants is limited, according to Robertson Parkman of the University of Southern California School of Medicine and the Children's Hospital of Los Angeles, by the lack of compatible donors. The bone marrow donor and recipient must be genetically matched so that the immune cells produced by the transplanted marrow do not recognize the recipient's tissues as foreign and mount an immune attack on them. Such an attack, which is called 'graft-versus-host disease', can cause widespread tissue damage and the death of the patient.

The opposite situation, that is rejection of a foreign marrow transplant by the recipient's immune system, is less of a problem. Most individuals who are to receive bone marrow transplants are first treated with radiation or drugs to destroy the blood-forming cells in their own marrow. This is done mainly to make room for the transplanted cells; otherwise they may be unable to establish themselves in the bone marrow because they cannot outcompete the resident cells. But the drug and radiation treatments also have the effect of suppressing the patient's immune responses.

Patients who have the ADA deficiency do not require the preliminary radiation treatment, however. Because of their severe immune deficiencies they cannot mount an effective rejection attack. Moreover, any transplanted cells with a functional ADA gene should have a strong selective advantage over the indigenous cells that lack the enzyme. All transplant patients are given immunosuppressive drugs after the transplant to diminish the potential for graft-versus-host diseases and for transplant rejection.

Most bone marrow donors are close relatives of the intended recipient, usually a brother or sister, whose genetic composition is very similar to that of the patient. No more than 30 per cent of patients who might benefit from bone marrow transplants have suitable donors, Parkman estimates, and even with a good donor-recipient match, between 10 and 20 per cent of individuals who receive the transplants die of graft-versus-host disease within a year. Graft-versus-host disease should not be a problem for patients who are given their own bone marrow cells after the cells have been 'cured' of their defect by gene transfer.

A third condition that is frequently mentioned as a possible early candidate for gene therapy is the Lesch–Nyhan syndrome, which is caused by lack of an enzyme called hypoxanthine–guanine phosphoribosyl transferase (HPRT). The human HPRT gene has also been cloned. Although this enzyme is part of the same synthetic pathway as ADA and PNP (see Fig. 15.1), HPRT deficiency causes a much wider range of symptoms than does the lack of the other two enzymes. Patients with Lesch–Nyhan syndrome have a high concentration of uric acid in the bloodstream, which results from the excess synthesis of purines and causes gout and kidney damage.

Conventional drug therapy can reduce the blood concentration of uric acid and prevent the kidney damage that formerly killed the patients at an early age. However, they also suffer serious and currently untreatable neurological problems, including mental retardation, cerebral palsy and, perhaps most distressing, a compulsion to mutilate themselves if they are not physically restrained.

The question that needs to be answered is whether introducing a good copy of the HPRT gene into the bone marrow will relieve the neurological symptoms, the causes of which are unknown. If the gene defect causes some intrinsic change in the anatomy or biochemistry of the brain neurons, having the active HPRT enzyme in the bone marrow is likely to be of little benefit. But if the neurological symptoms are caused by production of a toxic chemical, a functional HPRT enzyme, even if it is not present in the brain, might be able to keep the concentrations of the chemical low enough to alleviate the neurological problems.

An encouraging finding with regard to potential gene therapy for Lesch–Nyhan syndrome is the observation that individuals who are only partially deficient in HPRT have increased uric acid concentrations and gout but do not have the neurological symptoms. Less encouraging is the report by Parkman that there has been no clinical improvement in the one Lesch–Nyhan patient who has so far been given a bone marrow transplant, even though the patient's white blood cells produced normal amounts of HPRT after the transplant. The patient may have failed to respond because the effects of the enzyme did not extend to the brain. Another possibility is that the transplant was performed too late, after irreversible brain damage had already occurred. Because of its neurological ramifications, Lesch–Nyhan syndrome may be less amenable to gene therapy than the ADA or PNP deficiency.

At first glance, the hereditary anaemias, such as sickle cell anaemia or the thalassaemias, might appear to be excellent candidates for gene therapy. The defects, which affect the genes coding for the haemoglobin proteins, are well understood; the genes in question have been cloned; and the cell in which those genes are specifically expressed is made in the bone marrow. The problem is that regulation of the globin genes is much more complex than that of the ADA, PNP and HPRT genes.

ADA, PNP and HPRT are 'housekeeping' enzymes that are made all the time in all cells, which seem capable of tolerating some variation in how much is made. In contrast, the globin proteins are made only in a specific type of cell, the one that produces red blood cells, and at a particular time in the life-cycle of that cell. Moreover, the adult haemoglobin molecule contains two different proteins and the expression of the corresponding genes, which are located on different chromosomes, is coordinated so that the proteins are made in nearly equal amounts.

So far no one has been able to achieve completely normal regulation of globin genes that have been transferred into cultured cells. In particular, the transferred genes make very small amounts of product. Cure of the anaemias will require that haemoglobin be made in near-normal quantities. Not until the problem of globin gene regulation is solved will gene therapy for sickle cell anaemia and the other hereditary anaemias be feasible.

Still other diseases with genetic origins are not now candidates for gene therapy and may not be for many years, if ever. This is true for any condition for which the affected gene has not been identified and cloned. For example, Wiskott–Aldrich syndrome is marked by abnormalities in the lymphocytes and blood platelets can be cured by bone marrow transplantation, which indicates that gene therapy might work too – but the gene defect has not yet been pinpointed. Cystic fibrosis is another hereditary condition for which researchers have yet to nail down the gene affected, although they are getting very close (see Chapter 14). Nevertheless, gene therapy as it is now envisioned requires that having the appropriate gene active in bone-marrow-derived cells be sufficient to alleviate the disease symptoms. This will not work for cystic fibrosis, which affects many organs, but especially the lungs, liver and pancreas.

Gene transfer methods

Over the past several years investigators have developed a number of methods for introducing new genes into cells. In one of the earliest methods, and perhaps the simplest, cultured cells are incubated with a calcium phosphate precipitate of DNA. Frank Graham and Alex van der Eb of McMaster University in Hamilton, Ontario, originally observed that this DNA treatment facilitates uptake of the nucleic acid by cells. Later Michael Wigler, who is now at Cold Spring Harbor Laboratory on New York's Long Island, and Richard Axel and Saul Silverstein of the College of Physicians and Surgeons of Columbia University, modified the technique to allow it to be used for introducing individual, cloned genes into cells. The DNA thus taken up can become incorporated into the genome of the recipient cell and is passed on to the cell's progeny.

Very few of the cells incubated with the precipitated DNA actually take up the material, however. At best, just 1 in every 100 to 1000 cells does so, and a more typical efficiency is 1 in every 100 000 to 10 million cells. The low efficiency means that a selection method must be used to identify the rare cells that acquire the foreign DNA. This usually involves transferring a gene that will confer some kind of identifiable advantage on the cell that takes it up. Such a gene might code for resistance to an antibiotic, for example. Only those cells that acquire the gene will grow in the presence of the antibiotic while all the others will die. Investigators often want to transfer a gene that does not confer a selectable advantage and this can be done simply by transferring it together with one that does.

The calcium phosphate method has proved very valuable for studying control of gene expression. It has helped, for example, to identify the DNA segments that regulate gene transcription into messenger RNA. With the aid of recombinant DNA technology investigators can delete or alter specific DNA segments in and around a gene and then observe how the changes affect the gene's transcription after the altered sequences are transferred into cultured cells by the calcium phosphate method.

Despite the proven value of calcium-phosphate-mediated gene transfer in basic molecular biology, its low efficiency makes it unsuitable for use in gene therapy. Bone marrow consists of a mixed population of cells in varying stages of development, but for gene therapy to succeed the new gene must be introduced into a particular subpopulation – that of the primitive 'stem' cell. Stem cells have the potential to give rise to all the major cell lineages that originate in the bone marrow (Fig. 15.2). When the stem cells divide they form progeny that differentiate along several paths, with some of the descendants developing into red blood cells and others forming the various kinds of white blood cells. Differentiation is basically a one-way street; cells do not back-track as they move towards the mature state. They also lose their capacity to divide.

If a transferred gene enters only the mature cells of a particular lineage it will eventually be lost as those cells die off. However, if the gene is in the stem cells, they will continuously repopulate the bone marrow and the blood with cells bearing the transferred gene. The problem is that only about 1 in every 100 to 1000 bone marrow cells is a stem cell. W. French Anderson of the National Heart, Lung, and Blood Institute (NHLBI) in Bethesda, Maryland, calculates that just 10 to 100 stem cells in a typical bone marrow sample would acquire a gene by the calcium phosphate method if the transfer efficiency were 1 in a million. An individuals's total bone marrow contains from 100 million to 1 billion stem cells and reintroducing such a small number of genetically altered cells would be unlikely to have any noticeable effect unless they had a tremendous selective advantage over the unaltered population.

Researchers who are interested in developing gene therapy have therefore been concentrating their efforts on devising much more efficient means of transferring genes into cells than the calcium phosphate method. They have turned to viruses, which at least theoretically have the potential of infecting all the cells in a sample. In particular they are using retroviruses, which have RNA as their genetic material.

The life-cycle of the retroviruses makes them well-suited to

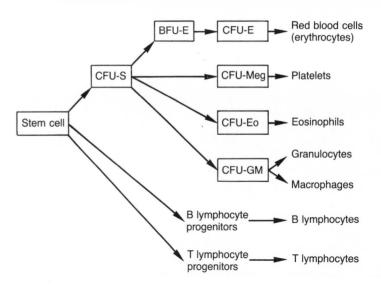

Fig. 15.2 The differentiation of bone marrow stem cells. An undifferentiated stem cell has the potential to produce all of the major types of blood cells including the red blood cells (erythrocytes); the megakaryotes that produce the platelets needed for blood clotting; the eosinophiles, white blood cells that contain granules that are stained by the red dye eosin; the granulocytes, another type of granule-containing white blood cells; the macrophages, which are so-called because they are the scavengers that engulf and 'eat' foreign antigens and abnormal or damaged cells; and the B and T lymphocytes. The designation CFU-S means colony-forming unit-spleen and refers to an immature type of cell that forms colonies in the spleen and gives rise to the various cell lines indicated. The BFU-Es, for burst-forming units-erythroid, are somewhat more differentiated cells that give rise only to red blood cells. The designations CFU-E, -Meg, -Eo and -GM refer to still more differentiated cell types that respectively produce erythrocytes, megakaryotes and platelets, eosinophils, and the granulocytes and macrophages.

being vehicles for gene transfer (Fig. 15.3). When one of them infects a cell the viral enzyme reverse transcriptase makes a DNA copy of the RNA genome of the virus. The DNA can insert itself into the host-cell genome, remain there permanently, and be transmitted to the progeny if the cell divides. The integrated retroviral DNA is bounded at each end by identical sequences called 'long terminal repeats' (LTRs), which contain the signals needed for transcribing the viral genes into RNA (Fig. 15.4). The RNAs thus produced serve both as messenger RNAs for making the viral proteins and also as genomes for the newly forming viral particles.

A foreign gene that has been inserted into the viral genome will also be transcribed into messenger RNA in infected cells under the influence of the LTRs, and several groups of investigators have developed retroviral vectors for gene transfer. The virus most frequently used for this purpose is Moloney murine leukaemia virus (MMLV), which as its name suggests causes leukaemia in mice. It does this only under very special condi-

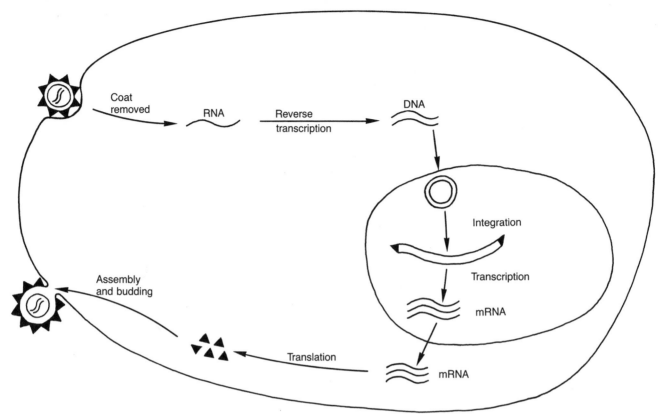

Fig. 15.3 Life-cycle of a retrovirus. The infecting virus is taken into the target cell where the viral coat is removed, thus releasing the free RNA genome. This is first transcribed into a DNA copy, which is itself copied to form a double-stranded DNA. After this molecule forms a circle, it is integrated into the cellular genome, where it is transcribed into messenger RNAs (mRNAs) together with the cell's other genes. The viral messenger RNAs are translated to produce the various proteins needed for making an infectious viral particle. The viral RNA and proteins are then assembled into a complete virus that buds from the cell to release free infectious virus.

LTR		gag	pol	env	LTR

Moloney murine
leukaemia virus

LTR		Cloned DNA	LTR

Vector

LTR	gag	pol	env	LTR

Helper virus

Fig. 15.4 Moloney mouse leukaemia virus and vector. The upper bar is a schematic representation of the normal genome of the Moloney mouse leukaemia virus. On the ends are the long terminal repeats (LTRs) which contain the sequences needed for integration of the DNA copy of the viral RNA into the host cell genome and also for controlling the transcription of the viral genes. The black bar represents the sequence needed for packaging the viral genome into complete infectious viral particles. The *gag*, *pol* and *env* regions respectively designate the genes coding for the viral core proteins, reverse transcriptase and envelope proteins. To make the vectors for gene transfer most of the viral sequences, with the exception of the LTRs and the packaging sequence, are replaced with the cloned gene or genes to be transferred. The helper virus, which makes the proteins needed for packaging the vector into infectious particles, retains the *gag*, *pol* and *env* genes; however, the packaging sequence is deleted, so that the helper genome can not be incorporated into infectious particles.

tions, however, when newborn animals are injected with large quantities of the agent. The animals do not develop the cancer if the virus is injected later in life, although they do become viral carriers.

How MMLV causes leukaemia is unclear. It does not contain any of the cancer-causing genes known as oncogenes. These genes are of cellular origin and in their normal form help to control cell growth and development. However, they can go awry, probably as a result of structural alterations that cause them to make abnormal products or to make their products at the wrong time in the cell-cycle. They can then produce the uncontrolled growth and other characteristics of malignant cells. Oncogenes were first discovered in certain cancer-causing animal viruses that had apparently picked them up from infected cells.

Clinicians would not want to give to human patients a retrovirus that is capable of replicating itself and spreading to other cells, especially if the virus in question has even a limited potential for causing cancer. Consequently, the MMLV genome used for the gene-transfer vectors has been modified so that the virus can no longer reproduce to make infectious particles but retains its ability to enter the target cells and insert its genetic information into the cellular genome.

The MMLV genome, like that of most other retroviruses, carries three principal genes: the *pol* gene, which codes for reverse transcriptase; the *env* gene, which codes for the proteins of the outer viral envelope; and the *gag* gene, which codes for the interior proteins of the viral core (Fig. 15.4). To make the vectors, recombinant DNA technology is used to cut out these three genes and replace them with whatever gene the investigator may wish to transfer – the ADA or HPRT gene, for example. A gene coding for resistance to an antibiotic or for some other selectable characteristic may also be incorporated in the vector genome to confer a selective advantage on the cells that contain it.

Bare viral DNAs or RNAs are not infectious. They have to be packaged in complete viral particles to enter cells efficiently. The retroviral vector DNA cannot package itself because it no longer contains the genes for making the necessary viral proteins. To produce the infectious vector the recombinant viral DNA is introduced by the calcium phosphate method into a line of cells that carries a 'helper' virus that does encode the *gag*, *pol* and *env* proteins. Ordinarily both the helper and vector viruses would be packaged into infectious particles under these conditions. This is undesirable because there is no way to separate the two and if both were to be introduced into bone marrow cells the viruses would replicate there – and spread.

The formation of helper virus can be prevented, however. The key development came in 1983 when David Mann, Richard Mulligan and David Baltimore of the Massachusetts Institute of Technology in Cambridge, Massachusetts, recognized that retroviral genomes carry a sequence that is located just between the left-handed LTR and the *gag* gene and is needed for packaging of the viral RNA into whole particles. If the packaging

sequence is removed from the helper virus RNA, it can no longer form intact viral particles. As a result, the only infectious material that comes out of the cells is the replication-defective vector virus.

Researchers from several laboratories have developed such vectors and used them to introduce a variety of foreign genes into cultured cells, including bone marrow cells. For example, Mulligan and his colleagues introduced a gene coding for resistance to the antibiotic neomycin into mouse bone marrow cells and Theodore Friedman and his colleagues at the Univerisy of California in San Diego, working with Inder Verma and A. Dusty Miller of the Salk Institute, La Jolla, California, put the human HPRT gene into mouse bone marrow cells. The transferred HPRT gene was expressed, although at that time it could not be determined whether it was active in stem cells.

Despite the successes with cultured cells, the results were more often than not disappointing when investigators went on to attempt to introduce genes into the bone marrow of living animals. In these experiments bone marrow cells were removed from the animals, usually mice, and incubated with the viral vector carrying the gene to be transferred (Fig. 15.5). Then the cells were re-injected into the mice, which had been exposed to a dose of X-irradiation sufficient to destroy their bone marrow. This is normally a fatal treatment, but the re-injected marrow cells replenish the animals' marrow and they survive. The protocol for these experiments is similar to that planned for human gene therapy trials, although clinicians hope to be able to avoid the step in which the patient's own bone marrow is destroyed. Nevertheless, this may be necessary if the genetically altered cells are unable to outcompete the indigenous cell population.

In any event, investigators could detect viral DNA in at least some of the cells produced by the reconstituted bone marrow of the mice, but when they looked for actual production of the enzymes specified by the transferred genes, they usually did not find it. There were some exceptions, however. Three independent groups introduced the neomycin-resistance gene into mouse bone marrow and found the gene product was made in relatively high concentrations in the various types of cells produced by the marrow. The production of the enzyme by diverse types of bone-marrow-derived cells showed that the neomycin-resistance gene entered the stem cells and remained active. The three groups who did the gene transfers were those of Alan Bernstein of the Mount Sinai Hospital Research Center in Toronto, Canada, who collaborates with Robert Phillips of that city's Hospital for Sick Children; Gordon Keller of the Basel Institute for Immunology, Switzerland, and Erwin Wagner of the European Molecular Biology Laboratory in Heidelberg, Federal Republic of Germany; and Anderson and his colleagues.

More recently, Anderson, Arthur Nienhius of NHLBI, and Richard O'Reilly of the Memorial Sloan-Kettering Cancer Institute in New York City, have also had success in introducing the human ADA gene into the bone marrow of living monkeys.

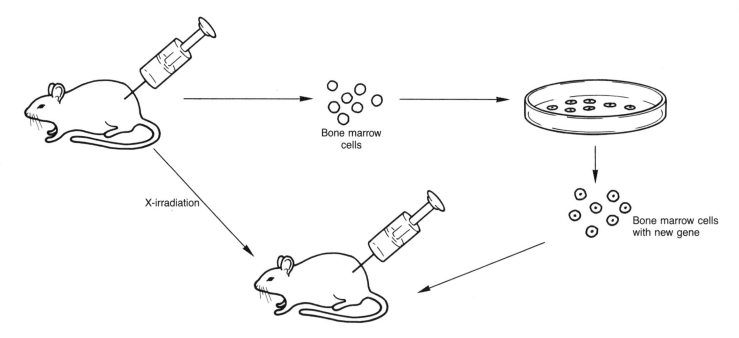

Fig. 15.5 Protocol for gene therapy. Bone marrow cells are withdrawn by syringe from the animal that is to receive the gene therapy. The cells are then cultured with cells that produce the vector virus to be used for introducing the new gene into the animal. The mouse is subjected to a dose of X-rays that will destroy its bone marrow, and then the bone marrow cells, which have acquired the new gene, are re-injected. A similar protocol will be used for therapy of human genetic diseases, although clinicians hope that the X-irradiation step can be avoided.

Bone-marrow-derived cells of five of the seven monkeys used produced the human enzyme. Although the levels of expression were only about 10 per cent of those thought necessary to overcome the effects of ADA deficiency, the results were nonetheless encouraging from the standpoint of eventually using the methods for human gene therapy. When Mulligan, with David Williams and Stuart Orkin of Harvard Medical School, attempted to perform a similar transfer of the human ADA gene into mice, they were unable to get expression of the gene in most of the animals, even though it was present in the animals' cells in intact and unrearranged form.

Nevertheless, both groups found that lymphocytes that were obtained from patients with the ADA deficiency and grown in culture can be 'cured' by introduction of a good copy of the ADA gene. The enzyme produced by the transferred gene protects the lymphocytes against the effects of the toxic chemical that would otherwise kill the cells. The results indicate that if the active ADA gene can be introduced into bone marrow stem cells, it will have the desired effect of protecting the lymphocytes against the lethal effects of the hereditary enzyme deficiency.

Why some investigators have obtained better results than others with gene transfers into living animals is unclear, although there are indications that the answer may reside in the vectors. Both the Anderson group and that of Keller and Wagner have been using a vector devised by Eli Gilboa and his colleagues at Princeton University. Bernstein and Phillips used a vector that they constructed themselves, but its design is similar to that of the Gilboa vector.

Miller, who is now at the Fred Hutchinson Cancer Research Center in Seattle, and Randy Hock, who is at the same institution, have recently reported another encouraging result with the Gilboa vector that may bring gene therapy for human hereditary diseases one step closer. Investigators have had trouble in getting expressions of genes transferred into human bone marrow cells in culture. But when Miller and Hock used the Gilboa vector to transfer either of two antibiotic-resistance genes into the human cells, between 5 and 20 per cent were infected and they became resistant to the appropriate antibiotic. The Seattle workers could not determine in these experiments whether the genes entered the stem cells, but their expression could be detected in the bone marrow 'progenitor' cells that while more advanced in development than true stem cells are still immature. Bernstein, Phillips and their colleagues have also been able to introduce an active antibiotic-resistance gene into progenitor cells from human bone marrow.

Researchers are interested in determining why the transferred genes have frequently worked so much better in cells growing in culture than in those of living animals. The problem might be a biological difference between the primitive stem cells and ordinary cultured cells. As mentioned previously, to obtain persistent expression of a transferred gene in living animals it must enter the bone marrow stem cells and retain its activity there. Bernstein, Phillips and their colleagues have found that transferred genes are active when they first enter stem cells, but are then turned off. Work from Peter Rigby's laboratory at the Imperial College of Science and Technology in London suggests that the LTRs of MMLV may be inactivated in stem cells and this may account for the poor expression there of genes transferred by vectors derived from that virus. Such inactivation would be consistent with the finding of Mulligan's group that the ADA gene is present in intact form in the mouse bone marrow cells but is nonetheless inactive.

The viral vectors clearly need more tinkering with before they can be used for human gene therapy trials. The new work is aimed not only at improving the expression of the genes transferred by the vectors but at eliminating a potential safety problem. Certain retroviruses cause cancer in animals by inserting in the cellular genome near one of the genes that has oncogenic potential. The potential oncogene then becomes activated by coming under the influence of the viral LTR, which is a strong promoter of gene expression. No one knows whether the viral vectors being considered for use in gene therapy might inadvertently activate an oncogene by inserting near it in the genome, but investigators, including Bernstein, Gilboa and Mulligan, are designing their newer vectors in such a way that the LTRs are lost when the viral DNAs integrate. That should minimize the likelihood of activating an oncogene.

Without the viral LTRs the transferred gene will not be expressed unless it is attached to other regulatory sequences that promote gene transcription. The gene could either be left attached to its own control sequences or it might be joined by recombinant DNA technology to other promoting and enhancing sequences.

Genes transferred by the current methods integrate randomly in the cellular genome. The investigator cannot direct them to any particular site and the chances that a gene will insert in its normal chromosomal location are vanishingly small. The influences of the chromosomal environment on gene regulation are poorly understood, but conceivably a transferred gene would be better regulated if it could be targeted to insert in the correct location. Such a development might improve the prospects of gene therapy for the hereditary anaemias, for example, or for any other condition for which accurate control of the transferred gene is critical. Perhaps best of all from the standpoint of gene therapy would be the ability to replace exactly a defective gene sequence with the corresponding good sequence. There is such a method for yeast, but it is not applicable to mammalian cells.

Oliver Smithies of the University of Wisconsin in Madison, Raju Kucherlapati of the University of Chicago and their colleagues have recently shown that it is possible to target a human beta-globin gene so that it inserts in its normal chromosomal location, although they have not achieved precise replacement of the endogenous beta-globin gene with the new one. The transferred gene was flanked both by vector DNA and by segments of the endogenous beta-globin gene. Moreover, the frequency of the targeted insertion is very low. Only about 1 cell in every 1000 that took up the DNA incorporated the beta-globin gene in the correct chromosomal location. This means that the method is not sufficiently efficient to use to insert genes into bone marrow stem cells for gene therapy. Nonetheless, Smithies, Kucherlapati and their colleagues have at least demonstrated the feasibility of targeted gene insertion and opened the way to developing more efficient methods that might be used for gene therapy.

Gene transfer into the germ line

The gene transfer methods that are currently under development for use in the therapy of human genetic diseases introduce new genes only into somatic cells, such as bone marrow cells. These genes will not be transmitted to future generations. A gene cannot be passed on to an individual's progeny unless it is present in the germ cells, which produce the sperm and eggs.

Over the past several years investigators have found that they can introduce new genes into the germ cells of experimental animals, including mice and fruit flies. Transfer into mice is accomplished by injecting a cloned gene with a very fine needle into the nuclei of newly fertilized eggs, while they are still in the single-cell stage (Fig. 15.6). The eggs are obtained by washing them out of the uteri of recently mated females. Those that survive the rather delicate injection procedure are then implanted in the uteri of foster mothers where they can develop normally until the time of birth.

If the injected gene integrates in the genome of the fertilized egg before it divides, all the cells of the resulting animal will contain the transferred gene, including the germ cells. Sometimes the integration does not occur before the first cell division of the embryo, however, and in that event the gene may not be inherited by all the body cells. In all cases the transferred gene integrates randomly in the mouse genome.

In the most experienced hands, 5 to 10 per cent of the injected mouse eggs survive and develop to term. As many as 30 per cent of the animals that are born carry the foreign gene and can transmit it to their offspring. The investigators who have pioneered the development of the mouse egg injection methods include Ralph Brinster of the University of Pennsylvania School of Veterinary Medicine in Philadelphia and Richard Palmiter of the University of Washington in Seattle; Frank Costantini of the College of Physicians and Surgeons of Columbia University in New York City and Elizabeth Lacy of the Memorial Sloan-Kettering Cancer Institute, also in New York City; Beatrice Mintz of the Fox Chase Cancer Center in Philadelphia; and Frank Ruddle of Yale University in New Haven, Connecticut.

Especially in the early experiments the transferred genes did not make any functional products in the recipient mice. This failure was often caused by rearrangement and consequent disruption of the genes, but faulty control also contributed to the expression problems. In 1982 Brinster, Palmiter and their col-

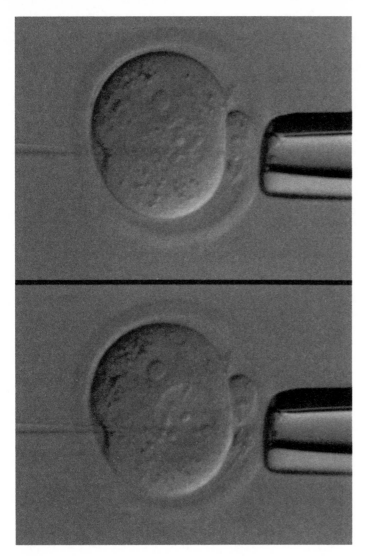

Fig. 15.6 Mouse egg injection. The upper picture shows a newly fertilized mouse egg that is being held on a large pipette. The male pronucleus (the nucleus of the sperm before its fuses with the egg nucleus) is being injected with foreign DNA by the small pipette, which has a diameter of about 1.5 micrometres. The lower picture shows that the injected pronucleus becomes swollen as a result. Also visible, just under the clear outer covering of the egg near the holding pipette, is the remnant of the fertilizing sperm. (From R. L. Brinster and R. D. Palmiter (1986) *Harvey Lectures, Series 80*, p. 1. Alan R. Liss, New York.)

leagues found that they could get good expression of transferred genes by attaching them to a regulatory sequence that is a strong promoter of transcription. They constructed hybrid genes in which the sequences coding for protein structure were tied to the promoter of the metallothionein gene. Metallothio-

nein is a protein that binds to potentially toxic metals such as cadmium and zinc, thereby removing them from circulation. The metallothionein promoter normally maintains the gene in the 'on' state, but also responds to high concentrations of the metals by increasing its transcription.

The Brinster–Palmiter group first showed that a hybrid gene containing the metallothionein promoter plus a viral gene for the enzyme thymidine kinase is expressed in mice. Expression was especially good when the animals were given sublethal doses of cadmium or zinc, two metals that ordinarily increase expression of the metallothionein gene. In subsequent experiments they hooked the metallothionein promoter to the gene for either rat or human growth hormone. Some of the mice that developed from eggs injected with these hybrid genes not only carried the foreign DNA in their cells, but also made the corresponding growth hormone and grew significantly faster than littermates that did not acquire the foreign gene (Fig. 15.7).

In the mice the hybrid genes were controlled more like the metallothionein gene than the growth hormone gene. Growth hormone is ordinarily made only in the pituitary gland at the base of the brain, but in these animals it was synthesized mainly in the liver and kidney, the sites where metallothionein is usually produced. Moreover, expression of the hybrid genes was stimulated by cadmium and zinc. Since these early experiments, several investigators have shown that genes introduced into mice by egg injection can be controlled normally. A recent example comes from Howard Goodman and his colleagues at Massachusetts General Hospital and Harvard Medical School in Boston. Insulin should be synthesized only in the beta cells of the pancreas and the production should be stimulated by increased concentrations of the blood sugar glucose. The Goodman group found that expression of the human insulin gene in mice followed this normal pattern.

Genes can also be introduced into the germ lines of fruit flies, by egg injection methods that were largely developed by Gerald Rubin and Allen Spradling of the Department of Embryology of the Carnegie Institution of Washington, which is located in Baltimore, Maryland. Rubin and Spradling discovered that 'transposable elements' make very good vectors for these transfers. Before that, gene transfer into fruit flies had not worked.

Transposable elements are naturally occurring pieces of DNA that can jump about the genome of the host species, sometimes landing in genes and thereby causing mutations. The elements are thought to occur in all species but have so far been isolated only from bacteria, yeast, maize and the fruit fly. In 1982 Rubin and Spradling isolated a particular kind of fruit fly transposable element, called the P element. They showed that genes that are incorporated into the P element and injected into very young fruit fly embryos transfer with very high efficiency – 50 per cent or more – into the germ lines of the resulting insects. The genes transferred by the P element insert in a variety of chromosomal locations, not the normal sites for the genes. But these transferred genes escape the rearrangements so frequently

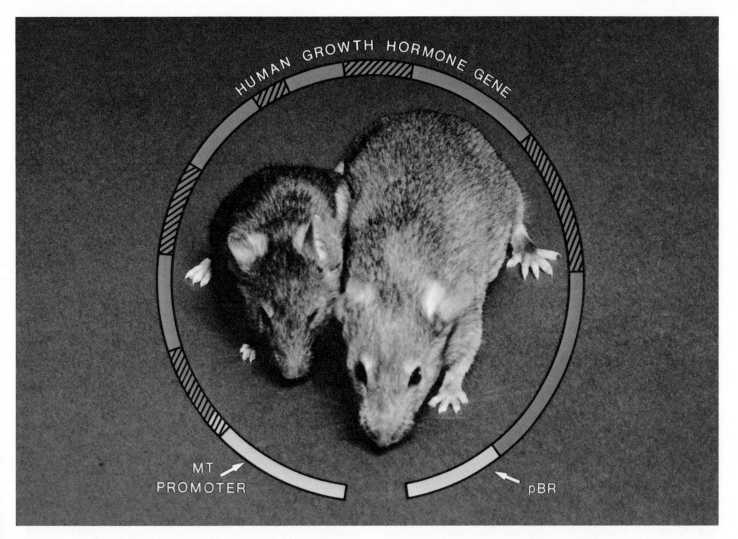

Fig. 15.7 'Mighty Mouse'. The two mice are sisters from the same litter and are approximately 24 weeks old. The animal on the right, which weighs twice as much as her normal littermate, contains a new hybrid gene consisting of the regulatory sequences from the metallothionein gene that have been linked to the protein-coding sequences of the human growth hormone gene. The gene was introduced by egg injection and became integrated into the DNA of the mouse that developed. The foreign gene is transmitted to the animal's progeny and those who acquire it also grow to about twice the normal size. The curved bar represents the structure of the transferred gene. The metallothionein sequence is in yellow and that of the human growth hormone sequence is in red. The region marked pBR is from the vector originally used to clone the hybrid gene. (From R. D. Palmiter, G. Norstedt, R. E. Gelinas, R. E. Hammer and R. L. Brinster (1983). *Science*, **222**, p. 809. By kind permission of the authors and publisher. © 1983 American Association for the Advancement of Science.)

seen in the genes that have been transferred into mice by egg injection without the benefit of a P element.

The high efficiency of integration and lack of rearrangement of genes carried by the P element has led many investigators to attempt to use it as a vector for gene transfer into mammalian cells. So far they have had no success. However, Rubin who is now at the University of California in Berkeley, and his Berkeley colleagues Frank Laski and Donald Rio, may have discovered the reason for the ineffectiveness of the P element as a vector in mammalian cells.

Movement of the element in the genome requires the activity of an enzyme called a transposase, which is encoded by the P element itself. In the fruit fly the transposase is active only in germ cells. Research by the Berkeley workers has now revealed why the enzyme is inactive in somatic cells. The cells lack the ability to remove one of the three intervening sequences that split the transposase gene into four segments. As a result, somatic cells cannot make an active messenger RNA for the enzyme.

Rubin and his colleagues went on to show that if they remove the intervening sequence from the transposase gene, the P element will work in fruit fly somatic cells as well as in germ cells. Possibly the same strategy can be used to adapt the P element as a vector for mammalian cells.

Although investigators developed the egg-injection methods primarily for studying how genes are turned on and off during embryonic development, the procedure may also be used to produce new strains of domestic animals. If extra copies of the growth hormone gene could be introduced into cows, for example, the animals might grow faster, as the mice did, and perhaps produce more milk. Brinster, among others, has been attempting to inject the eggs of domestic animals, including pigs, sheep and cows, with the growth hormone gene. These injections are much more difficult than those into mouse eggs, because the eggs of the domestic animals are more opaque, which makes it harder to see the nucleus. The first results indicate that a small percentage of the sheep and pigs that developed from eggs injected with the human growth hormone gene carried the foreign gene, but the animals did not grow any faster than controls.

Conceivably egg-injection might be used to introduce new genes into human embryos. The fertilized eggs could be obtained by the *in vitro* fertilization methods that have been developed to help couples who cannot conceive normally. Nevertheless, such experiments are not contemplated because of the ethical and practical problems that they raise. For one, egg injection has a high failure rate. According to Brinster, who may have more experience with the procedure than anyone else, only about 10 per cent of injected embryos actually produce living offspring, and no more than one-third of these – at best – carry the injected gene.

For another, there is currently no way to control where the new gene will integrate in the genome of the recipient cells. It could have deleterious effects if it disrupted an essential gene. Both Mintz and Brinster have evidence suggesting that this can happen, but the clearest example comes from Philip Leder's laboratory at Harvard Medical School. He and his colleagues apparently re-created a known developmental limb deformity in mice in one of their egg-injection experiments.

This type of insertional mutation would not be a problem if it occurred during gene transfer into bone marrow cells. In that kind of transfer many cells take up the new gene and integrate it in random chromosomal locations. If the inserted gene disrupted a gene in some of the cells they would simply be lost or perhaps not work the way they should, but the effect on the patient would probably be negligible. The possibility that an injected egg might produce an individual with a developmental deformity is a different situation entirely.

Moreover, Leder and Brinster have shown that they can create 'tumour-prone' mice by injecting genes into fertilized eggs. In both cases they used genes of known oncogenic potential, but foreign DNA may be carcinogenic even if it does not contain a known oncogene. As previously mentioned viral LTRs can cause certain animal cancers by inserting near and activating potential oncogenes in the cellular genome. Researchers are already working to diminish the possibility of this happening during gene transfer into somatic cells, but even if cancer did prove to be a rare outcome of gene therapy it would not necessarily preclude its use for hereditary diseases that already have early death as a certain outcome. Many of the drugs used to treat cancers are themselves carcinogenic, but the risk is deemed acceptable if they have a chance of saving patients from a currently life-threatening disease.

The same principle should also apply to the use of gene therapy to treat lethal hereditary diseases once researchers have overcome the problems that they have encountered in obtaining good expression of transferred genes. Meanwhile, the gene transfer techniques are proving to be valuable tools for studying how gene expression is controlled during development and may even prove useful for developing new strains of domestic animals.

Additional reading

Anderson, W. F. (1984). Prospects for human gene therapy. *Science*, **226**, 401.

Bernstein, A., Berger, S., Huszar, D. and Dick, J. (1985). Gene transfer with retrovirus vectors. In *Genetic Engineering: Principles and Methods*, vol. 7, ed. J. K. Setlow and A. Hollaender, pp. 235–61. Plenum Publishing, New York.

Brinster, R. L. and Palmiter, R. D. (1986). Introduction of genes into the germ line of animals. *The Harvey Lecture, Series 80*, p.1. Alan R. Liss, New York.

Friedman, T. (1985). Scientific and policy progress toward gene therapy. *BioEssays*, **3**, 40.

Friedman, T. (1985). HPRT transfer as a model for gene therapy. In *Genetic Engineering: Principles and Methods*, vol. 7, ed. J. K. Setlow and A. Hollaender, pp. 263–82. Plenum Publishing, New York.

Palmiter, R. D. and Brinster, R. L. (1985). Transgenic mice. *Cell*, **41**, 343.

Parkman, R. (1986). The application of bone marrow transplantation to the treatment of genetic diseases. *Science*, **232**, 1373.

Williams, D. A. and Orkin, S. H. (1986). Somatic gene therapy: current status and future prospects. *Journal of Clinical Investigation*, **77**, 1053.

16 Biotechnology, international competition and regulatory strategies

Joseph G. Perpich

The revolution in biotechnology is rapidly spreading across the globe. At the present time the key players are the United States, Japan and the nations of Western Europe. How each will fare in the competition over the development and marketing of the products of their biotechnology industries depends on many factors. These include the availability of private investment capital, the strengths and weaknesses of academic research institutions, the degree of cooperation between industry and academia, and the contributions of government.

Governmental contributions can include both the provision of direct support for biotechnological research and the fostering of a climate where the new biotechnology industries can thrive – or not thrive, as the case may be. A large part of that climate will depend on a government's policies for regulating the development of biotechnological products. An ideal regulatory strategy has to preserve the safety of mankind and the environment without hindering unduly the development of new products.

International strategies for fostering biotechnology

Biotechnology in Japan

Direct government involvement is characteristic of the Japanese effort to enhance the successful outcome of biotechnology development efforts in that country. In Japan, government programmes in biotechnology are the responsibility of the Science and Technology Agency, the Ministry of International Trade and Industry, and the Ministry of Agriculture, Forestry and Fisheries.

The policies of the Japanese government foster cooperation among industrial companies, including joint ventures involving two or more Japanese firms, and among the companies and academia. The government has targeted biotechnology for development, and working through the large, established Japanese firms has allotted significant resources for its potential commercial applications.

The Ministry of International Trade and Industry has selected three major biotechnology areas for a 10-year research and development programme. The programme will focus on improving mass cell culture techniques, recombinant DNA technology, and bioreactor development. Efficient bioreactors, which are the vessels in which biotechnological products are made by microorganisms or other cells, are essential for the successful transition from laboratory to commercial production of such products.

In addition, a foundation for biotechnology research and development, which consists of 14 chemical and pharmaceutical companies, has been created in Japan. The Ministry of Agriculture, Forestry and Fisheries has also announced the formation of a biotechnology centre to promote biotechnology research. Many of these initiatives are outgrowths of the Japanese Council for Science and Technology, which was chaired by Prime Minister Nakasone himself. The Council has proclaimed biotechnology to be a top priority of the Japanese government.

In contrast to the situation in the United States, where most government funding goes to basic research, government funding in Japan is primarily directed at applied research and is much more focused on specific projects. As a result, the United States has a large pool of scientists for basic research but lacks bioprocess engineers and the programme for training them. Bioprocess engineering is directed at solving the problems involved in scaling-up from the small amounts of materials that are handled in laboratory operations to the much larger amounts that must be handled for a commercial operation to be economically feasible.

Japan's problem is the opposite of that of the United States. Japan has a ready supply of bioprocess engineers, but insufficient numbers of people trained in molecular genetics. Consequently, Japan is now a world leader in fermentation technology and in the microbial manufacture of commodity chemicals such as amino acids, an area in which biotechnology may help to cut costs. But in the areas such as mammalian cell culture and protein engineering, Japan lags behind the United States and must increase its support for training and research and development if it is to catch up.

European approaches to fostering biotechnology

One result of the biotechnology revolution in the United States has been the establishment of numerous small companies that specifically aim to exploit the latest developments in molecular genetics and monoclonal antibody research. According to a 1984 survey by *Genetic Engineering News*, some 300 US companies, most of them new and small, have major biotechnology programmes. Europe, unlike the United States, suffers from a lack of the venture capital needed to establish such operations. As a result, funding for the European laboratories that do basic biotechnology research comes mainly from the traditional industrial corporations, financial institutions and the various governments.

The larger European companies, which operate on a multinational basis and have considerable financial resources, can coordinate their efforts with other groups around the world. For example, the West German chemical and pharmaceutical company Hoechst has donated $100 million to Harvard University and Massachusetts General Hospital in Boston to perform basic research in molecular biology and to train scientists for the company. Hoechst also has cooperative ventures in biotechnology with American, British and Japanese firms.

The European governments are developing their own strategies for promoting successful enterprises in biotechnology. These mainly include providing support for new biotechnology companies, for large corporation-based projects, for academic programmes in biotechnology, and for various industrial collaborations. The projects sponsored by the European governments, like those in Japan, tend to focus on applied rather than on basic research.

Public funding for biotechnology in the United Kingdom comes from several different sources, including the Department of Trade and Industry, the Science and Engineering Research Council and the Medical Research Council. The British Technology Group is a public corporation set up to help transfer basic research in biotechnology to commercial development. Among its other projects it has helped support Celltech, a small firm that specializes in monoclonal antibody technology.

In West Germany the Federal Ministry for Research and Technology supports basic biotechnology development and promotes the transfer of biotechnology to industry. The Ministry also provides grants to government laboratories, universities and institutes, including the Max Planck Institute. The Society for Biotechnological Research, a government laboratory that may be one of the best biotechnology research centres in Europe, is another institution that is engaged in transferring biotechnology into commercial application.

French government funding for biotechnology is provided through the Ministry of Research and Industry and government institutes such as the Institut de la Recherche et l'Industrie. In addition, the Centre National de la Recherche Scientifique and the Institut de la Santé et la Recherche Médicale conduct programmes in molecular biology that may prove applicable to the biotechnology industry.

The French government has biotechnology as a high priority, with the goal of capturing 10 per cent of the world biotechnology market by 1990. However, France may lag behind the United States, Japan and some of the other European countries in the overall commercial development of biotechnology.

The Netherlands has recently instituted a series of 'Innovation-Oriented Research Programmes' that are supported by two ministries: the Ministry of Economic Affairs and the Ministry of Education and Science. One of the programmes is devoted to biotechnology. It focuses on basic and applied research at universities and non-profit-making research institutes, and is administered and funded through the Ministry of Economic Affairs. The biotechnology programme has four subdivisions, one each concerned with manufacturing, health care, agriculture and the environment.

In universities in the Netherlands, biotechnology research is characterized by a multidisciplinary approach to applied projects, with some universities engaging in joint programmes. The research is supported by government grants and by individual contracts from government and industry. Universities have already set up several small commercial enterprises to market the products of their research. The universities also have plans to introduce a new educational programme in biotechnology. Industrial biotechnology activities are located in large multinational companies that generally emphasize the applications of the research to established product lines.

Switzerland, which has the highest per capita income among the industrial nations, does not have a government policy to promote biotechnology. Swiss science is highly decentralized, with most research and development taking place in industry. Swiss firms, in addition to having their own corporate programmes, support basic research in private institutes. The universities are encouraged to focus on basic research and education. The Swiss government supports two Federal Institutes of Research, one each in Zurich and Lausanne, but most of its research funding goes to the cantonial universities, which also receive appreciable research funding from the cantonial governments.

In addition to their individual initiatives the European countries are working together in a number of ways to compete more effectively in biotechnology with the United States and Japan. In 1982 the Commission of the European Communities initiated a 5-year Biomolecular Engineering Programme that has spent about $15 million on 100 contracts for specific research projects. Another Commission undertaking is the European Biotechnology Information Project, which facilitates the exchange of biotechnical information in Europe.

Other collaborative organizations with biotechnological interest include the European Federation of Biotechnology, which uses conferences and other mechanisms to further interdisciplinary activities in biotechnology, and the European Molecular Biology Organization, which promotes basic research and the transfer of information on molecular biology.

The European Economic Community (EEC) is taking steps to stimulate both basic and applied research in various technologies. A recent report from the EEC notes that practical applications of biotechnology are being delayed by a lack of insight into microbial physiology and identifies research areas on which future progress will depend. These areas include the physiology of organisms from unusual environments, the modification of proteins, cell-cycle regulation, and immobilized microorganisms for the production of genetically engineered proteins. The report concludes that there is a global shortage of well-qualified specialists to tackle these questions, the result of which may be the creation of a bottleneck in biotechnological innovation, especially in North America and, to a lesser extent, in Europe and Japan.

Despite the collaborative efforts of the EEC and the Commission of the European Communities, more will be required if Europe is to keep pace with Japan and the United States in the development and industrial application of biotechnology. For example, European management, labour and governments will have to reorganize to prepare for new technological developments.

A concerted effort will also be required to overcome a number of additional problems that could hinder biotechnological development in Europe. These include the insulation of European firms from competition, the high marginal tax rates that discourage capital investments, institutional rigidity, and excessive regulation. However, as Mark Dibner has noted in a review of European biotechnology efforts, Europe should not be underestimated. If the national and EEC efforts at coordination and support succeed, European industries will have a formidable presence in world biotechnology markets.

US strategies for biotechnology research and development

How can the United States, which has long occupied a strong position in biotechnology research and development, continue to achieve the technological innovations that are necessary for strong economic growth and development? Several recently completed studies specifically address this issue. The reports emphasize the importance of government investment in basic research and also call for increased attention by government, academia and industry to finding the best means of promoting technology transfer. The recommendations contained in the reports provide a strategy for fostering the industrial development of biotechnology and meeting the challenges of international competition.

President's Commission on Industrial Competitiveness

The President's Commission on Industrial Competitiveness, which consisted largely of chief executive officers of high technology companies, was established in June 1983 to advise US industry and government on government policies regarding international competition. About 18 months later the commission issued a report *Global Competition – The New Reality*, that

recommends increased government support for university research, including research on innovative manufacturing technologies. The report also specifically recommends a new tax credit to encourage more investment in university research by industry.

According to data compiled by the National Science Foundation (NSF), US spending for research and development (R&D) reached $122 billion in 1986, approximately $53 billion of which was provided by the federal government. Over the past 5 years increases in government support of R&D have been concentrated in the defence and basic research areas. Defence R&D received $37 billion in 1986 – 70 per cent of the total federal expenditures for R&D – and health research received $5.5 billion.

The presidential commission found that federally supported R&D in the civilian sector was largely fragmented and lacked adequate coordination. To combat these problems the commission report proposes that all civilian R&D agencies be combined in a single cabinet-level Department of Science and Technology. This proposal is unlikely to be implemented, however, because it would require a major reorganization of federal R&D programmes and congressional action would be required to change the relevant statutory authorities. The report also commends state and regional initiatives for fostering collaboration between government, universities and industry by means of fiscal and economic development programmes.

Congressional Budget Office

In June 1985 the Congressional Budget Office (CBO) issued a report entitled *Federal Financial Support for High Technology Industries*. This report stemmed from congressional concern over the increased international challenge, much of which is subsidized by foreign governments, to US technology. The report notes that the Congress has tried to lower the costs of private R&D by a combination of tax policy, direct spending and patent legislation.

The CBO report, like that of the presidential commission, presents options for tax credits to stimulate high technology research by industry. The overhaul of the tax codes passed by the US Congress in September 1986 continues the R&D tax credit for industry, although it reduces it from 25 per cent to 20 per cent on increases in R&D spending. In addition, the bill includes a new 20 per cent tax credit for new funds that industry spends on R&D at universities.

The report also suggests options for the direct federal support of industrial R&D. One of the proposed funding mechanisms would provide grants by agencies such as the NSF for applied research of general interest to high technology industries. Another would fund specific projects directly. A third option would have the federal government purchase R&D, as the Department of Defence has done to help the US semiconductor industry.

Office of Technology Assessment

The US Congress created the Office of Technology Assessment (OTA) in 1972 to provide it with advice on science and technology. A recent OTA report, *Commercial Biotechnology – An International Analysis*, suggests that government support for research and for training in the scientific disciplines that are related to biotechnology has been very important in maintaining the US lead in biotechnology.

The OTA report emphasizes the need to fund research in fermentation technologies and to train students in bioprocess engineering, areas in which, as mentioned previously, the United States is deficient, especially with regard to Japan. The report further recommends that the government should fund applied research on bioreactor design and on the biology of microorganisms that are useful for making industrial products. Increased support for more basic studies in molecular biology, especially the molecular biology of plants, is another recommendation in the OTA report.

The National Academy of Sciences

The National Academy of Sciences (NAS), with a membership consisting of outstanding US scientists, serves as an independent advisor to the federal government. In 1984 the NAS issued the *Report of the Research Briefing Panel on Chemical and Process Engineering for Biotechnology*, which also points out the need to strengthen the knowledge base for bioprocess engineering in the United States. Without such a base, the report states, the United States will be unable to maintain its leading role in biotechnology.

Despite such urgings, fewer than 20 US universities currently operate relevant programmes in biochemical engineering. The programmes produce no more than 60 graduates with doctoral or master's degrees annually, even though the United States needs two to three times that number. In contrast, West Germany, Japan and Great Britain have federally supported biotechnology institutes with impressive operating budgets that bring together academic and industrial investigators for cross-disciplinary research. No similar institutes exist yet in the United States.

Consequently, the objectives recommended by the NAS panel include increasing engineering capabilities for the design of systems for growing large quantities of bacterial, plant and animal cells; an expansion in the knowledge base required for the large-scale recovery and purification of biological molecules; and the training of the next generation of biochemical engineers.

Similarly, the National Academy of Engineering called in 1984 for the establishment of 25 engineering research centres at a cost of $25 million per centre, to be borne by the NSF. Eleven centres, including a Biotechnology Process Engineering Research Center at the Massachusetts Institute of Technology (MIT) in Cambridge, have been established to date. The Biotechnology Center, which is under the direction of Daniel E. Wang, combines basic research in molecular and cell biology with applied research in chemical engineering.

In addition, an NAS Panel on Biotechnology in Agriculture has emphasized the importance of genetic engineering to agriculture. The *Report of the Research Briefing Panel on Biotechnology* underscores the need for research on a wide range of topics, ranging from increasing the quality and efficiency of food production to improving protection against pest organisms. The panel's recommendations include increasing the funding for agricultural research at existing agencies, among them the US Department of Agriculture, the NSF, the National Institutes of Health and the Department of Energy.

The report also calls for special short-term grants of 5 to 10 years' duration to help integrate interdisciplinary molecular methods into basic and applied agricultural biology. Moreover, the panel suggests expanding training grants for young scientists, increasing career development awards for outstanding young investigators, funding grants to provide answers to physiological and biochemical regulatory questions, and providing resources to maintain plant germ plasms. The need for dynamic leadership in applying biotechnology to agriculture could be met, the panel notes, by establishing a full-time position of science director in the Department of Agriculture's Office of Grants and Program Systems.

Office of Science and Technology Policy

The Office of Science and Technology Policy (OSTP), which was established in 1962 to advise the executive branch of government on science and technology, has made its own recommendations on strategies for fostering biotechnology, which were based on the OTA and NAS reports.

Former OSTP director, George A. Keyworth, II, was instrumental in setting up the NSF's engineering research centres. In 1985 Keyworth and Erich Bloch, the director of the NSF, proposed that $500 million be appropriated over several years for the research centres. Six centres, one of which is devoted to biotechnology research, were established in 1985. Five more were created in 1986, although none is for biotechnology, and an additional three or four centres will be set up in 1987.

Moreover, OSTP is concerned with applying biotechnology to the expanding needs of agriculture. The office has proposed that the NSF and the Departments of Agriculture and Energy fund several federal centres for doing research on agricultural biotechnology. The projects of the centres would be coordinated through interdisciplinary centres at universities. Programme support, which would be available for 5 years, would add as much as $250 million to the competitive grant pool for plant biotechnology. The research done at the centres would focus on areas ranging from the biological control of agricultural pests to the maintenance of water quality. Funds for training postdoctoral scientists in interdisciplinary plant research would also become available.

A commitment to these programmes remains despite the

budget reductions that are mandated by the Gramm-Rudman-Hollings Act, which was passed by the US Congress in 1985 in an effort to control the soaring deficits. John P.McTague, then the acting director of OSTP, testified before Congress in 1986 that 'we are aggressively pursuing an initiative to establish university-based, multidisciplinary, problem-focused science and technology centres around critical areas of broad national needs and relevant to industrial technology'.

McTague considers the NSF Engineering Research Centres Program as a good working model for the proposed research centres for agricultural biotechnology. He notes that the initiation of the plans for the engineering centres 'is a good beginning – but only a beginning. This concept deserves emulation across the broad spectrum of interdisciplinary science, mathematics, and engineering'. New interdisciplinary centres in biotechnology, which will be concerned with bioprocess engineering, agriculture, energy and the environment, will be developed over the next 5 years. They should help to bring government,

university and industry together to foster the industrial applications of biotechnology.

The National Institutes of Health

The proportion of funds contributed by the federal government to support biomedical research has declined slightly during the past decade, as has that contributed by non-profit-making organizations (Fig. 16.1). Meanwhile, the research conducted

Fig. 16.1 National support for health R&D in the United States from 1975 to 1985. Figures in bars are percentages. During this decade the proportion of funds contributed by the National Institutes of Health (NIH) and other federal agencies declined slightly, as did that contributed by foundations and other non-profit-making organizations ('Other'). Meanwhile industry expanded its support for health R&D. (Data are from the 1985 *NIH Data Book*.)

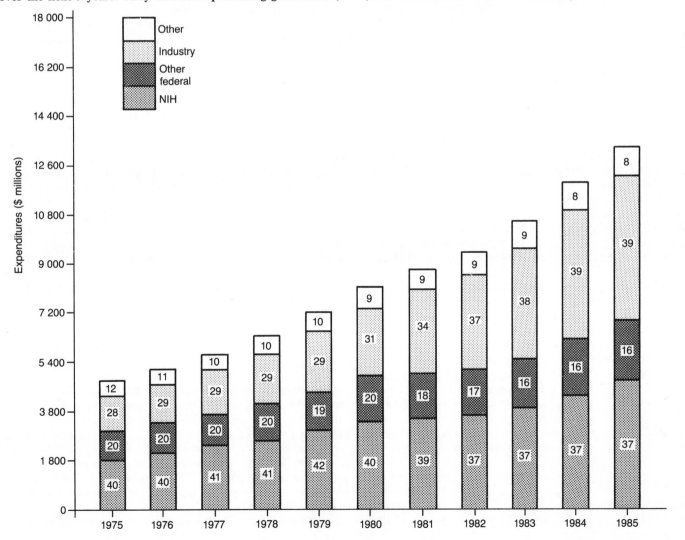

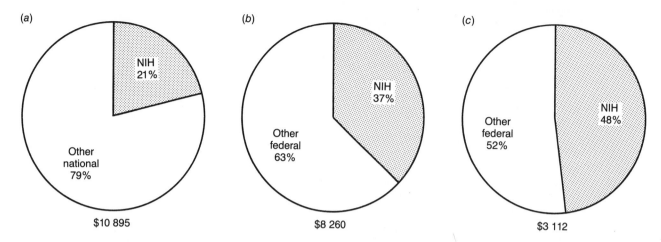

Fig. 16.2 Basic research support by the US National Institutes of Health (NIH) in fiscal year 1983. The NIH is the US leader in providing funds for basic research, contributing 21 per cent of the total US funds expended for that purpose in fiscal year 1983 (*a*) and 37 per cent of the monies provided by the federal government (*b*). Moreover, nearly half of the federal support for basic research in colleges and universities came from the NIH (*c*). The figures under the circles give the total expenditures in those categories in millions of dollars. (Data are from the 1985 *NIH Data Book*).

by industry has become increasingly important. Industry's contributions now exceed those of the National Institutes of Health (NIH), the government agency with the major responsibility for performing and funding biomedical research.

Nevertheless, the NIH continues to be the US leader in support for basic research. In fiscal year 1983, the NIH contributed 21 per cent of all national support for basic research and 48 per cent of the federal funding for basic research at universities and colleges (Fig. 16.2). The NIH also has by far the largest federal role in the support of biotechnology R&D (Table 16.1).

The NIH has initiated a review of its mission to determine whether new directions are needed in its contributions to biotechnology R&D in industry and academia. James Wyngaarden, the Institute's director, noted that the NIH will examine its role in sponsoring the type of applied research that he defines as falling between the basic research that has been traditionally supported by the NIH and the proprietary research that is undertaken by industry in the development of specific products. The task force that is carrying out this broad review of the NIH activities in biotechnology will also focus on the research and training that are needed to increase the nation's expertise in all major phases of industrial biotechnology.

The role of industry

By the end of 1985 the cumulative investment of the private sector in biotechnology totalled slightly more than $4 billion

dollars, with 65 per cent derived from equity purchases, 15 per cent from contract research and joint ventures, and 14 per cent from R&D limited partnerships. The health area has seen the greatest growth in industry investment (Fig. 16.3). Originally most investments went into recombinant DNA technology, but during the past 2 years investment in monoclonal antibody technology has increased dramatically (Fig. 16.4).

Industry's biotechnology investments go to universities as well as to the more traditional profit-making operations. In 1984 industry provided approximately $120 million to university research in biotechnology. Although these funds represent only a small portion of industry's total commitment to biotechnology R&D, the investment is productive. University-based research has generated almost one-quarter of all biotechnology patent applications over the past 5 years. This may reflect the fact that the university research that is supported by industry tends to be more applied than that funded by NIH grants.

Industry funding to universities is proving successful in promoting the transfer of basic research findings to biotechnological applications in pharmaceuticals, chemicals and agriculture. However, getting the resulting products to the marketplace depends on the other side of the biotechnology ledger – that dealing with the regulatory oversight of product development.

US biotechnology regulatory policies

In February 1975 an international group of scientists took the unprecedented step of convening a meeting to consider the ethical and safety implications of the then newly discovered recombinant DNA technology. The ability to combine the DNA of different species and to introduce foreign genes into organisms had raised concerns that scientists might inadvertently create new pathogens, which could cause devastating epidemics, or otherwise upset the delicate balance of nature.

One outcome of the meeting, which was held at the Asilomar Conference Center in Pacific Grove, California, was the develop-

Table 16.1. *Biotechnology*[a] *research funded by the US federal government in the fiscal years 1984 to 1987*

	Amount of funding (millions of dollars)[b]			
	1984	1985	1986 (est.)	1987 (budget est.)
National Institutes of Health	1633	1839	1836	1801
Directly related[c]	521	639	636	621
Broader science base[d]	1112	1200	1200	1180
National Science Foundation	61	82	89	109
Department of Agriculture	39	77	74	79
Department of Energy	38	43	43	44
Department of Defense	33	34	43	49
Army	24	22	28	31
Office of Naval Research	9	12	15	18
Department of Commerce/ National Bureau of Standards	1	1	3	4
Environmental Protection Agency	0	2	6	8
Total	1805	2078	2094	2094

[a]For the purposes of this analysis the Office of Technology Assessment's definition of biotechnology is used: Biotechnology includes any technique that uses living organisms (or parts of organisms) to make or modify products, to improve plants or animals, or to develop microorganisms for specific uses.

[b]Rounded to the nearest million.

[c]NIH defines directly related research as that research involving genetic manipulation, cloning of DNA, use of special techniques to isolate and detect DNA, creation of hybridomas and production of monoclonal antibodies, and computer methods used to analyse DNA and protein sequences.

[d]Broader science base research refers to that research underlying the new biotechnology and includes free-ranging investigations in genetics and molecular biology, cell biology and immunology.

Source: Data were obtained from the White House Office of Science and Technology Policy and biotechnology program officials in each of the cited Departments and agencies (J. G. Perpich).

ment of a series of guidelines that regulated and, in some cases prohibited, recombinant DNA experiments. In the United States the guidelines were produced under the aegis of the NIH. The guidelines required, among other things, that all NIH-funded experiments that involve recombinant DNA be submitted to the NIH Recombinant DNA Advisory Committee for approval. Other countries developed their own guidelines, although these were often similar to those promulgated by NIH.

The NAS celebrated the tenth anniversary of the Asilomar conference by holding a symposium on biotechnology in Feb-

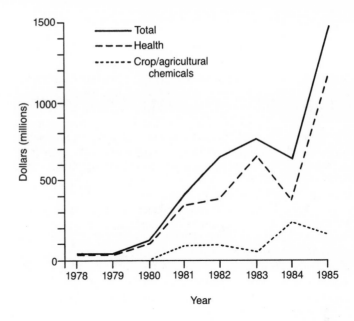

Fig. 16.3 Private support of biotechnology research. Between 1978 and 1985, private support of biotechnology research increased from nearly nothing to about $1500 million per year, with the cumulative total now being about $4 billion. Health-related research received the lion's share of the support, with much less going to research on agricultural biotechnology. The latter's share has increased somewhat since 1983, however. (After J. R. Murray (1986). The first $4 billion is the hardest. *Bio/Technology*, **4**, 293–6.)

Fig. 16.4 Cumulative investment in specific biotechnologies since 1978. Although early investment in recombinant DNA technology far surpassed that in other forms of biotechnology, monoclonal antibody investment has surged ahead since 1984.

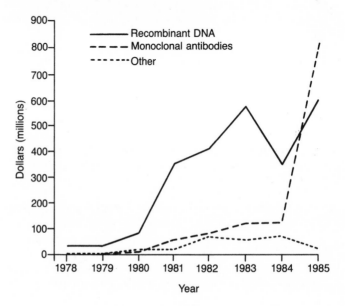

ruary 1985. Alexander Rich of MIT noted at the symposium that hundreds of thousands of recombinant DNA experiments had been performed in the previous 10 years and that many of the concerns over the potential risks of this technology, which had been raised at the Asilomar conference, had been laid to rest.

Today in the United States the focus of public concern has shifted from regulation of recombinant DNA research to the nature and scope of the government's role in regulating the industrial development of biotechnology products. The NIH guidelines for recombinant DNA research, which were developed between 1976 and 1978, have been relaxed as experience has shown little risk from the routine types of laboratory experiment. The guidelines nonetheless stand as the national standards for governing federally supported research, which is primarily done in university laboratories. Compliance of industry with the guidelines has been voluntary, however, and as commercial manufacture of the products of recombinant DNA technology becomes a reality, the need for the coordinated regulation of the production and release of those products has increased.

The Biotechnology Science Coordinating Committee

Confusion often reigned over the regulation of the first biotechnology products, especially in those cases in which the products were genetically altered microorganisms that were to be deliberately released into the environment. The requirements of regulation – which products should be subject to oversight, the types of regulation required, and the identity of the government agency that would have responsibility for a particular product – were initially clouded by uncertainty.

A Cabinet Council Working Group on Biotechnology, with Keyworth as chairman, was created in 1984 to resolve these issues. The Working Group reviewed all federal policies that pertained to biotechnology regulation and issued a report for public comment on 31 December 1984. In response to the comments that were subsequently received by the OSTP and other federal agencies, the OSTP announced the establishment of the Biotechnology Science Coordinating Committee (BSCC) on 14 November 1985.

The committee is composed of senior representatives from the agencies that sponsor or regulate biotechnology research or products. These include the NIH, NSF, the Department Agriculture, the Environmental Protection Agency and the Food and Drug Administration. It is chaired alternately by the Assistant Director for Biological, Behavioral and Social Sciences of the NSF and the Director of the NIH.

A major initial responsibility of the committee was to assist the Domestic Policy Council Working Group on Biotechnology in the development of a coordinated framework for regulating biotechnology. This framework, which was announced on 16 June 1986, outlines the jurisdictions of the various agencies. The three agencies with principal regulatory authority are the

Food and Drug Administration (FDA), the US Department of Agriculture (USDA) and the Environmental Protection Agency (EPA). The committee defined the statutory authorities for each agency and then specified which had the lead in regulating specific products (Table 16.2).

The BSCC has also proposed definitions for the organisms that require regulation to provide a common scientific basis for regulatory oversight by the individual federal agencies. Two categories of organisms would be subject to regulation, according to the committee's recommendations. The first includes genetically engineered organisms that are formed by the deliberate combination of genetic material from different species. The second category includes microorganisms that are themselves pathogens or have been genetically altered with genes from pathogens. Pathogen is defined as a virus or microorganism that has the ability to cause disease in other living

Table 16.2. *Agencies of the US federal government that are to have responsibility for the approval of various commercial biotechnology products, according to the coordinated framework of the Biotechnology Science Coordinating Committee*

Subject	Responsible agency(ies)
Foods/food additives	FDA[a], FSIS[b]
Human drugs, medical devices and biological products	FDA
Animal drugs	FDA
Animal biological products	APHIS
Other contained uses	EPA
Plants and animals	APHIS,[a] FSIS,[b] FDA[c]
Pesticide microorganisms released in the environment (all)	EPA,[a] APHIS,[d]
Other uses (microorganisms):	
1. Intergeneric combination	EPA,[a] APHIS[d]
2. Intrageneric combination:	
(a) Pathogenic source organism:	
agricultural use	APHIS
non-agricultural use	EPA,[a,e] APHIS,[d]
(b) No pathogenic source organisms	EPA Report
Non-engineered pathogens	
(i) Agricultural use	APHIS
(ii) Non-agricultural use	EPA,[e] APHIS[d]
Non-engineered non-pathogens	EPA Report

[a] Lead agency.
[b] FSIS (Food Safety and Inspection Service), under the Assistant Secretary of Agriculture for Marketing and Inspection Services, is responsible for food use.
[c] FDA is involved when in relation to a food use.
[d] APHIS (Animal and Plant Health Inspection Service) is involved when the microorganism is a plant pest, animal pathogen or regulated article requiring a permit.
[e] EPA requirements will only apply to environmental release under a 'significant new use rule' that EPA intends to propose.

organisms, whether human, animal, plant or another microorganism.

An organism from either category could be exempted from review if the combined genetic material that it contains is not considered to pose an increased risk to human health or the environment. Examples would include genetically engineered organisms that are developed by transferring a well-characterized DNA sequence that does not code for a gene product, even if that DNA originally came from a pathogen. Such sequences include those involved in regulating gene expression.

Now that the committee has classified organisms for the purpose of regulatory review, its next task is to define what constitutes 'release into the environment'. This definition is important because it sets the stage for EPA and USDA regulation of biotechnology products that are to be used for agricultural and chemical applications outside the laboratory or manufacturing plant. The committee is therefore establishing a working group for developing scientific recommendations on greenhouse containment and on how to conduct the small-scale field trials that must be carried out before more extensive environmental releases can be contemplated.

The BSCC, in its comprehensive framework, also takes into account the broad goals that were proposed by an Ad Hoc Group of Government Experts that was convened by the Organization for Economic Cooperation and Development (OECD). In 1986 the Ad Hoc Group issued a report entitled *Recombinant DNA Safety Considerations: Safety Considerations for Industrial, Agricultural, and Environmental Applications of Organisms Derived by Recombinant DNA Techniques*. The report urges consistent international approaches to product regulation.

To assure data exchange and minimize trade barriers, the OECD report recommends further R&D support for testing methods, equipment design, and studies of microbial taxonomy to produce information for use by such national and international organizations as the World Health Organization and the Commission of European Communities. The report also recommends that the large-scale industrial applications of biotechnology should whenever possible make use of microorganisms that are intrinsically of low risk and that such microorganisms be handled under the conditions of 'good industrial large-scale practice'. Further research is called for to improve techniques for monitoring and controlling the non-intentional release of genetically engineered organisms that are used in large-scale industrial applications.

The Food and Drug Administration

The FDA has the authority to regulate the manufacture of foods, food additives, drugs and cosmetics. Virtually all biotechnology companies are involved with the FDA because many biotechnology products are in the health area. To date, the FDA has been able to regulate biotechnology products without major changes in its current statues. FDA policy states that 'the use of a given biotechnological technique does not require a different administrative process. Regulation by FDA must be based on the rational scientific evaluation of products and not on any a priori assumptions about certain processes'.

The FDA uses documents entitled 'Points to Consider' to provide guidance on how FDA regulations apply to biotechnology products. This procedure is flexible enough to adapt to rapidly changing scientific and product developments. Five such documents have already been issued. They deal with the manufacture of monoclonal antibodies for in vitro diagnosis; the production and testing of interferon for investigational use in humans; the manufacture of monoclonal antibodies for therapeutic or diagnostic use in humans; the production and testing of new drugs and biological materials that are produced by recombinant DNA technology; and the characterization of cell lines used to make biological products.

These documents request specific information on the origins and characteristics of the cell lines that have been used to make the product in question. They ask, for example, whether the cells have oncogenic potential. The documents also spell out quality control requirements for the assessment of the product's identity and purity. A number of tests are suggested for showing that the product is free from contaminants, whether they be infectious agents, DNA or protein.

The National Institutes of Health

The NIH is collaborating with the FDA on the production of guidelines concerning the future use of cloned genes for gene therapy in humans. Clinical protocols for gene therapy trials that are funded by NIH will have to be approved by the NIH Recombinant DNA Advisory Committee, which is otherwise known as RAC. In preparation for this, the NIH has issued a 'Points to Consider' document on the design and submission of protocols for introducing new genes into the somatic cells of humans. Estimates are that such clinical experiments may begin by 1988 (also see Chapter 15).

Congressional and public scrutiny of the ethical implications of gene therapy will no doubt continue, especially over any attempts to transfer genes unto the germline, which would effect a permanent hereditary change in an individual who had undergone such a manipulation. Germline gene transfers into humans are not being contemplated at this time, however.

In 1986 the Boston-based Committee for Responsible Genetics asked the RAC to amend the NIH guidelines on recombinant DNA research to limit somatic cell gene therapy to 'life-threatening or severely disabling conditions' and to forbid germline therapy. The RAC, acting on the recommendations of its gene therapy subcommittee that no new amendments to the guidelines were required, did not make the suggested changes. The NIH committee did not wish to adopt limitations that would diminish its flexibility in providing advice concerning gene therapy experiments.

The NIH is likely to become involved in an additional area concerning patient rights. Patients have sued to establish their right of ownership to genes and cells obtained during clinical experiments and that have been shown to have possible commercial value. This situation is likely to lead the NIH to require that local Institutional Review Boards, which have the responsibility for determining whether the clinical experiments proposed by investigators at the institution are ethical, make certain that patients are informed about the potential commercial use of their body materials before they participate in clinical research. The OTA is also reviewing this subject.

The Environmental Protection Agency

The EPA has regulatory authority over activities, including chemical manufacturing and pesticide use, that might cause pollution of the land, air or water. The agency has recently begun regulating industrial biotechnology under the Federal Insecticide, Fungicide and Rodenticide Act (FIFRA) and the Toxic Substances Control Act (TSCA).

The EPA's regulatory concerns arise from the possible deleterious consequences of agricultural and chemical products that have been produced by biotechnology being released outside the laboratory. The concern extends to the field-testing of genetically modified organisms. The first such case – and one that illustrates some of the early regulatory confusion – arose in 1983 when the NIH's RAC approved an experiment in which a genetically modified strain of a bacterium was to be field-tested for controlling frost damage in plants.

The experiment was to be conducted by Steven Lindow and Nickolas Panopoulos of the University of California at Berkeley. These investigators had genetically modified the bacteria, which occur naturally in the environment, by removing a gene that enables the bacterial cells to act as nucleating centres for frost-crystal formation. The idea was that with the gene no longer present, the bacteria would be less capable of initiating frost formation on plants.

After the NIH committee approved the experiment, Jeremy Rifkin of the Foundation on Economic Trends in Washington, DC, filed suit to block the proposed test on the grounds that the NIH had failed to assess the environmental impact of releasing the altered microbe, as required by the National Environmental Policies Act. Rifkin won the suit, and the NIH set out to perform the necessary environmental assessment.

The situation was complicated, however, because Advanced Genetics Sciences, a new biotechnology company in California, wanted to perform essentially the same experiment. Although the recombinant DNA guidelines do not require the RAC's approval for experiments that are not supported by government funds, Advanced Genetics Sciences voluntarily submitted its proposed experiment to the committee.

Meanwhile, the EPA began to develop its own plans to regulate genetically engineered microbes that are intended for release into the environment. This action would have the advantage of including organisms developed by commercial firms that are not covered by the NIH guidelines. Advanced Genetics Sciences then submitted its proposed experiment to the EPA, as did other companies that were planning experiments involving the environmental release of genetically engineered plants or bacteria.

The NIH eventually amended its guidelines for recombinant DNA research in November 1985, and agreed to defer to another federal agency to which an experiment had been submitted, if reviews by both agencies would overlap. In the case of the frost-resistant microbe the NIH determined that a review by the EPA of the potential environment impact of the modified bacteria would be sufficient. Consequently, Lindow and Panopoulos also submitted their application to the EPA.

Although the EPA approved the request of Advanced Genetics Sciences for a permit for field testing of the bacteria in November 1985, the company acknowledged in January 1986 that it had conducted an outdoor test of the microbe – on the roof of a company building – a year before the EPA had granted its permission for environmental release. As a result, the EPA announced that the company had violated agency rules, suspended the permit, and imposed a fine on the company.

In an agreement announced in June 1986 the EPA withdrew its charges that Advanced Genetic Sciences had falsified data in its application to the agency. The company plans to apply for a permit. Meanwhile, in May the EPA had approved the request of Lindow and Panopoulos for a permit for their experiment. However, community protests in the region where the test was to be conducted and further legal challenges prevented the experiment from taking place during the 1986 growing season.

This lengthy regulatory saga has an ironic conclusion in that the frost-resisting bacterium, which was produced by a gene deletion, would now be subject to less stringent regulation, according to the new coordinated framework from the BSCC. The NIH's RAC, at its September 1986 meeting, agreed to consider a proposal to have the NIH guidelines cover only organisms that have been genetically modified by the addition of foreign DNA and thereby exempt gene deletion experiments from review.

The current guidelines exempt such experiments in the laboratory, but not under conditions of large-scale production or when environmental release is involved. All in all, the framework should provide better guidance concerning regulatory requirements both for the various government agencies and for the biotechnology industry, and consequently strengthen the industry's compliance with those requirements.

The EPA currently defines its regulatory boundaries by declaring that the Departments of Agriculture and the Interior control plants and animals and that the FDA's jurisdiction extends to food, food additives, drugs and cosmetics. The breadth and scope of the EPA's authority under FIFRA and TSCA to regulate the manufacture of pesticides and chemicals nevertheless brings a large segment of the biotechnology industry under the EPA's product review.

For its regulations, the EPA has adopted the scientific principles and criteria that have been set forth by the BSCC for classifying genetically engineered microorganisms. Microbial products that are made from a genetically engineered microorganism that is itself a pathogen or contains genetic material from a pathogen must undergo regulatory review before they can be released into the environment – unless the product is intended solely for non-pesticidal agricultural use. In that case it is to be reviewed instead by the Department of Agriculture.

The EPA will also regulate those microorganisms that have been formed by genetic engineering with materials that do not come from pathogens. Microbial products, whether pathogenic or non-pathogenic, will be subject to review under FIFRA (for pesticides) or under TSCA (for other chemicals) before any environmental release is permitted, including small-scale field testing and other research on the environmental impact of a product.

Although previous EPA regulations for environmental release exempt field tests on plots smaller than 10 acres, this exemption does not apply to genetically engineered microorganisms. In addition, non-indigenous microorganisms that are used as pesticides will be subject to regulatory review under FIFRA, but the review will be abbreviated for those organisms that are not genetically engineered.

Congressional oversight and legislative activities on the environmental release of biotechnology products were prominent in 1985 and 1986 and will undoubtedly continue. Bills were introduced in Congress to establish separate authorities within the EPA for the regulation of biotechnology research, field testing and product marketing.

Sound regulation in the environmental area depends on having accurate methods for assessing the risk of releasing a product. Much more research needs to be done on several aspects of risk assessment. More needs to be learned, for example, about how readily the genetic material that has been inserted into an organism is transferred to other organisms in the surrounding environment. The taxonomy and natural histories of bacteria in soil and water habitats also require more investigation.

US Department of Agriculture

The US Department of Agriculture (USDA) regulates agricultural products, including plant, meat and poultry products. In March 1986 the US Government Accounting Office (GAO) issued a report that pointed up inadequacies in the USDA's regulation of biotechnology. The GAO report noted that the department did not have a well-defined regulatory structure and that its Recombinant DNA Research Committee lacked the authority and direction to oversee biotechnology research and regulation. The report urged that the USDA's regulatory structure be more clearly defined. The BSCC's proposed coordinated framework should help address these concerns.

The USDA guidelines for regulating agricultural biotechnology apply to all areas of research funded by the department. The USDA guidelines on recombinant DNA research, which are partly based on the NIH guidelines on recombinant DNA research, encompass all phases of research. They include research performed under contained, laboratory conditions and specialized isolation research, as may be required, for example, in developing vaccines for serious diseases such as foot-and-mouth disease. The environmental release of genetically altered organisms also comes under the USDA regulatory purview.

The USDA identifies the organisms that it intends to regulate according to the definitions supplied by the BSCC. Among the products to be regulated by the department are vaccines for veterinary use. For veterinary biological products the USDA provides regulatory guidance on how to conduct field trials involving live, genetically engineered organisms and on what data are needed to support applications for product licences.

The USDA also identifies for regulation products made by the genetic engineering of plant pests, which are defined as all organisms that can cause disease or damage to plants. The agency maintains that its traditional statutory and regulatory authorities concerning seeds and meat and poultry products apply equally well to biotechnology products in these categories and that no changes are needed at this time.

One potential biotechnology application has been drawing a great deal of attention. This is the development of new strains of domestic animals by the introduction of new genes into the animals' germlines. As described in Chapter 15, investigators are attempting to breed larger, faster-growing pigs and sheep by transferring human growth hormone genes into the animals.

In 1985 Rifkin filed suit to halt these gene transfer experiments on the grounds that the USDA had not filed the required environmental impact statement on the work. He further asked the court to establish the principle that there should be no crossing of species barriers in mammals. The District Court in Washington, DC, dismissed the suit in April 1986 and noted that no environmental impact statement was required because the research had not progressed far enough for such an assessment to be made.

The Occupational Safety and Health Administration

The mission of the Occupational Safety and Health Administration (OSHA) is to make the workplace safe for employees. According to a recent statement by the agency, its current regulations already provide 'an adequate and enforceable' basis for protecting employee safety in the biotechnology industry. The agency statement concludes, 'No additional regulation of workplaces using biotechnology appears to be needed at this time since no hazard or hazards from biotechnology, per se, have been identified'.

The Commerce and Defense Departments

Export controls, another area of regulatory oversight, fall under the purview of the Commerce Department. Steps are under way at the Commerce Department to draft a set of export regulations governing biotechnology. The issue is how to balance the interests at stake. Export regulations are largely driven by national security conditions.

The Department of Defense, which advises the Commerce Department on national security matters, wants to restrain biotechnology exports that might help potential adversaries wage chemical or biological warfare. Fermentation technology and high-capacity separatory devices rank high on its list of exports that might have this effect. Nevertheless, although export controls may be necessary to protect the national security, if they are too extensive or stringent they might harm the international exchanges upon which academic and industrial biotechnology research depends.

The President's Commission on Industrial Competitiveness, which was mentioned previously, has noted that the national security, as well as a successful foreign policy, depends on maintaining US industrial competitiveness and that export restrictions are generally antithetical to competitiveness. It has made some excellent recommendations on the subject of the federal role in developing international trade policies, including export regulations.

Among these are the establishment of a cabinet-level Department of Trade to improve policy-making in that area and a more effective coordinating mechanism for balancing domestic and international policies. The Commission also called for revisions in domestic trade laws and a reform of US anti-trust policy to include exemptions for mergers that promote national objectives. Additional recommendations suggest streamlining and automating the existing licensing process for all controlled exports. Such changes are necessary to preserve competitive responses to other nations. Finally, the President's Commission suggested several measures for financing exports and concluded with a call for strengthening the multilateral trading system.

Within the United States, some groups have expressed concern that biotechnology might be used for the purposes of biological warfare. Rifkin, for example, submitted a proposal to the RAC that called for a working group to examine these potential uses. The proposal was turned down. Central to this decision was a statement from the Department of Defense noting that the proposed working group is unnecessary because the department adheres fully to national policy as spelled out in the Biological Weapons Convention of 1972, which prohibits research on biological weapons. The Defense Department also emphasized that all of its research programmes that involve recombinant DNA are unclassified and conducted in full compliance with the NIH guidelines.

At a review of the operations of the Biological Weapons Convention held in Geneva in September 1986 a major topic was the potential misuse of biotechnology for biological warfare. Agreement was reached on a number of steps to exchange information and data on biotechnological R&D to ensure that it is not used to develop biological weapons.

Rifkin has also sued the Defense Department to halt its plans to construct new laboratories at the Dugway Proving Grounds in Utah for testing highly infectious and lethal biological aerosols. The suit alleged that the department had failed to prepare an Environmental Impact Statement on its biological research programme. The Court enjoined the construction of the laboratory until the Army could conduct the required assessment of potential environmental problems, which the Army has agreed to do. As this case and the proposal to the RAC demonstrate, there is likely to be continuing public debate on the potential misuse of biotechnology for biological warfare.

Conclusions

National concern over the future competitiveness of domestic biotechnology industries remains high in the United States, Japan, the European countries, and elsewhere. As these countries strive to remain competitive by assisting the vigorous growth of these industries, it is well to remember how the United States achieved its position of leadership in biotechnology.

Frank Press, the current president of the NAS, has identified five key elements that contributed to this situation. These are: a quarter century of strong federal support for research in basic biology; a powerful research system in the universities that is driven by a catalytic blending of research and teaching; the presence of clinicians in research laboratories who knew that a better understanding of gene regulation would have an enormous impact on the diagnosis and treatment of disease; the scientific community's acceptance of its responsibility to alert the public to the potential risks of recombinant DNA technology and the need for appropriate oversight of the research by the NIH and its advisory groups; and the ability to conduct biotechnology research without enormous financial resources.

Today in the United States there is a growing partnership – between government, academia and industry – that is priming the flow of basic science discoveries and their applications in biotechnology. Nevertheless, cognizant of concern that biotechnology might result in harm to man and the environment, David Bazelon, retired Senior Circuit Judge of the US Court of Appeals for the District of Columbia, points out that society must be informed about 'what is known, what is feared, what is hoped, and what is yet to be learned'. Science must provide an assessment of any risk posed by biotechnology, but it is society that must ultimately decide whether that risk is acceptable.

Future leadership in biotechnology in the United States and other countries depends on government financial support and tax policies to foster technological innovation, as well as on government investment in biotechnology R&D programmes. Government, university and industry, working in collaboration under enlightened public oversight, will develop a sound data base for risk assessment. As the data base grows, speedy and effective regulatory review of biotechnology products will permit the promise of industrial biotechnology to be realized internationally, to the benefit both of the industrial world and of the less developed countries.

Additional reading

Bloch, E. (1986). Basic research in economic health: the coming challenge. *Science*, **232**, 595–9.

Blumenthal, D., Gluck, M., Louis, K. S. and Wise, D. (1986). Industrial support of university research in biotechnology. *Science*, **231**, 242–6.

Cape, R. E. (1986). Future prospects in biotechnology: a challenge to United States leadership. In *Biotechnology in Society: Private Initiatives and Public Oversight*, ed. J. G. Perpich, pp. 5–9. Pergamon Press, New York.

Congressional Budget Office (1985). *Federal Financial Support for High Technology Industries*. The Congress of the United States, Congressional Budget Office, Washington, DC (June 1985).

Dibner, M. D. (1986). Biotechnology in pharmaceuticals: the Japanese challenge, *Science*, **229**, 1230–5.

Dibner, M. D. (1986). Biotechnology in Europe. *Science*, **232**, 1367–72.

Marcum, J. M. (1986). The technology gap: Europe at the 'Crossroads'. *Issues in Science and Technology* (Summer 1986), 28–37. National Academy of Sciences, Washington, DC.

Milewski, E. A. (1986). Discussion on human gene therapy. *NIH Recombinant DNA Technical Bulletin*, **9**, 88–130.

Office of Science and Technology Policy (OSTP) (1985). Coordinated framework for regulation of biotechnology and establishment of the Biotechnology Science Coordinating Committee: notice. *Federal Register*, **50**, 47174.

Office of Science and Technology Policy (OSTP) (1986). Coordinated framework for regulation of biotechnology: announcement of policy and notice for public comment. *Federal Register*, **51**, 23302.

Olson, S. (1986). *Biotechnology: An Industry Comes of Age*. National Academy Press, Washington, DC.

Perpich, J. G. (ed.) (1986). *Biotechnology in Society: Private Initiatives and Public Oversight*. Pergamon Press, New York.

Perpich, J. G. (1986). A federal strategy for international industrial competitiveness. *Bio/Technology*, **4**, 522–5.

President's Commission on Industrial Competitiveness (1985). *Global Competition: The New Reality*, vol. 1. Report of the President's Commission on Industrial Competitiveness. US Government Printing Office, Washington, DC (January 1985).

Prestowitz, C. V. (1986). Japanese biotechnology industry development. In *Biotechnology: Industrial Competitiveness and International Trade Policies*, ed. J. G. Perpich. *Technology in Society*, **8**, 219.

Glossary

Words in *italic type* are cross-references to other entries in the Glossary.

Adenosine deaminase deficiency: A *genetic disease* in which a lack of the *enzyme* adenosine deaminase results in a severe loss of immune responses that usually proves fatal in the first few years of life. The disease is a likely candidate for the first human *gene therapy* trials.

Agrobacterium tumefaciens: A species of *bacteria* that produces *crown gall tumours* in the plants it infects. The tumours are caused by the acquisition of *DNA* transferred from the *Ti plasmid* of the bacteria into the plant *cells* during infection. *A. tumefaciens* has been adapted for introducing new genes into crop plants.

Amino acid: The building block molcules that are linked together to make *peptides* and *proteins*.

Amniocentesis: A procedure in which fluid containing fetal cells is withdawn from the uterus of a pregnant women by a needle inserted through the abdominal wall. The cells can be analysed for indications of the presence of certain *genetic diseases* in the fetus.

Antibiotic: Any of a group of structurally diverse chemicals produced by microorganisms, including *bacteria* and moulds, that kill other microorganisms. Antibiotics are used to treat bacterial diseases and some of the agents have proved valuable in cancer chemotherapy.

Antibody: Antibodies are *proteins* produced by the *B lymphocytes* of the immune system. They help to defend the body against foreign invaders, including pathogenic *viruses* and *bacteria*, by searching out and binding to foreign molecules.

Antigen: Strictly speaking, any molecule that is recognized and bound by an *antibody*. Usually this is the molecule used to elicit production of the antibody.

Antigenicity: The ability to be recognized and bound by an *antibody*.

Anti-idiotype antibody: An *antibody* made in response to, and recognizing the *iodiotype* of, another antibody. Anti-idiotype antibodies are under investigation as possible *vaccines*.

Autosome: Any *chromosome* other than those that determine the sex of an organism. The human species, for example, has 46 chromosomes, two of which (the X and Y chromosomes) determine the sex of an individual and the rest of which are autosomes.

Autotroph: An organism that can grow on carbon dioxide as its sole source of carbon.

Bacillus thuringiensis: A species of *bacteria* that infects insects and is used as a biological insecticide.

Bacteria: Small, *prokaryotic* organisms that lack a *nucleus* and usually have rigid cell walls. Although some bacterial species cause disease in man, most are useful both in nature (for example those that carry out *nitrogen fixation* or break down organic matter) and in the *biotechnology* industry.

Bacteroid: The form taken by the *bacteria* of the genus *Rhizobium* in the nitrogen-fixing nodules on the roots of *legume* plants.

Baculovirus: A type of *virus* that infects insects and is used as a biological insecticide.

Base-pairing: In the *DNA double helix* the two strands are held together by weak bonds that join the bases protruding from one chain to those of the other chain. The bases are paired so that adenine joins with thymine, and guanine with cytosine. Thus, the sequence of one DNA strand specifies the sequence of the second.

Bioleaching: The use of solutions containing *bacteria* to dissolve and recover valuable metals from ores.

Biomining: The use of *bacteria* to recover valuable metals from ores. As the biomining methods are currently constituted, they are generally applied to ores of such poor quality that metal recovery by more traditional operations is not economically feasible.

Bioreactor: A vessel or tank in which *fermentation* reactions are carried out. Bioreactors are equipped with a variety of instruments for monitoring and adjusting the conditions in the tank so that the reaction proceeds efficiently.

Biotechnology: The application of living organisms or of substances made by living organisms to make products of value to man.

B lymphocyte: A type of white blood *cell* that produces *antibodies* when appropriately activated. 'B' refers to the bursa of fabricius, the organ where the B cells mature in the chicken. The comparable organ in mammals is unknown.

Callus: An undifferentiated mass of *cells* that forms on plant wounds and is also produced by plant tissues growing in *cell culture*.

Carbohydrate: Biological compounds containing carbon, hydrogen and oxygen, usually with two hydrogens and one oxygen for every carbon atom. The carbohydrates include the simple *sugars*, such as glucose, and more complex molecules, such as starch and cellulose, in which many sugars have been joined together to form large *polymers*.

Carrier: An individual who carries and can transmit a *gene* for a *hereditary disease*, but who does not have the disease. Many genetic diseases are caused by *recessive genes*, two copies of which must be inherited for the condition to develop; a carrier, who has only one copy, will not show any symptoms.

Cell: The fundamental unit of which living organisms are composed. Some organisms, such as *bacteria*, consist only of one cell, whereas others, such as man, contain billions.

Cell culture: Growing cells or tissues on appropriate nutrients in laboratory dishes.

Chimeric antibodies: *Antibodies* in which the individual *protein* chains are composed of segments from two different species. The most common combination today is mouse with human.

Chloroplast: A type of membranous particle that occurs in plant *cells* and is the site of *photosynthesis*.

Chorionic villus biopsy: A procedure in which a small sample of tissue is removed from the chorion (one of the membranes of the fetus) and examined for fetal abnormalities, especially those of genetic origin.

Chromatin: The complex of *protein* and *DNA* of which the *chromosomes* are composed.

Chromosomes: The thread-like structures in the *cell nucleus* that contain the *genes* in a linear array. Each *somatic cell* in a *eukaryote* has a double set of chromosomes.

Clone: A group of genetically identical *cells* or organisms that are all descended asexually from the same individual. (See also *gene cloning*).

Codon: A sequence of three successive *nucleotides* in *DNA* or *RNA* that either specifies a single *amino acid* in a *protein* or serves as a stop signal marking the end of a *gene*.

Cofactor: A small molecule necessary for the catalytic activity of an *enzyme*.

Cross-protection: A situation in which infection of a cell with one viral strain protects the cell against infection by another, related strain.

Crown gall: A tumour formed on *dicotyledonous plants* as a result of infection by *Agrobacterium tumefaciens*. The tumour is caused by transfer of *DNA* from the *Ti plasmid* of the *bacterium* to the plant cell.

Culture: See *cell culture*.

Deoxyribonucleic acid (DNA): One of the two types of *nucleic acid* and the material of which genes are composed. A strand of DNA is formed by linking together many individual building blocks called *nucleotides* in a large molecule. The linear sequence of the nucleotides specifies the linear sequence of *amino acids* in the *protein* of a product *gene*.

Deoxyribose: A sugar occurring in *nucleic acids*, specifically in *deoxyribonucleic acids*.

Dicotyledonous plant (dicot): One of the two major subclasses of plants. The embryos of dicots have two seed leaves (cotyledons). Trees, shrubs, and many common crop plants such as the tomato, potato and tobacco are dicots. (See also *monocotyledonous plants*.)

DNA: See *deoxyribonucleic acid*.

DNA polymerase: The *enzyme* that copies the *DNA* of the genetic material when it duplicates before cell division.

Dominant gene: One in which inheritance of single copy of the *gene* is sufficient to confer the trait specified.

Double helix: The three-dimensional structure of *DNA* in which the two strands of the molecule are twined together in a regular helical structure.

Enantiomers: Compounds that are mirror images of one another. They can occur in any compound in which four different substituent groups are attached to the same carbon atom.

Enhancer: A regulatory *DNA* sequence associated with a *gene* that increases *transcription* of that gene into *messenger RNA*.

Enzyme: Proteins that act as biological catalysts, speeding up the rates of chemical reactions in living organisms. Most enzymes are extremely specific and act on a very limited range of molecules.

Escherichia coli (E. coli): A common *bacterium* that lives, among other places, in the human intestinal tract. Much of the fundamental work in understanding the biochemical basis of life was performed on *E. coli*. Laboratory strains of the bacterium are now used in *biotechnology* for *gene cloning* and making *recombinant proteins*.

Eukaryote: Any organism in which the *cells* have a true *nucleus* enclosed by a membrane. The *chromosomes* are contained within the cell nucleus. Eukaryotes are generally considered to be more advanced forms of life than *prokaryotes*.

Evolution: The development of new forms of life from pre-existing species.

Fatty acid: Organic acids found in fats and also as intermediates in such energy-producing pathways as the *tricarboxylic acid cycle*. Fatty acids have a number of diverse applications. They can be used, for example, as flavouring agents and lubricants.

Feedstock chemicals: Chemicals used in large quantities as starting materials for the manufacture of other chemicals.

Fermentation: The use of microorganisms, or in some cases of *enzymes* extracted from microorganisms, to carry out any of a wide variety of chemical reactions. The classic example of fermentation is the conversion of sugar to alcohol by yeast.

Fermentor: The tank or other apparatus in which *fermentation* reactions are carried out. The fermentor is usually equipped with devices for monitoring and adjusting the conditions of the reaction it contains.

Gene: The basic unit of heredity, a gene encodes a specific trait that can be passed on from one generation to the next. In molecular terms, a gene is a segment of *nucleic acid*, usually *deoxyribonucleic acid*, that specifies the structure of a single *protein* chain.

Gene cloning: The use of *recombinant DNA* technology to isolate and produce large quantities of an individual *gene*.

Gene library: A set of *cell clones*, now usually clones of *bacterial* cells, each of which contains a *DNA* fragment from a particular source. A human gene library, for example, consists of cell clones containing fragments of human DNA.

Gene therapy: Treatment of a human *genetic disease* by introducing a new *gene* into the patient to take over the function of the defective gene that is causing the disease.

Genetic disease: A disease that is caused by a defective *gene* and can be transmitted from one generation to the next.

Genetic engineering: The application of *recombinant DNA* methods to confer new traits on organisms by introducing new *genes* into their *cells*.

Genome: The complete *gene* composition of an organism.

Germ cells: The cells that give rise to the sperm in males and the eggs in females.

Glycolysis: The series of reactions that begins the breakdown of the sugar glucose and results in the release of a small quantity of energy for use by the *cell*. The *fermentation* of sugar to alcohol in yeast occurs through the glycolytic pathway.

Graft-versus-host disease: A condition that occurs in persons who have received bone marrow transplants. It is caused by the transplanted marrow (the graft) producing immune *cells* that attack the recipient's tissues as foreign.

Haemoglobin: The oxygen-carrying molecule of red blood *cells*. Haemoglobin consists of four *protein* chains, two each of the protein alpha-globin and two each of beta-globin, plus the red-coloured pigment haem.

Hereditary disease: A disease that can be transmitted from one generation to the next and is the result of a defective *gene*.

Heterotroph: An organism that cannot grow on carbon dioxide but must have more complex carbon sources such as *sugars*.

Heterozygote: *Genes* are inherited in pairs, with one member of the pair coming from the mother and one from the father. A heterozygote is an individual in which the pair consists of two different variants of the gene in question. For example, a normal beta-globin gene might be paired with a *sickle cell* gene.

Homozygote: An individual in which the two members of a particular *gene* pair are the same.

Hormone: A biologically active substance produced in one part of the body and carried by the blood to another part where it exerts its effects.

Hybridization: 1. In molecular biology, a procedure in which single-stranded *nucleic acid* molecules, from either *DNA* or *RNA*, are mixed. If the molecules have similar structures they will hybridize, that is, bind to one another by the *base-pairing* rules. The procedure can be used to give an indication of the degree of relatedness of two *genes*, or to detect RNAs or DNAs using a specific nucleic acid probe. 2. The crossing of two species to form hybrids that have some of the characteristics of each parent.

Hybridoma: A hybrid cell made by fusing an antibody-producing *B lymphocyte* with a *myeloma cell*. Hybridomas are the source of *monoclonal antibodies*.

Hypoxanthine–guanine phosphoribosyl transferase (HPRT): An *enzyme* needed for the synthesis of the building blocks of *nucleic acids*. An HPRT deficiency that results from a *gene* defect is the cause of *Lesch–Nyhan disease*.

Idiotype: Region characteristic of a particular *antibody*. The idiotype, which is located in or near the *antigen*-binding site of the antibody, can itself stimulate antibody production and may have a shape complementary to that of the antigen recognized by the antibody.

Immunogenicity: The ability to elicit an immune response.

Interferons: A group of related *proteins* having a variety of activities in the body. These include fighting viral infections and stimulating the cell-killing abilities of certain types of immune cells. The interferons are being tested in therapies for cancer and acquired immune deficiency syndrome (AIDS).

Interleukins: A group of *proteins* that transmit signals for growth and development among immune cells and are necessary for mounting normal immune responses. Interleukin-2, for example, is undergoing clinical trials for treating cancer and acquired immune deficiency syndrome (AIDS).

Leghaemoglobin: A type of *haemoglobin* found in the *nitrogen-fixing* nodules on the roots of *legume* plants. Leghaemoglobin tightly binds oxygen, which would otherwise poison the nitrogen fixing enzymes.

Legume: Member of a group of plants that includes peas, beans and alfalfa. Legumes can form *symbiotic* relationships with *nitrogen-fixing bacteria* of the genus *Rhizobium*.

Lesch–Nyhan disease: A *genetic disease* caused by a defective *gene* for the *enzyme hypoxanthine–guanine phosphoribosyl transferase*. The condition is characterized by increased concentrations of uric acid in the blood and neurological symptoms, including mental retardation and a compulsion for self-mutilation.

Linkage analysis: The association of a particular structural variation in *DNA* with the presence of a mutant *gene* that can cause a *genetic disease*. Linkage analysis can be used for diagnosing genetic diseases and detecting the gene *carriers*.

Long terminal repeats (LTRs): Duplicated *DNA* sequences found at the ends of the *genomes* of certain *viruses*. They carry signals needed for the expression of the viral *genes*.

Meiosis: Nuclear division in the *germ cells*, which results in the halving of the number of *chromosomes* so that the egg cells or sperm that are produced receive just one set of chromosomes instead of the double set found in each *somatic cell*.

Messenger RNA (mRNA): A form of *RNA* that is copied from the *DNA* of the *genes* so that its structure contains the 'message' encoded in the gene. The messenger RNA carries that message to the *ribosomes* where it directs *protein* synthesis.

Metabolism: The sum total of all the biochemical reactions in the *cell*.

Mitochondrion: A membrane-enclosed particle that is the site of the principal energy-yielding reactions of the *cell*, including those of the *tricarboxylic acid cycle*.

Mitosis: Nuclear division in the *somatic cells*. Before the *nucleus* divides, the *chromosomes* are duplicated so that each daughter cell that will be formed receives a complete double set of chromosomes.

Monoclonal antibody: An *antibody* produced by a *hybridoma clone*. Because all the cells in a clone are identical, they all produce the same antibody molecule. Monoclonal antibody technology provides a way of producing essentially unlimited quantities of a pure antibody.

Monocotyledonous plant (monocot): One of the two major classes of plants. Monocots have embryos with just one seed leaf (cotyledon). The grasses and major cereal crops, including wheat, corn and rice, belong to this class. (See also *dicotyledonous plant*.)

Mutagenesis: The introduction of changes into an organism's *genes*.

Mutant: An organism that has undergone some change as a result of a *gene mutation*.

Mutation: Any change in the *DNA* of a *gene* that produces some alteration in the organism carrying that gene.

Myeloma: A tumour of the *antibody*-producing *B lymphocytes*. Myeloma cells are used to make *hybridomas* for the production of *monoclonal antibodies*.

nif genes: A complex of *genes* needed for *nitrogen fixation*. Only *prokaryotes*, including certain *bacteria* and blue-green algae, carry *nif* genes, but some of these organisms can form close *symbiotic* relationships with green plants, especially *legumes*.

Nitrogenase: An enzyme complex needed for the reduction of atmospheric nitrogen to ammonia during biological *nitrogen fixation*.

Nitrogen fixation: The reduction of atmospheric nitrogen to ammonia. The reaction, which requires a great deal of energy, can be carried out chemically or biologically.

Nodulation: The formation of *nitrogen-fixing* nodules on the roots of plants by *bacteria*, such as those of the genus *Rhizobium* that form *symbiotic* relationships with *legumes*.

Nucleic acid: Molecules consisting of a backbone of alternating *sugar* and phosphate groups from which protrude certain organic bases, which are attached to the sugars. The nucleic acids are so called because they were originally isolated from *cell nuclei*. They include *deoxyribonucleic acid*, which forms the genetic material, and the *ribonucleic acids*.

Nucleotide: A molecule consisting of a *sugar* to which is attached a phosphate group and an organic base. Nucleotides are the building blocks of the *nucleic acids*.

Nucleus: The most prominent structure within the *cell* of *eukaryotes*, the nucleus is bounded by a membrane and contains the *chromosomes*.

Oncogene: A *gene* that can cause cells to undergo cancerous *transformation*. Oncogenes are aberrant forms of normal cellular genes that are needed for growth and differentiation.

Pathogen: Any microorganism or *virus* that causes disease.

Peptide: A molecule formed by the linking together of two to perhaps 100 *amino acids*. (See also *protein*.)

Photosynthesis: The conversion of carbon dioxide and water to *carbohydrates* by plants, using light as the energy source.

Plasmid: A small, usually circular, molecule of *DNA* that occurs in *bacteria* and is not part of the bacterial *chromosome*. Plasmids, which carry small numbers of *genes*, have been adapted as *vectors* for transferring genes between species.

Polymer: A large molecule formed by linking together many simple building blocks. *Proteins, nucleic acids* and *polysaccharides* are examples of naturally occurring polymers.

Polysaccharide: Large *carbohydrates* formed by linking together many *sugar* molecules. Starch and cellulose are examples.

Prokaryote: Organisms, including the *bacteria*, that consist of usually a single cell that does not have a *nucleus*. Prokaryotes are usually considered to be less advanced forms of life than *eukaryotes*.

Promoter: A regulatory *DNA* sequence that is associated with a *gene* and is necessary for the accurate initiation of the gene's *transcription*.

Protein: A biological *polymer* formed by linking together many *amino acids*. The principal distinction between *peptides* and proteins is size, proteins usually having molecular weights greater than 10 000 whereas the molecular weights of peptides are less than 10 000. The categories may overlap, however. Proteins serve as *enzymes* and also help to form cellular structures.

Protoplast: A plant *cell* that has been denuded of its rigid cellulose wall. Protoplasts are used to produce *somatic cell hybrids* and also as targets for *gene* transfer in plant *genetic engineering*.

Receptor: A molecule, usually a *protein*, that is present on a cell's outer membrane. A receptor recognizes and binds a specific intercellular messenger such as a *hormone* or growth factor, thereby initiating the messenger's effects within the cell.

Recessive gene: One encoding a trait that will not be evident unless two copies of the *gene* are present. Many *genetic diseases* are caused by recessive genes.

Recombinant DNA: A *DNA* molecule formed by joining DNA segments from two or more sources.

Recombinant protein: A *protein* made, usually in *bacteria*, yeast or other *cultured cells*, as the product of a *gene* in a *recombinant DNA* molecule. Many human proteins of potential clinical importance are now made as recombinant proteins.

Restriction enzyme: A bacterial *enzyme* that cuts *DNA* at a specific *nucleotide* sequence. Restriction enzymes are used for making *recombinant DNA*, for *gene* mapping, and in the diagnosis of *genetic diseases*.

Restriction fragment length polymorphism (RFLP): A change in the size of the fragments produced when a *DNA* sample is digested by a *restriction enzyme* that is the result of a change in the DNA sequence. RFLPs are used in diagnosing *genetic diseases*.

Restriction mapping: The mapping of a *DNA* molecule by digesting it with *restriction enzymes*. Each DNA molecule will have its own characteristic pattern of sites recognized by the enzymes, and the maps can therefore be used to compare the relatedness of DNAs from different sources.

Retrovirus: Any of a group of *viruses* that have *RNA*, instead of *DNA*, as their genetic material and the life-cycles of which include a step in which the viral RNA is copied into DNA. Retroviruses are being developed as *vectors* for *gene* transfer.

Rhizobium: A genus of *bacteria* that includes the *nitrogen-fixing* species that form *symbiotic* relationships with *legume* plants.

Ribonucleic acid (RNA): One of the two major types of *nucleic acid*. RNA is similar in structure to *DNA*, but contains the *sugar* ribose instead of deoxyribose. The RNAs include *messenger RNA*, *ribosomal RNA* and *transfer RNA*.

Ribose: A *sugar* occurring in the *ribonucleic acids*.

Ribosomal RNA (rRNA): The type of RNA that combines with several *proteins* to form the *ribosomes*.

Ribosome: An intracellular particle composed of *proteins* and *ribosomal RNA*. Ribosomes are the sites of *protein* synthesis.

RNA: See *ribonucleic acid*.

Sickle cell anaemia: A hereditary form of anaemia caused by a defect in the *gene* coding for beta-globin, one of the two *proteins* of *haemoglobin*.

Single-cell protein: *Proteins* made by single-celled organisms such as *bacteria*, fungi or algae for use in animal or human foods.

Somaclonal variation: Genetic variability occurring in plants that have been regenerated from *cells* maintained in culture. The variability can arise even in cells that originally came from the same plant and would be expected to have identical genetic compositions.

Somatic cell: Any of the cells of an organism except the *germ cells*.

Somatic cell hybrid: A hybrid produced by fusing *somatic cells* from two different species, especially plant species.

Stem cell: An immature form of bone marrow *cell*. It divides to form progeny some of which serve to maintain the stem cell population while others give rise to all the major cell lineages originating in the bone marrow.

Steroid: A member of a class of compounds that have a core structure of one five-carbon and three six-carbon rings. Cholesterol is a steroid, as are the male and female sex *hormones* and a number of additional hormones that regulate salt and water balance and metabolism.

Substrate: A compound that is acted on by an *enzyme* and thus converted to another compound.

Sugar: A simple, usually sweet, form of *carbohydrate*. Some sugars, such as glucose, are energy sources for *cells*. They are also building blocks for *polysaccharides* such as starch and cellulose.

Symbiosis: A relationship in which individuals of two different species live together to their mutual benefit. The relationship between *nitrogen-fixing Rhizobium bacteria* and *legume* plants is an example.

T lymphocyte: A type of white blood *cell* that performs several functions needed for normal immune responses. Some T lymphocytes are killers that destroy cells recognized as foreign (virus-infected cells for example). Other T cells act as regulators – either helpers or suppressors – of other immune cells, including the *antibody*-producing *B lymphocytes*. 'T' refers to the thymus gland, the organ in which T lymphocytes mature.

Thalassaemia: A type of hereditary anaemia in which a *gene* defect results in a deficiency or complete absence of one of the two *proteins* of the *haemoglobin* molecule.

Thiobacillus ferrooxidans: A species of *bacteria* that is used for *biomining*.

Ti plasmid: A *plasmid* carried by the *bacterium Agrobacterium tumefaciens*. The bacteria transfer the plasmid into plant cells where it induces the formation of *crown galls*. Modified forms of the Ti plasmid are now being used for the *genetic enginering* of plants.

Tissue plasminogen activator (TPA): An *enzyme* that converts the inactive enzyme precursor plasminogen to the active enzyme plasmin, which acts to dissolve blood clots. TPA that is made by *recombinant DNA* technology is being tested clinically for dissolving the blood clots in the coronary arteries of heart attack victims.

Transcription: The copying of the *DNA* of *genes* into *messenger RNA*, which is the first step in gene expression.

Transfer RNA (tRNA): A type of *RNA* that picks up *amino acids*, carries them to the *ribosomes* where *protein* synthesis is taking place, and lines them up in the proper position on the *messenger RNA* so that they can be joined together to form a protein.

Transformation: 1. The introduction of foreign genes into cells, which acquire some new characteristic as a result. 2. The conversion of normal cells to cancer cells.

Transit peptide: A *peptide* containing from about 15 to about 25 *amino acids* that is attached to a *protein*. A transit peptide directs the protein to its final destination inside or outside the cell and is then removed.

Translation: The conversion of the *nucleotide* sequence of a *messenger RNA* into the *amino acid* sequence of a protein during protein synthesis.

Transposable element: A piece of *DNA* that can move about the *genome* and can cause *mutations* if it inserts into a *gene*. Transposable elements, which are also called transposons, are being investigated as potential *vectors* for gene transfer.

Tricarboxylic acid (TCA) cycle: One of the central pathways of *metabolism* in the *cell*, the TCA cycle yields much of the energy released during the breakdown of sugars and fats. The cycle is named after the first compound in the pathway.

Vaccine: A preparation introduced into the body to stimulate immunity against a *pathogen*, usually a *bacterium* or *virus*. Most vaccines are made from the pathogen itself, which is killed or attenuated so that it no longer causes disease.

Vector: In *biotechnology*, a *DNA* molecule, usually a *plasmid* or viral DNA, used to transfer *genes* into cells.

Virus: Particles consisting of a *nucleic acid* core, which may be either *DNA* or *RNA*, surrounded by a *protein* coat. Viruses are parasites that reproduce by entering cells and taking over the cells' own synthetic machinery.

Index

Page numbers in **bold** refer to illustrations and figures.

Abdullah, Ruslan 124
acetic acid fermentation 33
Acetobacter fermentation **59**
acetone 28
   fermentation production 31–2, **33**
acetyl CoA 31–3, 34
Acinetobacter sp. H01–N 69
acquired immune deficiency syndrome *see* AIDS
adenosine deaminase (ADA) deficiency 186–8, 192
adenosine triphosphate *see* ATP
adriamycin 57
Aerobacter aerogenes 137
aflatoxin 61
agricultural chemicals 1
Agrobacterium 115, 127, 133
   herbicide tolerance 135
Agrobacterium tumefaciens 130–2, **136**
agrochemicals 126
AIDS
   molecular mechanisms 47
   recombinant interleukin-2 in 51
   vaccine development 49
   virus 49
alanine 17
alcohol 30
aldolase 23–4
Alexander, Hannah 168
alfalfa 125
algae
   in sewage treatment 73
   in single-cell protein production 72–3
amino acids 9–10
   for animal feed 29
   microbial fermentation 60–1
   nutritional supplements 61
   opines 131
   synthesis 20, 64
amniocentesis 174
ampicillin 64
Amrhein, Nicholas 136
amylase 19
   recovery 37–8

Anabaena **108**, 116–17
Anderer, F. 159
Anfinsen, Christian 161
animal feed
   fermentation waste 39
   single-cell protein 78, 80
antibiotics 1
   microbial products 29, 56, 57, 60
   synthesis by immobilization techniques 64
antibodies
   anti-idiotype 156, **157**
   binding 160
   chimaeric 153, 156, 157
   enzyme-like behaviour 168
   network theory of regulation 156
   peptide elicited and medical applications 168–70
   protein reactive 160
   reacting with viral particle **162**
   structure of molecule **153**, **154**, **159**
   to myohaemerythrin 167
anticodon 10
anti-depressants 61
antigenicity 160, 161, 164, 167
anti-inflammatory compounds 61
alpha-1-antitrypsin deficiency 182
arachidonic acid 61
Arnon, Ruth 161
aroma chemicals **63**
asparaginase 22
Aspergillus **58**
Aspergillus niger 34, 36
Atassi, M. 161
ATP 16, 23
autoimmune conditions 147
auxins 119
Avery, Oswald 7
Axel, Richard 188
Azolla 105
Azospirillum 110, 117

B cell lymphoma 151–2
B lymphocytes 145, 147
Bacillus thuringiensis 95, 141–2, **143**
bacitracin 64
bacteria
    carbon cycling 96–8
    citric acid production 34
    discovery 28
    gene transfer in environment 100
    genetic improvement of leaching 87
    high protein content 71
    leaching 87
    metal leaching 82–4
    mining and 98
    nitrogen fixing 104
    in non-photosynthetic single-cell protein production 73–4
    release into environment 94
    in single-cell protein production 72, 73, **74**
baculovirus 95
Baltimore, David 191
barnacles 40
Bast, Robert 149
Beachy, Roger 140
Beadle, George 5–6
bean phaseolin genes 135
beer haze formation 20
Berg, Paul 11
beverage industry 20
Bingham, John 127
bioleaching 82, 84
    chemical reactions 84
    dump 84, 86
    in situ 86
    operations 84, 86–8
biological control 94–5
biosorption 98
biosynthesis, enzymes as catalysts 56–7
biotechnology industry 197
Biotechnology Science Coordinating Committee 204–5
bloat 125
blood
    clots 21, 51, 61
    fetal analysis 174
blue-green algae 105, 107, 116
bombesin 157
bone marrow
    gene transfer into cells 186–8, 189
    stem cells 189
    transplant 152
Boyer, H. 12
Bradyrhizobium 111, 113
Brassica napus 124–5
Brill, Winston 107
Broadbalk Experiment 103

Browne, Christine 120
Brownlee, George 183
butanol 28
    fermentation production 31–2, **33**

cachectin 47–8
callus 133
    cells 119
cancer
    breast **150**
    colorectal 148
    development stages 48
    diagnosis **150**
    diagnosis with monoclonal antibodies 148
    genetics 48
    immunity 156
    lung 149, 151
    molecular mechanisms 47
    ovarian 149
    virus 48
cancer therapy 2, 47
    antibiotics 57
    monoclonal antibodies 145, 149–52, 157
    recombinant interleukin-2 in 51
    recombinant leucocyte interferon 50
    steroids 67
Candida utilis 71, 77, 78
Cannon, Frank 112
carboxypeptidase A **15**
carcinoembryonic antigen (CEA) 148
carcinogens, chemical 48
carrot 133
casein 20
Catharanthus roseus 126
cattle breeding 1
cauliflower mosaic virus (CaMV) 132–3, 134
caulimovirus 132
cell culture, plant
    clones 120
    hybrid production by protoplast fusion 124–5
    limitations 123
    methods 119–20
    synthesis 126
    transformation 127–8
cells 4
    animal 5
    communication 47
    fluorescence-activated sorting 125
cellulose as raw material for fermentation 30
Chaetomium cellulolyticum 80
Chakrabarty, Ananda 97
chalcone synthase gene 135
Chaleff, Roy 123
Chargaff, Erwin 8

cheese
    production 20
    wrapping 21
chemical
    industry and enzymes 27
    synthesizing 20
chemotherapy 57, 67
chillhaze 20
chill-proofing 20
Chlamydia trachomatis 147–8
Chlorella 72, 73
chlorobenzoic acid degradation 97, **98**
chloroplasts
    genomes 135
    transit peptide 135
cholesterol 67
chorionic villus biopsy 174–6
chromatin 4
chromosomes 4–5
    abnormalities in cancer cells 48
Chung, Annie 12
citric acid
    contaminants in production 36
    fermentation 33–4
citrus tristeza virus 140
Claviceps purpura 61
clones 12, **13**
    banks 44
    in plant propagation 120
Clostridium acetobutylicum 31
Cocking, Edward 124, 125
codons 9
cofactors
    iron–molybdenum of nitrogenase 116
    regeneration schemes 23
Cohen, Stanley 12
Colmer, Arthur 82
Comai, Luca 139
co-metabolism 96–7
Commerce and Defense Departments 207–8
Commercial Biotechnology – An International Analysis 200
Congressional Budget Office 199
copper 98
    bacterial leaching 82, 84, **86, 88, 89**
corn
    starch source for fuel alcohol 30–1
    wet-milling process **31**
corn syrup, high fructose 19, 26
    see also glucose syrup, high fructose
cosmetic industry, plant products 126
Crick, Francis 8
crop improvement 1, 2, 130
cross-protection 140
crown gall 131

cryopreservation 120
Cuttitta, Frank 149
cystic fibrosis 172, 183, 186
cytokine 42, 47
cytokinins 119

2,4-D 97
dairy industry 20
Darwin, Charles 2–3
Davies, Earl 182
DDT 96, 98
de la Pena, Alicia 133–4
degradation enzymes as catalysts 56–7
degradation, microbial 96, 97–8
della-Cioppa, Guy 139
deoxyribonucleic acid *see* DNA
detergents, enzyme 20
developing nations 1
diabetics 21, 22
disease
    autosomal dominant 181
    counselling in genetic 174
    diagnosis with monoclonal antibodies 145, 147–8
    ethics of prenatal diagnosis 183–4
    genetic 50, 172–85
    genetics of 47
    human hereditary 186–95
    immunodiagnosis 168
    mechanisms of resistance 140
    resistant plants 130
    X chromosome linked 181, 182
DNA 1, 5–14
    direct uptake by plant cells 133–4
    double helix 8
    injection into germ cells 133
    mutagenesis 25
    polymorphism 177–8
    precipitation 188
    prenatal analysis 176
    probes 100, 182
    rapid sequencing 45
    recombinant *see* recombinant DNA technology
    replication **9**, 25
    retroviral vector 191
    structure **7**, 8
    transfer in crop improvement 132
    transposable elements 194–5
Doolittle, Russell 164
Down's syndrome 174
Dreesman, Gordon 156
Duchenne muscular dystrophy 183
Dusty Miller, A. 191
dwarfism 50, 186

EcoRI 11, **13**
ecosystem, effect of new microorganism on 99
eicosapentaenoic acid 61
elite trees 119
enantiomers 17, 23
environmental contamination 94
Environmental Protection Agency 206–7
enzymes 15–18
    assays 22
    in chemical industry 27
    cofactors 23, 25
    conditions for action 17
    deficiency diseases 50
    in detergents 20
    diagnostic tests 22
    economics 26–7
    in food and beverage manufacture 19–20, 27
    immobilization 17–18
    industrial uses 16, 17, 19
    large volume transformation of chemicals 20
    medicinal applications 21–2
    microbial production 56–7
    modification 24–5
    multi-enzyme systems 25–6
    in nitrogen fixation 106
    in pharmaceutical industry 26
    restriction 11, 176, 177, 178–9
    specificity 16
    substrate ranges 24
    in water treatment 26
EPSP synthase 137, 139
ergot 62
Escherichia coli
    cloning of *B. thuringiensis* toxin 95
    gene expression 45
    growth rate for fermentation 40
    human protein production 45
    potential hazards of DNA recombinant technology 14
    protein synthesis signals 45
    and recombinant molecules 11
    recombinant protein production 45
ethanol
    anaerobic metabolism 30
    ethylene synthesis 30
    fermentation 29, 30
    for fuel 30
    sugar fermentation **32**
    yeast fermentation 30
eukaryotes 4
European collaboration 198–9
Evans, David 122
evolution 2–3

factor VIII 46, 50, 182
fatty acids 61–2

Federal Financial Support for High Technology Industries 199
feedstock chemicals 28
fermentation
    alcoholic 28, **29**
    contamination 38
    continuous 36
    cycle time 37, 40
    economics 36–9
    heat generation 40
    microbial 57–63
    organic acids 33–6
    oxygen availability 38–9
    production costs 36–8
    products **29**
    range of compounds produced 61–3
    role of microorganisms 28–9
    viscosity of broth 39
    waste disposal 39
    yield 37
fermentors 76, **77**
Fersht, Alan 25
fertilizers 1, 103–4
fixed bed reactor 19
flavanones 115, **116**
flavones 115
flavonoids 135
flavour chemicals 63
Fleming, Alexander 29, 57
food
    packaging 20
    processing 19–20, 27
food and drink industry 19–20, 27, 126
Food and Drug Administration 205
foot-and-mouth disease 169, 170
Fraley, Robert 137
France 198
Frankia 110, 112
Franklin, Rosalind 8
French Anderson, W. 189
Friedman, Theodore 191
frost resistance 96
fruit juice 20
Fujita, Yasuhiro 126
fungi, in single-cell protein production 79–80
Fusarium graminearum 79

galactose 20
Gaucher's disease 185
Gautheret, Robert 119
geminivirus 132
gene cloning 12, **13**, 147
    in crop improvement 130
    human **44**
    library 44
    for probes 100

and protein biochemistry 42–3
  protein production 43
  regulatory protein production 47
  of *T. ferrooxidans* 91–2
  technology 43–4
gene transfer 2
  of beta-globin 193
  by calcium phosphate 189
  of herbicide tolerance 137, 139
  into germ line 193–4, 196
  into mouse eggs 193, **194, 195**
  into protoplasts 133
  methods 188–96
  of selectable marker by injection 133–4
  vectors 130–3
  for virus-resistant plants 140–1
genes 2
  amplification systems 132–3
  antibody **155**
  bacterial 10
  bank 43, 44
  characterization 45
  chimaeric 132, 133, 134, 135
  control of expression 12, 189
  in crop improvement 2, 130
  env coding region 164
  expression of plant 134–5
  expression of *T. ferrooxidans* 91–2
  factor VIII cloning 182
  globin 180
  beta-globin 180
  heat-shock 135
  hereditary diseases 2
  identification of clone 44
  isolated 130
  lesions in disease 172
  LHCP 134
  libraries 43, 44
  light-sensitive 134–5
  in livestock improvement 2
  manipulation 45
  mutations 5–6, 130
  nif 104, **106,** 107–10, 118
  nod 113, 114, 115–16
  patatin 135
  phenylalanine hydroxylase cloning 182
  plant viruses as vectors 132–3
  point mutation 179–80
  rbcS 134
  regulation of expresion 134
  regulatory sequences 132
  sickle cell 178
  transfer systems for *T. ferrooxidans* 90–1
  therapy with cloned normal genes 185
  therapy for human hereditary diseases 186–95
  transfer among bacteria 100
  transmission in plant transformation 128
  transposable elements as vectors 134
  vectors 90
  viral 48
  virulence 131
genetic
  code **8,** 9
  counselling 174
genetic engineering 25
  environmental impact of microorganisms 98–101
  in fermentation processes 40
  glyphosate tolerant plants 139
  insect resistant plants **143**
  plant disease resistance 141
  protein production 45–6
Genetic Engineering News 198
Gerin, John 169
Getzoff, Elizabeth 165
gibberellic acid 65
Global Competition – The New Reality 199
glucoamylase 19
gluconic acid 36
glucose 20
  conversion by microorganisms 30
  isomerase 19
  oxidase 20
glucose syrup, high fructose 65
glutamine 111
glycolysis 15
glyphosate 135–7
  tolerance and genetically engineered plants 139
  tolerant callus **138**
Golgi complex **5**
gonorrhoea 147
Goodman, Howard 137, 194
governments and the biotechnology industry 197
graft-versus-host disease 152
Graham, Frank 188
Green, Nicola 164
Green Revolution 1, 130
Griffiths, Fred 7
growth hormone 2, 186
  injection of gene into eggs 196
  production by recombinant DNA technology 30
  production from recombinant *E. coli* clones 45
  recombinant human 50
  releasing factor (GRF) 50, 51
Guillemin, Roger 50
Gusella, James 183
gypsy moth 99, 141

Haber-Bosch process 103
Haberlandt, Gottleib 119
haemagglutinin protein 164, 165

haemoglobin 6, 172
haemophilia 50, 181, 182
halogen pollution 96–7
heart attack 21
Helling, R. 12
Hellstrom, Ingegerd 156
Hellstrom, Karl Erik 156
heparinase 21
hepatitis B 169–70
herbicides 1
    plant resistance 130
    tolerance 135, 137
hereditary diseases 2
heredity 3–4
herpes virus 148
hexokinase 15
Hinkle, M. E. 82
Hoechst 198
Hood, Leroy 12
Hozumi, Nobumichi 153
Hunkapillar, Michael 12
Huntington's disease 172, 181
    ethics of prenatal diagnosis 184
    genetic marker 183
hyaluronic acid 26
hybrid plant production 124–5
hybridoma 146, 152
    technology 100
hydrops fetalis 173
hypotensives 61
hypoxanthine guanine phosphoribosyl transferase (HPRT)
        deficiency 188

ice damage 95
immobilized cells
    for biochemical synthesis 63–5
    techniques 64
immune
    suppression 145, 147
    system 145, 147
immunoaffinity chromatography 167
immunogenicity 159, 161, 164
    of intact protein 165
immunogens 164
immunological catalysis 168
immunomodulators 61
industrial investment 202, **202**
influenza virus 164
inheritance
    in peas 3–4
    physical traits 1
insects
    microbial control 94
    resistant plants 141–2

insulin 2
    porcine 21
    production by recombinant DNA technology 30
    production from recombinant *E. coli* clones 45
    recombinant human 50
interferon 2
    in cancer therapy 47
    from recombinant *E. coli* clones 45
    gene cloning 47
    production 45
    recombinant 47
    recombinant leucocyte 50
interleukin 2, 42
    antagonists 53, **54**, 55
interleukin-1 47
interleukin-2 26, 47
    recombinant 51, **53**
introduction of organisms to new environment 99
iodine-131 149, 151
iron contamination in citric acid production 36
iso-propyl alcohol 29
itaconic acid 36

Janda, Kim 168
Japan 197
Jerne 156
Johnston, William 149
Jones, Laurie 119
Journal of the Brno Society of Natural Science 4

Kaiser, E. 25
Kan, Yuet Wai 176
kanamycin resistance 134
Kennedy, Ronald 156
Khorana, Har Kobind 9
kidney transplantation 145, 147
Kishore, Ganesh 139
Klebsiella pneumoniae **106**, 107–9
Klee, Harry 137
Klug, Aaron 167
Knackmuss, Hans 97
Köhler, Georges 145
Koprowski, Hilary 148–9
Kucherlapati, Raju 193

lactase 20
lactic acid fermentation 36
lactosamine 26
lactose 20
    raw material for fermentation 30
large white butterfly 141
Lawrence, D. 25
leghaemoglobin 112, **113**
legumes 103, 104, 105, 110

bloat-safe 125
seed inoculation 117
Lerner, Richard 164, 165, 167, 168, 169
Lesch–Nyhan syndrome 185, 188
lettuce somaclonal variation 120–2
leukaemia 191
leukotrienes 61
Levy, Ronald 151
Lindow, Steven 95
linkage analysis 177–8
lipase 23, **24**
Lithospermum erythrorhizon 126, **128**
livestock breeding 2
LSD 62
Lucas, John 120
Lymanthria dispar 99
lymphokine 47
lymphoma 152
lysozyme, antigenic structure 161, 164

McCarty, Maclyn 7
McClintock, Barbara 120
McKinney H. H. 140
MacLeod, Colin 7
Madagascar periwinkle 126
maize 133
introduction of new genes into protoplasts 133
transposable element 120, 134
Manduca sexta 141, 142
Maniatis, Tom 43
Mann, David 191
Medicago sativa 125
Meischer, Friedrich 5
Mendel, Gregor 3–4
messenger RNA 9
point mutations 179–80
metabolite production of plants 125–7
metal leaching, bacterial 82–4
metallothionen promoter 194
methanol 76
methotrexate 151
Methylophilus methylotropus 75, 76
Micractinium 73
microbe counting in environmental samples 99–100
microcosm 99
milk production 1
Miller, Richard 151
Milstein, Cesar 145
mining
bacteria and 98
microbes in 82, 88
Minna, John 149
mitochondria 5
genomes 135

mitosis 4
Moloney *see* murine Moloney leukaemia virus
monoclonal antibodies 100, 145–58
applications 147
in cancer diagnosis 148–9, **150**
in cancer therapy 149–52
discovery 145–7
future 156–7
limitations on therapy 152–3
peptide induced antibodies 165
production 146
side effects 147, 149
morphine 126
Morrison, Sherie 153
mosquito control 95, 141
moulds
citric acid production 34
in food processing 80
in single-cell protein production 79–80
Mulshine, James 148
multi-enzyme systems 25–6
multinational companies 198
multiple sclerosis 147
murine Moloney leukaemia virus 164, 167, 189, 191
Murray, H. C. 67
muscular dystrophy 183
mushroom mycelium 80
mutagenesis 25
myasthaenia gravis 147
Mycobacterium tuberculosis 168–9
mycoprotein 79, 80
myeloma 146
myohaemerythrin 165–7

NAD 23
Nadler, Lee 152
National Academy of Sciences 200
National Institutes of Health 14, 201–2, 205–6
Neïsseria gonorrhoeae 174–8
Netherlands 198
Neuberger, Michael 153
Neurath, Robert 170
Nicotiana tabacum 125
Niman, Henry 165
Nirenberg, Marshall 9
nitrogen fertilizer 103–4
nitrogen fixation 103–18
amount fixed 103
bacteria 104
genetics 107–8
introduction of *nif* genes to plants 118
non-legume crops 118
regulation of *nif* gene expression 108–10
symbiotic 110–12

nitrogenase
  complex 106
  effect of oxygen 106–7
  iron–molybdenum cofactor 116
  oxygen sensitivity 116
nodulins 112
nucleic acid 5
  segments 26
nucleotides 5, **6**

Occupational Safety and Health Administration 207
Ochoa, Severo 9
Office of Science and Technology Policy 200–1
Office of Technology Assessment 200
Oi, Vernon 153
oil, price of 29, 40
oil seed rape 124–5
oligonucleotide
  in DNA replication 25
  probes 181
On the Origin of Species by Means of Natural Selection 3
oncogene 48
one-grained spelt 133
opines 131
  genes coding for synthesis 134
Order, Stanley 151
Orthoclone OKT-3 147
Orton, Thomas 123
oxygen 75

Paecilomyces varioti 80
palm-oil trees 119
pancreatin 21
papain 20, 25
Papaver 126
papaya ringspot virus 140
paraffin hydrocarbons 77, 78
Pasteur, Louis 28
pathogen-free plant strains 120
Pauling, Linus 6
peas
  Afghanistan variety 117
  inheritance in 3–4
  light-sensitive genes 134–5
pectin 20
'Pekilo' process 80
penicillin 1
  commercial production 36
  enzyme conversion 20, **21**
  G synthesis by immobilization techniques 64
  microbial fermentation 29
Penicillium **59**
pepsin 20

peptide immunogens
  analysis for specific proteins 167
  immunogens 164–5
  medical applications 168–70
  potential applications 167–70
  reaction with small region of protein 167–8
peptides
  antibodies to 164
  chemically synthesized immunogens 165
  conformation 165
  eliciting antibodies 164–5
  eliciting protein-reactive antibodies 159
  modified natural as pharmaceutical 52
  myohaemerythrin 166–7
pest control, microbial 94–5
pesticides 1
Peterson, D. H. 67
petroleum 29
petunia
  glyphosate-tolerant 136, **139**
  introduction of new genes into protoplasts 133
  plant transformation 127
Petunia hybrida 136
phage contamination of fermentation processes 38
pharmaceuticals
  chemicals 49–55
  industry and enzymes 26
  industry and microbial fermentation 29
  modified natural protein, 50, 51–2
  plant products 126
  synthesis 20
phenylalanine 182
phenylketonuria 181–2, 184
Phillips, Donald 117
Pieris brassicae 141
plant
  breeding 1
  propagation by cell culture 119
plasma membrane 5
plasmids 12
  Ti 128, 132
  transfer between *Pseudomonas* species 97
  as vectors for gene transfer 90, 91
plasmin 51, 52
poliovirus **163**
polychlorinated biphenyls (PCB) 96
polymorphic restriction sites 180
polysaccharides 57
poppy 126
Porphyridium centrum 61
Postgate, John 107
potato
  patatin gene 135
  somaclonal variation 122
Power, Brian 120

pregnancy diagnosis 145, 147
President's Commission on Industrial Competitiveness 199
prokaryotes 4
pro-urokinase 21
prostaglandins 61
protease 18
protein
  adhesive 40
  engineering 40, 65–6
  human 42
  immunogenic sites 167
  immunology 160–1, 164
  intercellular signalling 47
  made by plant nodule 112
  mimetics of natural bioregulatory 52–3
  modified natural as pharmaceuticals 51–2
  peptide immunogen analysis 167
  production from gene cloning 43
  production of recombinant 45, **46**
  recombinant natural as pharmaceuticals 50
  regulatory 47
  sequence analysis 12
  synthesis 5, 9–10
  synthesis of foreign by microorganisms 65–6
  three-dimensional structure 52–3
protoplast
  fusion 124–5, 127
  and somatic cell hybrids **121**
Pruteen 75, 76
Pseudomonas fluorescens 95
Pseudomonas, halogen degradation 97
Pseudomonas syringae 95–6
purine nucleoside phosphorylase (PNP) deficiency 186–8

radiation 151
recombinant DNA technology 11–12, 24
  agricultural applications 95–6
  in AIDS 49
  antibody synthesis 153
  and bioleaching bacteria 87, 90–1, 92
  in crop improvement 130
  economics 26, 27
  and fermentation 39–40
  in gene cloning 43
  growth hormone production 30
  in human genetic disease 172
  human insulin production 30
  and monoclonal antibodies 157
  movement of *B. thuringiensis* toxin gene 95
  plant transformation 127–8
  in rennet production 20
  restrictions on 12, 14
Reineke, Walter 97
rennet production 20
rennin 20

Report of the Research Briefing Panel on Chemical and Process
  Engineering for Biotechnology 200
retrovirus 189, **190**
Rhizobium 103, **104**, 105
  genetic analysis 112–15
  interactions with legumes 110–11
  nif genes 112, 114
  nod genes 113, 114, 115–16
  plasmids **114**
  regulation of *nod* gene expression 115–16
Rhodopseudomonas 73
Rhodopseudomonas capsulata **101**
ribonucleic acid *see* RNA
ribosomal RNA analysis, 100
ribosomes **5**
rice
  improvement 124
  paddies 105
  regeneration of protoplasts 133
ricin 151
RNA 9–10; *see also* messenger RNA
  genomes in plant viruses 140
Rogers, Stephen 137
Rosen, Steven 151
Rosenberg, Stephen 51
Roundup 135–6
Roux, Wilhelm 4
rubber, synthetic 28–9

Saccharomyces **66**
Saccharomyces cerevisiae 30, 31
  baker's yeast production 71
  human protein production 45
St Vitus' dance 62
Scenedesmus 72, 73
Schleiden, Matthias 4
Schlom, Jeffrey 149
Schulz, Peter 168
Schwann, Theodor 4
Sears, Henry 151
Sela, Michael 161
selective culture media 100
self-pollination 124
sewage treatment 73
Shah, Dilip 137
Shaked, Zeev 23
Sharp, Roderick 122
shikonin 126, **128**
Shulman, Marc 153
sickle cell anaemia 6, 172–3, 174
  crises 186
  evolution of gene 178
  prenatal diagnosis 177, 184
silk protein engineering 40
Silverstein, Saul 188

Simberloff, Daniel 99
simian virus 40 11
single-cell protein 71–81
    amino acid supplementation 80
    bacteria in production 73–6
    continuous production **74**
    developments in production **72**
    economics 81
    fermentor design 75
    fungi in production 79–80
    moulds in production 79–80
    non-photosynthetic production 73–4
    photosynthetic organisms in production 72–3
    yeast in production 76–8
    yield 76
site-directed antibodies 160
Skoog, Folke 119
small cell carcinoma of the lung 149
Smith-Gill, Sandra 164
Smithies, Oliver 193
somaclonal variation 120–3
somatic
    embryogenesis 124
    hybridization 127
Southern blot analysis 176, 181
spina bifida 174
Spirulina 71, 72, 73
Stalker, David 139
Steplewski, Zenon 148–9
steroids
    pharmaceutical uses 67
    synthesis 67–8
    transformation by microorganisms 66–8
Streptococcus pneumoniae 7
streptokinase 21, 51
streptomycin
    resistant tobacco plants 125
    synthesis by immobilization techniques 64
substrate composition for microorganisms 69
sulphite waste 77, 80
sweet manufacture 20
Switzerland 198
synthetic seeds 120
systemic lupus erythematosus 147

2,4,5-T 97
T lymphocytes 147, 168
T-cell receptor 47
Tabata, Mamoru 126
Tainer, John 165
Tatum, Edward 5–6
Tay-Sach's disease 174, 186
termites 104

thalassaemia 172–4, 176–7
    ethics of prenatal diagnosis 183–4
    oligonucleotide probes 181
    point mutations 179–80
thalidomide 17
Thiobacillus ferrooxidans 82, 84
    gene transfer systems 90–1
    genetic improvement 87
    selectable markers 87, 90
    vectors for new genes 90
Thompson, John 124
Thy-1 antigen 168
tissue culture *see* cell culture, plant
tissue plasminogen activator (TPA) 15, 21, 26, 51
    directed protein modification 52
    production 46
tobacco 125
    callus 133
    chalcone synthase gene 135
    hornworm 141, 142
    introduction of cloned toxin gene 142
    introduction of new genes into protoplasts 133
    leaf-specific gene 135
    mosaic virus (TMV) 140–1
    plant transformation 127
    transposition of maize transposable element 134
    tumour induction **132**
    virus-resistant plants **142**
tomato **121**
    somaclonal variation 122–3
torula yeast 71, 77
Tramantano, Alfonson 168
transposase 195–6
tricarboxylic acid cycle 34
Trichoderma harzianum 80
Trichosporon pullulans 79
Trifolium repens 125
Triticum monococcum 133
Tsui, Lap Chi 183
tuberculosis 168–9
tumour
    -prone mice 196
    diagnosis 148
    formation in plants 131
    necrosis factor 47–8
tyrosyl-tRNA synthetase 25

Uhr, Jonathan 151
United Kingdom 198
United States 197
    biotechnology regulatory policies 202–4
    Department of Agriculture 207

strategies for biotechnology research and development 199–202
uranium 98
  bacterial leaching 82, 87
urokinase 21, 51

vaccines
  anti-idiotype antibodies 156
  peptide immunogens 169
Vale, Wylie 50
van der Eb, Alex 188
van Leeuwenhoek, Antony 28
van Regenmortel, M. H. V. 167
venereal disease 147–8
Verma, Inder 191
Vibrio cholerae **101**
viral vectors 191, 193
virus
  cancer causing 48
  plant as gene vector 132–3
  protein coding region 164
  resistant plants 140–1
  serotypes 169
Vitetta, Ellen 151

water treatment 26
Watson, James D. 8
wax esters 69

West Germany 198
wheat
  dwarf virus 133
  introduction of new genes into protoplasts 133
whey 20, 77
  for single-cell protein production 78
white clover 125
Whitesides, George 23
Wigler, Michael 188
Wiley, Donald 164
wine 20
Wiskott–Aldrich syndrome 188
Woo, Savio 182
wood pulp 80
world population 1

yeast
  baker's 76
  citric acid production 34
  commercial production 71
  fermentation 30–1
  food grade 71
  single-cell protein production 76–8
  substrates 76
  torula 71, 77
Youle, Richard 151

Zea mays 13